PROTEIN TYROSINE KINASES

CANCER DRUG DISCOVERY AND DEVELOPMENT

Beverly A. Teicher, Series Editor

Protein Tyrosine Kinases: *From Inhibitors to Useful Drugs*, edited by *Doriano Fabbro and Frank McCormick*, 2005

Biomarkers in Breast Cancer: *Molecular Diagnostics for Predicting and Monitoring Therapeutic Effect*, edited by *Giampietro Gasparini and Daniel F. Hayes*, 2005

Death Receptors in Cancer Therapy, edited by *Wafik S. El-Deiry*, 2005

Bone Metastasis: *Experimental and Clinical Therapeutics*, edited by *Gurmit Singh and Shafaat A. Rabbani*, 2005

The Oncogenomics Handbook, edited by *William J. LaRochelle and Richard A. Shimkets*, 2005

Camptothecins in Cancer Therapy, edited by *Thomas G. Burke and Val R. Adams*, 2005

Combination Cancer Therapy: *Modulators and Potentiators*, edited by *Gary K. Schwartz*, 2005

Cancer Chemoprevention, Volume 2: *Strategies for Cancer Chemoprevention*, edited by *Gary J. Kelloff, Ernest T. Hawk, and Caroline C. Sigman*, 2005

Proteasome Inhibitors in Cancer Therapy, edited by *Julian Adams*, 2004

Nucleic Acid Therapeutics in Cancer, edited by *Alan M. Gewirtz*, 2004

Cancer Chemoprevention, Volume 1: *Promising Cancer Chemopreventive Agents*, edited by *Gary J. Kelloff, Ernest T. Hawk, and Caroline C. Sigman*, 2004

DNA Repair in Cancer Therapy, edited by *Lawrence C. Panasci and Moulay A. Alaoui-Jamali*, 2004

Hematopoietic Growth Factors in Oncology: *Basic Science and Clinical Therapeutics,* edited by *George Morstyn, MaryAnn Foote, and Graham J. Lieschke*, 2004

Handbook of Anticancer Pharmacokinetics and Pharmacodynamics, edited by *William D. Figg and Howard L. McLeod*, 2004

Anticancer Drug Development Guide: *Preclinical Screening, Clinical Trials, and Approval, Second Edition*, edited by *Beverly A. Teicher and Paul A. Andrews*, 2004

Handbook of Cancer Vaccines, edited by *Michael A. Morse, Timothy M. Clay, and Kim H. Lyerly*, 2004

Drug Delivery Systems in Cancer Therapy, edited by *Dennis M. Brown*, 2003

Oncogene-Directed Therapies, edited by *Janusz Rak*, 2003

Cell Cycle Inhibitors in Cancer Therapy: *Current Strategies*, edited by *Antonio Giordano and Kenneth J. Soprano*, 2003

Fluoropyrimidines in Cancer Therapy, edited by *Youcef M. Rustum*, 2003

Chemoradiation in Cancer Therapy, edited by *Hak Choy*, 2003

Targets for Cancer Chemotherapy: *Transcription Factors and Other Nuclear Proteins*, edited by *Nicholas B. La Thangue and Lan R. Bandara*, 2002

Tumor Targeting in Cancer Therapy, edited by *Michel Pagé*, 2002

Hormone Therapy in Breast and Prostate Cancer, edited by *V. Craig Jordan and Barrington J. A. Furr*, 2002

Tumor Models in Cancer Research, edited by *Beverly A. Teicher*, 2002

Tumor Suppressor Genes in Human Cancer, edited by *David E. Fisher*, 2001

Matrix Metalloproteinase Inhibitors in Cancer Therapy, edited by *Neil J. Clendeninn and Krzysztof Appelt*, 2001

Farnesyltransferase Inhibitors in Cancer, edited by *Saïd M. Sebti and Andrew D. Hamilton*, 2001

Platinum-Based Drugs in Cancer Therapy, edited by *Lloyd R. Kelland and Nicholas P. Farrell*, 2000

Signaling Networks and Cell Cycle Control: *The Molecular Basis of Cancer and Other Diseases*, edited by *J. Silvio Gutkind*, 1999

Apoptosis and Cancer Chemotherapy, edited by *John A. Hickman and Caroline Dive*, 1999

Antifolate Drugs in Cancer Therapy, edited by *Ann L. Jackman*, 1999

Antiangiogenic Agents in Cancer Therapy, edited by *Beverly A. Teicher*, 1999

Anticancer Drug Development Guide: *Preclinical Screening, Clinical Trials, and Approval*, edited by *Beverly A. Teicher*, 1997

Cancer Therapeutics: *Experimental and Clinical Agents*, edited by *Beverly A. Teicher*, 1997

PROTEIN TYROSINE KINASES

FROM INHIBITORS TO USEFUL DRUGS

Edited by

DORIANO FABBRO, PhD

Novartis Institutes for BioMedical Research Basel,
Novartis Pharma AG,
Basel, Switzerland

and

FRANK MCCORMICK, MD, PhD

UCSF Comprehensive Cancer Center,
University of California, San Francisco,
San Francisco, CA

HUMANA PRESS
TOTOWA, NEW JERSEY

This publication is printed on acid-free paper.♾
ANSI Z39.48-1984 (American National Standards Institute) Permanence of Paper for Printed Library Materials.

Production Editor: Amy Thau

Cover design by Patricia F. Cleary

Cover Illustration: Figure 9 from Chapter 9, "Structural Biology of Protein Tyrosine Kinases," by Sandra W. Cowan-Jacob, Paul Ramage, Wilhelm Stark, Gabriele Fendrich, and Wolfgang Jahnke.

For additional copies, pricing for bulk purchases, and/or information about other Humana titles, contact Humana at the above address or at any of the following numbers: Tel.: 973-256-1699; Fax: 973-256-8341; E-mail: orders@humanapr.com or visit our Website at www.humanapress.com

Printed in the United States of America. 10 9 8 7 6 5 4 3 2 1
Library of Congress Cataloging-in-Publication Data
eISBN: 1-59259-962-1
Protein tyrosine kinases : from inhibitors to useful drugs / edited by
Doriano Fabbro and Frank McCormick.
   p. cm. -- (Cancer drug discovery and development)
 Includes bibliographical references and index.
 ISBN 1-58829-384-X (alk. paper)
 1. Protein-tyrosine kinase--Inhibitors--Therapeutic use. 2.
Cancer--Chemotherapy. I. Fabbro, D. II. McCormick, Frank, 1950- III.
Series.
 RC271.P76P76 2005
 616.99'4061--dc22
                      2005006248

PREFACE

Protein kinases function as components of signal transduction pathways, playing a central role in diverse biological processes, such as control of cell growth, metabolism, differentiation, and apoptosis. The development of selective protein kinase inhibitors that can block or modulate diseases with abnormalities in these signaling pathways is considered a promising approach for drug development. The function of many protein kinases is deregulated in human cancers. Deregulation, whether as a result of deletion, mutation, or amplification, is manifested as aberrant activation, prime examples of which are kinases including Bcr-Abl, EGFR family members, Flt-3, Met, etc., as well as kinases involved in the neovascularization of tumors like KDR. A decade ago, these protein kinases were considered prime targets for the development of selective inhibitors. Currently, over 20 different kinases—the majority being receptor protein tyrosine kinases (RPTKs)—are being considered as potential therapeutic targets in oncology.

Although the success of agents such as Glivec® (Imatinib mesylate, Glivec/Gleevec®) and Iressa™ (Gefitinib) has provided a proof of concept that such agents can be therapeutically effective and retain an acceptable safety profile, the clinical experience with other tyrosine kinase inhibitors is still limited.

A comprehensive overview of the drug discovery processes aimed at generating inhibitors for the treatment of malignancies believed to be dependent on the gain of function of protein tyrosine kinases (PTKs) has to contain a summary of those drug discovery programs that have devoted their efforts to generating low molecular-weight (LMW) inhibitors directed against either the adenosine triphosphatase (ATP)-binding site (summarized in Chapter 1) or the Src homology 2 (SH2) domain, an important noncatalytic module that recognizes a short phosphotyrosine-containing sequence in other proteins. A review on the advances made targeting this critical SH2-binding event, which would result in the inactivation of undesirable signal transduction networks, is found in Chapter 2.

Epistatically, PTKs are located either upstream and/or downstream of tumor suppressor genes or oncogenes and have been demonstrated to play central roles in apoptosis, proliferation, invasion, and differentiation. The signal transduction pathways of PTKs, in particular receptor PTKs, is intimately linked to the phosphoinositide 3-kinase (PI3-K) pathway as activation of cells by a wide variety of stimuli leads to rapid changes in 3-phosphorylated inositol lipids through the action of a family of enzymes known as PI3-Ks. The dissection of PI3-K signaling pathways has been greatly aided by genetic approaches and by the availability of two pharmacological tools, wortmannin and LY294002. In Chapter 3, a comprehensive summary is given to explain why the PI3-Ks represent a reasonable target for pharmaceutical intervention. All the reasoning for the activation of PI3-K as target is central to the coordinated control of multiple cell-signaling pathways leading to cell growth, cell proliferation, cell survival, and cell migration.

Aberrant activation of tyrosine kinases, owing to mutation or overexpression, is sufficient for them to become transforming in cellular and animal models. The majority of targets are RPTKs. Deregulating mutations of over half of the known RPTKs have been associated with different human malignancies. To illustrate the rationale and the progress made towards generating "selective" LMW kinase inhibitors, a few selected examples have been chosen that include the targets of Glivec (platelet-derived growth factor receptor, Kit, Bcr-Abl), FLT-3, JAKs, as well as Src. A special chapter has also been devoted to the normal function, role in disease, and application of platelet-derived growth factor antagonists. All of these efforts illustrate the tremendous biological complexity that is encountered by targeting these kinases and render a conclusion about the actual level of understanding of the molecular epidemiology and pathophysiology, as well as disease relevance of these kinases. In particular, the success story of Glivec has taught the academic, as well as pharmaceutical fields, some lessons regarding the inhibition of these kinases from the points of view of therapeutics and biology (Chapters 4–8).

A successful development of protein kinase inhibitors is based primarily on solid epidemiology allowing the identification and validation of the target along with the knowledge of the structure of the kinase. The structural understanding of protein kinases has significantly progressed as structures of kinases both in phosphorylated or nonphosphorylated forms, active or inactive states, unliganded or complexed to substrate analogs or inhibitors, and with only the catalytic domain present or in a multidomain construct including SH3 and SH2 domains have become available. All of this knowledge is being used for structure-based design and has been summarized in Chapter 9.

Robust predictive preclinical in vitro and, in particular, in vivo screening model systems that allow rapid optimization of lead compounds are key to a successful drug discovery effort as they are crucial for determining the safety of kinase inhibitors. Animal models as used in cancer can be divided conveniently into models designed to understand the natural history of cancer and models that are useful for the testing, selection, and profiling of new anticancer treatment modalities. The advantages and disadvantages of in vivo preclinical models for testing protein kinase inhibitors with antitumor activity have been summarized in Chapter 10.

Finally, one of the most important steps in the drug discovery process leading to kinase inhibitors is to determine "on-target" vs "off-target" effects by demonstrating that the protein kinase inhibitor downregulates the function of the target in vitro and in vivo with all the expected consequences (downregulation of given pathway[s] and growth arrest). Therefore, phosphoprofiling or phosphoproteomics that include the large-scale determination of protein phosphorylation in cells and tissue is one approach that can be used to characterize biological states, including therapeutic responses to provide a comprehensive picture of cellular states. The utility of these methods in drug discovery and development is discussed in Chapter 11.

Understanding the role of a potential target in cancer development and progression is as relevant as the efficient optimization of an inhibitory compound's potency, toxicity, and pharmacokinetic profile. To be a valid target, a kinase should play a fundamental role in the pathogenesis of a disease and the rationally designed LMW compounds, which are almost exclusively directed against the ATP-binding site of the kinase, should be able to revert the effects of the disease-causing kinase in preclinical models, and should be translatable to clinical settings.

Doriano Fabbro, PhD
Frank McCormick, MD, PhD

CONTENTS

Preface v

Contributors xi

1. Protein Tyrosine Kinases as Targets for Cancer
   and Other Indications 1
   Mark Pearson, Carlos García-Echeverría, and Doriano Fabbro

2. Inhibitors of Signaling Interfaces: *Targeting Src Homology 2*
   Domains in Drug Discovery 31
   Carlos García-Echeverría

3. PI3-Kinase Inhibition: *A Target for Therapeutic Intervention* 53
   Peter M. Finan and Stephen G. Ward

4. Src as a Target for Pharmaceutical Intervention:
   Potential and Limitations 71
   Mira Šuša, Martin Missbach, Rainer Gamse, Michaela Kneissel,
   Thomas Buhl, Jürg A. Gasser, Markus Glatt, Terence O'Reilly,
   Anna Teti, and Jonathan Green

5. Activated FLT3 Receptor Tyrosine Kinase as a Therapeutic Target
   in Leukemia 93
   Blanca Scheijen and James D. Griffin

6. JAK Kinases in Leukemias, Lymphomas, and Multiple Myeloma ... 115
   Renate Burger and Martin Gramatzki

7. Glivec® (Gleevec®, Imatinib, STI571):
   A Targeted Therapy for Chronic Myelogenous Leukemia 145
   Elisabeth Buchdunger and Renaud Capedeville

8. Platelet-Derived Growth Factor: *Normal Function, Role in Disease,*
   and Application of PDGF Antagonists 161
   Tobias Sjöblom, Kristian Pietras, Arne Östman,
   and Carl-Henrik Heldin

9. Structural Biology of Protein Tyrosine Kinases 187
   Sandra W. Cowan-Jacob, Paul Ramage, Wilhelm Stark,
   Gabriele Fendrich, and Wolfgang Jahnke

10. Testing of Signal Transduction Inhibitors
    in Animal Models of Cancer 231
    Terence O'Reilly and Robert Cozens

11. Phosphoproteomics in Drug Discovery and Development 265
     Michel F. Moran, Jarrod A. Marto, Cynthia J. Brame,
        Olga Ornatsky, Mark M. Ross, Leticia M. Toledo-Sherman,
        Alfredo C. Castro, Brett Larsen, Henry Duewel,
        Christopher Hosfield, Christopher Orsi, Thodoros Topaloglou,
        Daniel Figeys, Jennifer A. Caldwell-Busby, and David R. Stover

Index 279

CONTRIBUTORS

CYNTHIA J. BRAME, PhD • *MDS Proteomics Inc., Toronto, Canada and Charlottesville, VA*

ELISABETH BUCHDUNGER, PhD • *Senior Research Investigator I, Oncology Research, Novartis Institutes for BioMedical Research Basel, Novartis Pharma AG, Basel, Switzerland*

THOMAS BUHL, PhD • *Senior Research Investigator I, Musculoskeletal Diseases (MSD), Novartis Institutes for BioMedical Research Basel, Novartis Pharma AG, Basel, Switzerland*

RENATE BURGER, PhD • *Jerome Lipper Multiple Myeloma Center, Department of Medical Oncology, Dana-Farber Cancer Institute, Harvard Medical School, Boston, MA*

JENNIFER A. CALDWELL-BUSBY, PhD • *MDS Proteomics Inc., Toronto, Canada and Charlottesville, VA*

RENAUD CAPDEVILLE, MD • *Group Leader, Clinical Research, Novartis Oncology, Novartis Pharma AG, Basel, Switzerland*

ALFREDO C. CASTRO, PhD • *Syntonix Pharmaceuticals, Waltham, MA*

SANDRA W. COWAN-JACOB, PhD • *Group Leader, Discovery Technologies, Protein Structure Unit (DT/PSU), Novartis Institutes for BioMedical Research Basel, Novartis Pharma AG, Basel, Switzerland*

ROBERT COZENS, PhD • *Unit Head, Drug Discovery Pharmacology (DDP), Novartis Institutes for BioMedical Research Basel, Novartis Pharma AG, Basel, Switzerland*

HENRY DUEWEL, PhD • *MDS Proteomics Inc., Toronto, Canada and Charlottesville, VA*

DORIANO FABBRO, PhD • *Unit Head, In Vitro Profiling (IVP), and FIP Head of Signaling Pathways/Oncology Research, Novartis Institutes for BioMedical Research Basel, Novartis Pharma AG, Basel, Switzerland*

GABRIELE FENDRICH, PhD • *Research Investigator I, Discovery Technologies, Protein Structure Unit (DT/PSU), Novartis Institutes for BioMedical Research Basel, Novartis Pharma AG, Basel, Switzerland*

DANIEL FIGEYS, PhD • *MDS Proteomics Inc., Toronto, Canada and Charlottesville, VA*

PETER M. FINAN, PhD • *Novartis Horsham Research Centre, Horsham, UK*

RAINER GAMSE, MD • *DA Project Management Director, MSD Management, Novartis Institutes for BioMedical Research Basel, Novartis Pharma AG, Basel, Switzerland*

CARLOS GARCÍA-ECHEVERRÍA, PhD • *Global Discovery Chemistry-Oncology, Novartis Institutes for BioMedical Research Basel, Novartis Pharma AG, Basel, Switzerland*

JÜRG A. GASSER, PhD • *Senior Research Investigator I, Musculoskeletal Diseases (MSD), Novartis Institutes for BioMedical Research Basel, Novartis Pharma AG, Basel, Switzerland*

MARKUS GLATT, PhD • *Senior Research Investigator I, Musculoskeletal Diseases (MSD), Novartis Institutes for BioMedical Research Basel, Novartis Pharma AG, Basel, Switzerland*

MARTIN GRAMATZKI, MD • *Division of Stem Cell and Immunotherapy II, and Department of Medicine, University of Kiel, Kiel, Germany*

JONATHAN GREEN, PhD • *Senior Research Investigator II/NDS, Musculoskeletal Diseases (MSD), Novartis Institutes for BioMedical Research Basel, Novartis Pharma AG, Basel, Switzerland*

JAMES D. GRIFFIN, MD • *Department of Medical Oncology, Dana-Farber Cancer Institute; Department of Medicine, Brigham and Women's Hospital, Harvard Medical School, Boston, MA*

CARL-HENRIK HELDIN, PhD • *Ludwig Institute for Cancer Research, Uppsala, Sweden*

CHRISTOPHER HOSFIELD, PhD • *MDS Proteomics Inc., Toronto, Canada and Charlottesville, VA*

WOLFGANG JAHNKE, PhD • *Senior Research Investigator I/NLS, Discovery Technologies, Protein Structure Unit (DT/PSU), Novartis Institutes for BioMedical Research Basel, Novartis Pharma AG, Basel, Switzerland*

MICHAELA KNEISSEL, PhD • *Research Investigator II, Musculoskeletal Diseases (MSD), Novartis Institutes for BioMedical Research Basel, Novartis Pharma AG, Basel, Switzerland*

BRETT LARSEN, MSc • *MDS Proteomics Inc., Toronto, Canada and Charlottesville, VA*

JARROD A. MARTO, PhD • *MDS Proteomics Inc., Toronto, Canada and Charlottesville, VA*

FRANK MCCORMICK, MD, PhD • *UCSF Comprehensive Cancer Center, University of California at San Francisco, San Francisco, CA*

MARTIN MISSBACH, PhD • *Unit Head, Disease Area Bone, Muscle, and Gastrointestinal (BMG), Novartis Institutes for BioMedical Research Basel, Novartis Pharma AG, Basel, Switzerland*

MICHEL F. MORAN, PhD • *McLaughlin Centre for Molecular Medicine, Hospital for Sick Children, University of Toronto, Toronto, Canada*

OLGA ORNATSKY, PhD • *MDS Proteomics Inc., Toronto, Canada and Charlottesville, VA*

TERENCE O'REILLY, PhD • *Senior Research Investigator I/NLS, Oncology, Novartis Institutes for BioMedical Research Basel, Novartis Pharma AG, Basel, Switzerland*

CHRISTOPHER ORSI, MEng • *MDS Proteomics Inc., Toronto, Canada and Charlottesville, VA*

ARNE ÖSTMAN, PhD • *Cancer Center Karolinska, Karolinska Institute, Stockholm, Sweden*

MARK PEARSON, PhD • *Research Investigator II, Oncology In Vitro Profiling (IVP), Novartis Institutes for BioMedical Research Basel, Novartis Pharma AG, Basel, Switzerland*

KRISTIAN PIETRAS, PhD • *Ludwig Institute for Cancer Research, Stockholm, Sweden*

PAUL RAMAGE, PhD • *Group Leader, Discovery Technologies, Protein Structure Unit (DT/PSU), Novartis Institutes for BioMedical Research Basel, Novartis Pharma AG, Basel, Switzerland*

MARK M. ROSS, PhD • *MDS Proteomics Inc., Toronto, Canada and Charlottesville, VA*

BLANCA SCHEIJEN, PhD • *Department of Medical Oncology, Dana-Farber Cancer Institute; Department of Medicine, Brigham and Women's Hospital, Harvard Medical School, Boston, MA*

TOBIAS SJÖBLOM, PhD • *The Sydney Kimmel Comprehensive Cancer Center, Johns Hopkins University, Baltimore, MD*

MIRA ŠUŠA, PhD • *Senior Research Investigator I/NLS, Musculoskeletal Diseases (MSD), Novartis Institutes for BioMedical Research Basel, Novartis Pharma AG, Basel, Switzerland*

WILHELM STARK, PhD • *Research Investigator I, Discovery Technologies, Protein Structure Unit (DT/PSU), Novartis Institutes for BioMedical Research Basel, Novartis Pharma AG, Basel, Switzerland*

DAVID R. STOVER, PhD • *Oncology Antibody Lab I, Novartis Institutes for BioMedical Research Basel, Novartis Pharma AG, Basel, Switzerland*

ANNA TETI, PhD • *University L'Aquilla, L'Aquilla, Italy*

LETICIA M. TOLEDO-SHERMAN, PhD • *MDS Proteomics Inc., Toronto, Canada and Charlottesville, VA*

THODOROS TOPALOGLOU, PhD • *MDS Proteomics Inc., Toronto, Canada and Charlottesville, VA*

STEPHEN G. WARD, PhD • *Department of Pharmacy and Pharmacology, University of Bath, Bath, UK*

1 Protein Tyrosine Kinases as Targets for Cancer and Other Indications

Mark Pearson, PhD,
Carlos García-Echeverría, PhD,
and Doriano Fabbro, PhD

CONTENTS

INTRODUCTION
ABL, C-KIT, AND PLATELET-DERIVED GROWTH FACTOR RECEPTOR
INDICATIONS FOR IMATINIB OTHER THAN CANCER
FLT3
EPIDERMAL GROWTH FACTOR RECEPTOR
SMALL-MOLECULE INHIBITORS OF EGFR AND HER2
VASCULAR ENDOTHELIAL GROWTH FACTOR RECEPTOR
NONCANCER VEGF-TARGETED THERAPIES
NOVEL TYROSINE KINASE TARGETS
IGF-IR
C-MET
JAK FAMILY MEMBERS
CLINICAL QUESTIONS
CONCLUSIONS AND PERSPECTIVES

1. INTRODUCTION

The identification and characterization of the members of individual signal transduction cascades, and advances in understanding how these signals are integrated in normal and pathological conditions have provided new strategies for therapeutic intervention. Rapid progress has occurred in last few years in the development of inhibitors that target protein tyrosine kinases (PTKs), enzymes that transfer the γ-phosphate group of adenosine triphosphate (ATP) to the

From: *Cancer Drug Discovery and Development:*
Protein Tyrosine Kinases: From Inhibitors to Useful Drugs
Edited by: D. Fabbro and F. McCormick © Humana Press Inc., Totowa, NJ

Table 1
Tyrosine Kinase Targets In Human Cancer

Target	Genetic lesions	Associated cancers
EGFR family	Overexpression (amplification)	Breast, lung, prostate, colon
EGFR	Extracellular domain deletions	Breast, lung, glioma
HER-2/ErbB2	Overexpression (amplification)	Breast, ovarian, colon, lung, gastric
VEGFR family		
VEGFR-1	None	Solid tumors angiogenic/ metastatic
VEGFR-2	None/Overexpression	Solid tumors angiogenic/ metastatic
Flt-4		
Flt3	Internal tandem duplication	AML
c-Kit	Activating point mutations	GIST, seminoma, mastocytosis
Abl	Translocation (Bcr-Abl)	CML
JAK	Activation translocation Tel-JAK2	CML, T-ALL solid cancers
PDGFR family		
PDGFR-α	Overexpression	Glioma, glioblastoma, ovarian
	Translocation Fip1L1-PDGFR-α	HES
PDGFR-β	Translocation Tel-PDGFR-β	CMML
	Translocation ColIAI-PDGF	dermatofibrosarcoma protuberans
	Overexpression	Glioma
IGF-IR	Overexpression	Colorectal, pancreatic, breast, ovarian, MM
c-Met	Overexpression	Hepatocellular carcinoma
	Activating point mutations	Renal carcinoma
		HNSCC metastases
FGFR family		
FGFR1	Translocations BCR-; FOP-; ZNF198-; CEP110-FGFR1	CML Stem cell myeloproliferative disorder
FGFR3	Translocations	Multiple myeloma
	Activating point mutations	
FGFR4	Overexpression (amplification)	Breast, ovary
c-Ret	Translocations	Thyroid carcinoma
	GOF point mutations	MEN2A, MEN2B, FTMC familial and sporadic

hydroxyl group of tyrosine residues on target proteins. Although PTKs represent a small percentage of the total number of kinases in the "kinome," 90 of 518, a disproportional number of inhibitors currently in clinical trials are directed against them; e.g., more than 20 different tyrosine kinases are being evaluated as potential targets in oncology. There are a number of reasons why tyrosine kinases have been considered to be good targets. Epistatically, PTKs are located upstream and downstream of tumor suppressor genes or oncogenes and have been demonstrated to play central roles in apoptosis, proliferation, invasion, and differentiation *(1)*. Aberrant activation of tyrosine kinases, owing to mutation or overexpression, is sufficient for them to become transforming in cellular and animal models. The majority of targets are receptor protein tyrosine kinases (RPTKs), as deregulating mutations of over half of the known RPTKs have been associated with different human malignancies; *see* Table 1 for examples. Finally, and equally as important as the epidemiological and biochemical data, the prevalence of PTKs as targets is because of the fact that they are considered druggable.

Several approaches have been adopted to inhibit or decrease kinase activity. The clinical efficacy of antibodies that inhibit ligand-mediated activation of RPTKs (e.g., Herceptin™, Erbitux™, or Avastin™) provides a proof-of-concept of the therapeutic validity of targeting this family of enzymes. Low molecular mass compounds currently in the clinic are, almost without exception, directed against the kinase ATP binding site. Co-crystallization of kinases with ATP analogs or inhibitors and homology modeling studies have been used to identify structural features around the ATP binding site that have allowed the design of an inhibitor specific for one or several kinases *(2)*. Despite the lack of complete specificity—i.e., one inhibitor, one kinase—therapeutically beneficial clinical use, in the absence of limiting toxicities, has been demonstrated for small molecular mass inhibitors such as imatinib and gefitinib. Indeed, once registered, the activity of imatinib against other kinases has been successfully exploited in the treatment of other indications *(3)*.

This chapter will provide a brief summary of the clinical progress in targeting certain classes of tyrosine kinases. Although mainly focused on targeting kinases involved in cancer, the use of inhibitors in nonmalignant conditions will also be addressed where appropriate. The reader is directed to the subsequent chapters for a more detailed description of the role of specific kinases in disease and inhibitors that have been developed against them.

2. ABL, c-KIT, AND PLATELET-DERIVED GROWTH FACTOR RECEPTOR

Preclinical studies with imatinib mesylate (1, Fig. 1; Glivec®/Gleevec®, Novartis Pharma AG) demonstrated potent in vitro inhibition of Abl as well as the platelet-derived growth factor receptor (PDGFR) and c-Kit *(4,5)*. The Bcr-Abl

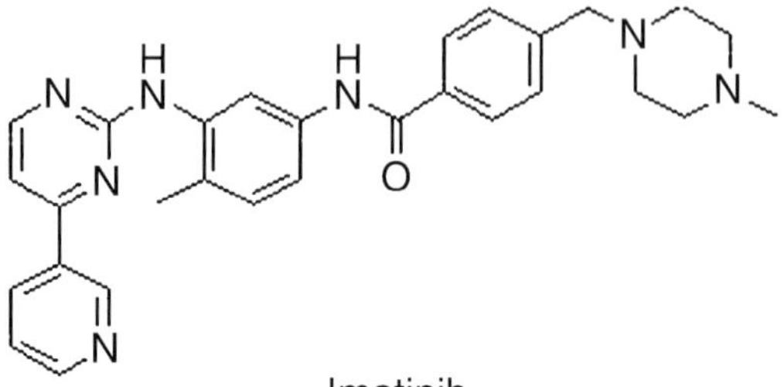

Imatinib
1

Fig. 1. Structure of imatinib.

Table 2
Representative Examples of Small-Molecule Inhibitors of Tyrosine Kinases
In Clinical Development

Target	Compound	Company	Status	Indication
EGFR	Gefitinib	AstraZeneca	Registered	NSCLC
EGFR	Erlotinib	OSI/Genentech/ Roche	Phase III	NSCLC
EGFR, HER-2, ErbB3, ErbB4	Canertinib	Pfizer	Phase II	Solid tumors
EGFR, HER-2	EKB659	Wyeth-Ayerst	Phase I	Solid tumors
EGFR, HER-2	Lapitinib	GlaxoSmithKline	Phase I	Solid tumors
HER-2	Mubritinib	Takeda	Phase I	Solid tumors
VEGFR	Valatinib	Novartis	Phase II-III	Solid tumors
VEGFR	SU11248	Pfizer	Phase III	Solid tumors
c-Kit			Phase I	GIST
Flt-3			Phase II	AML
VEGFR	SU6668		Phase II	Solid tumors
ABL	Imatinib	Novartis	Registered	CML
PDGFR				CMML, HES
c-KIT				GIST
VEGFR	ZD-6474	AstraZeneca	Phase III	Solid tumors
Flt3	Midostaurin	Novartis	Phase II	AML
c-KIT			Phase I	Solid tumors
PDGFR				
Flt3	CEP-701	Cephalon	Phase I	AML
Flt3	MLN-518	Millenium	Phase II	AML

fusion protein is constitutively active and is present in 95% of patients with chronic myeloid leukemia (CML) and 15–30% of patients with acute lymphoblastic leukaemia (ALL). The CML-like disease, observed on expression of Bcr-Abl in murine bone marrow, established this fusion protein as a causative event in CML *(6)*. Imatinib was shown to inhibit Bcr-Abl-mediated proliferation of established cell lines and freshly isolated leukemic cells from CML and ALL patients *(4)*. In clinical trials, imatinib was well tolerated and induced a response in patients at all stages of CML, resulting in its approval by the US Food and Drug Administration (FDA) for treatment of this indication *(7)*. Whereas a stable response was observed in patients treated in chronic phase, most patients in accelerated or blast crisis, after an initial response, developed resistance to imatinib and relapsed *(8)*. Multiple mechanisms appear to underlie the development of resistance; however, it appears that at least half of relapsed patients display point mutations in the ATP-binding domain of Bcr-Abl, rendering kinase activity insensitive to imatinib *(9,10)*. Subsequently, imatinib was also approved for treatment of unresectable and/or metastatic gastrointestinal stromal tumors (GISTs), a rare, chemoresistant sarcoma *(11,12)*. A significant proportion of GISTs express constitutively activated c-kit, as a result of point mutations in the juxtamembrane or kinase domains. Patient response to imatinib could be stratified according to expression of mutated c-kit, those responding best expressing the activated kinase *(13)*. Of note, a small fraction of the tumors, which responded to imatinib despite containing wild-type c-kit, were shown to contain mutations in PDGFR-α, resulting in its constitutive activation *(14)*. Subsequently, other malignancies that display constitutive activation of the PDGFR, resulting from either the presence of aberrantly expressed ligand, the COL1A1-PDGFB fusion in dermatofibrosarcoma protuberans *(15,16)*, or fusion of the receptor, TEL-PDGFRβ in chronic myelomonocytic leukemia (CMML) *(17)*, have responded to treatment with imatinib. An interstitial chromosomal deletion leading to expression of Fip1L1-PDGFRα was recently discovered in patients with hypereosinophilic (HES) syndrome *(18)*. In a small study, imatinib treatment led to disease remission in 9 of 11 HES patients. Five of these patients were found to express the Fip1L1-PDGFRα fusion product. Notably, relapse in one patient was associated with the appearance of a T674I mutation, homologous with T315I imatinib-resistant mutation observed in Bcr-Abl. Imatinib is additionally being considered for treatment of other neoplasms that contain c-Kit mutants, such as seminoma *(19)* and systemic mast-cell disease *(20)*, as well as Kit/PDGF hypersensitive indications such as polycythemia vera and myelofibrosis with myeloid metaplasia *(3,21)*.

Finally, preclinical studies have indicated imatinib inhibition of PDGFR underlies its ability to reduce the interstitial tumor pressure in orthotopically implanted colon carcinomas, suggesting additional use in combination with chemotherapeutic agents. In an analogous manner, CDP-860 (Celltech/Zymogenetics), a humanized antibody directed against PDGFR-β, is being assessed in combination with chemotherapeutic agents in Phase II clinical trials *(22)*.

3. INDICATIONS FOR IMATINIB OTHER THAN CANCER

The well-tolerated toxicity profile observed in patients receiving imatinib has prompted studies investigating potential uses of the drug in indications other than cancer. These have principally focused on vascular wall proliferative diseases linked to pathological overactivation of the PDGF system including restenosis and allograft arteriosclerosis (chronic rejection). In different animal models, treatment with imatinib resulted in almost complete inhibition of allograft associated ateriosclerosis. A recent study indicated that imatinib was additionally effective in a model of diabetes associated atherosclerosis *(25,26)*.

4. FLT3

Cytogenetic analysis indicates that the Bcr-Abl fusion protein is the only abnormality that occurs during chronic-phase CML, whereas the acceleration and blast phases are accompanied by the appearance of additional lesions, typically in known tumor suppressor genes *(27–29)*. Although these secondary mutations may be required for disease progression, the fact that even blast-phase patients initially respond to imatinib treatment indicates that the continued activity of the Bcr-Abl kinase is a prerequisite for CML. Point mutations and nonrandom chromosomal translocations that induce the activation of specific oncogenes or create novel chimeric genes are characteristic of many leukemias and lymphomas. Expression of the mutated allele in murine bone marrow frequently leads to partial or complete recapitulation of the chronic phase of the disease, suggesting that targeting the aberrant activity may represent a generally valid therapeutic approach.

Approximately 30% of acute myeloid leukemia (AML) patients have a hyperactivating gain of function mutation in the RPTK Flt3 *(30,31)*. The presence of this mutation is associated with a poor clinical prognosis and expression of Flt3 mutants in murine bone marrow results in development of a myelodysplastic syndrome (MDS) *(32)*. Translocations that result in the expression of AML1/ETO, CBFB/SMMHC, or PML/RARα fusion proteins have been identified in Flt3-expressing leukemias *(33,34)*. Full-blown AML is experimentally recapitulated when Flt3 mutations were introduced into transgenic mice expressing one of the fusion proteins, indicating cooperation occurs between the two mutations *(35)*. At least four different inhibitors are currently in early-stage clinical trials with AML patients harboring Flt3 mutations; CEP-701 (2, Fig. 2; Cephalon Inc/Kyowa Hakko Kogyo Co. Ltd.) *(35)*, Midostaurin/PKC412 (3, Fig. 2;Novartis AG) *(36,37)*, SU-11248 (4, Fig. 2; Pfizer) *(38,39)*, and MLN-518 (5, Fig. 2; Millenium Pharmaceuticals) *(40)*. All these compounds induce cytostatic or cytotoxic effects in cells expressing Flt3 mutant receptor, but do not affect cells that express the wild-type receptor. Administration

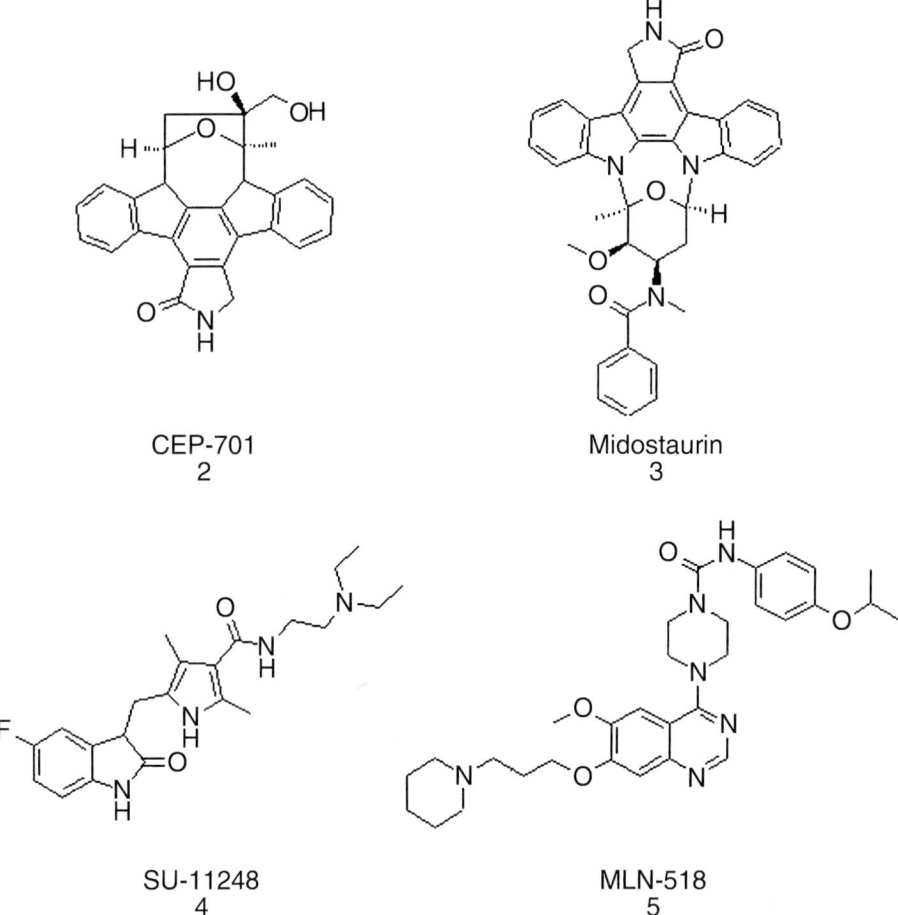

CEP-701
2

Midostaurin
3

SU-11248
4

MLN-518
5

Fig. 2. Flt-3 kinase inhibitors.

of Flt3 inhibitors to mice suffering from MDS resulted in an increase in overall survival *(35,37,40,41)*. Subsequent clinical trials, albeit small, have suggested that treatment of patients, selected for the presence of the mutation, can result in disease remission *(41–43)*.

5. EPIDERMAL GROWTH FACTOR RECEPTOR

The epidermal growth factor receptor (EGFR) family of receptors consists of four structurally related receptors, for which a variety of different ligands have been characterized *(44–47)*. Ligand binding has been shown to induce receptor hetero- and homodimerization. ErbB2/Her2 is the preferred partner for all

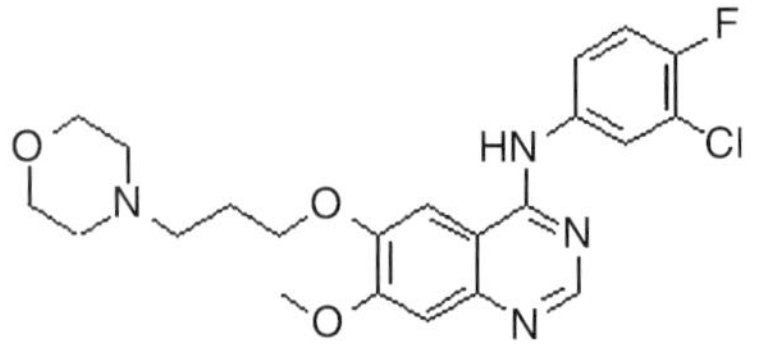

Gefitinib
6

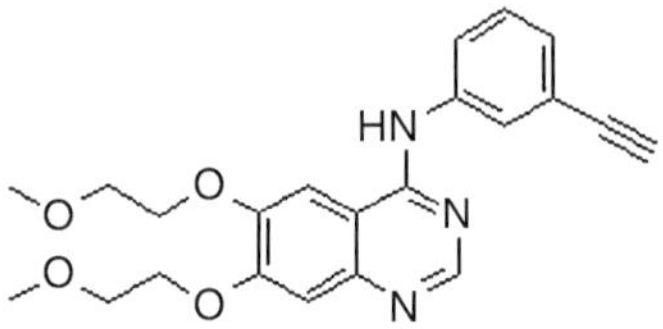

Erlotinib
7

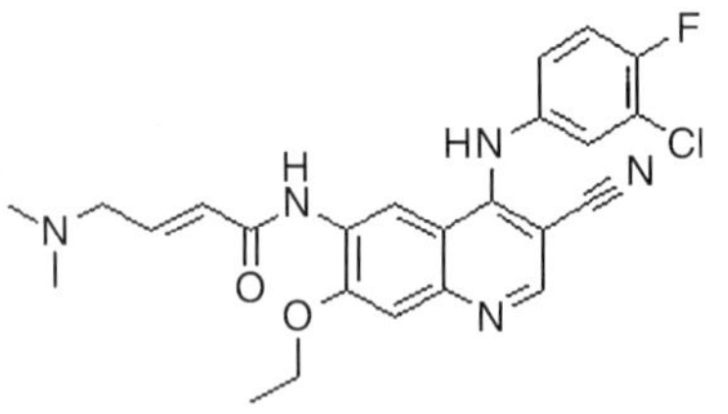

EKB-569
8

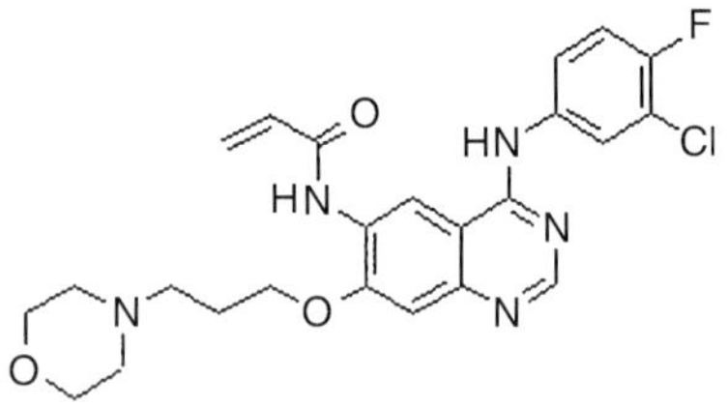

Canertinib
9

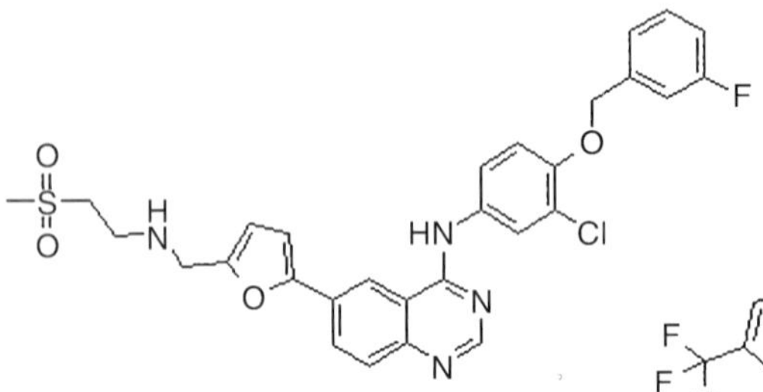

Lapatinib
10

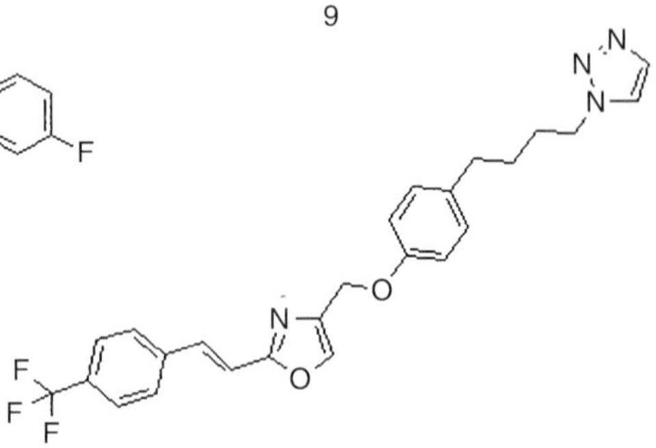

Mubritinib
11

Fig. 3. Representative examples of kinase inhibitors of the EGFR system.

other receptors, however the dimerization partner also depends on both ligand and cell type. Receptor activation leads to phosphorylation of the intracellular cytoplasmic domain and recruitment of docking proteins that initiate downstream signaling cascades such as the PI3K/PKB and MAPK pathways *(46)*. Overexpression of EGFR family members, particularly EGFR or Her2, is frequently observed in solid tumors *(48)*. In addition, EGFRvIII variants, bearing a mutation in the extracellular domain that leads to constitutive kinase activation, have been observed in glioblastoma and non-small-cell lung cancer (NSCLC) *(49,50)*. EGFR mutation or high levels of expression are correlated with poor prognosis and more aggressive disease *(47)*. The frequent involvement in

Table 3
Antibodies Currently In Clinical Development

Agent phase	Company	Structure	Target	Clinical
IMC-225 (Erbitux™, cetuximab	ImClone/ Merck KgaA/Bristol-Meyers Squibb	Chimeric Monoclonal	EGFR	Phase II–III
ABX-EGF	Abgenix Inc./ Amgen Inc.	Humanized monoclonal	EGFR	Phase II
h-R3	York Medical Bioscience Inc.	Monoclonal	EGFR	Phase II
MDX-447	Medarex/Merck KgaA	Humanized Bivalent Monoclonal	EGFR	Phase I–II
Trastuzumab (Herceptin™)	Genentech Inc.	Monoclonal	HER-2	Registered
MDX-210	Medarex/Novartis	Humanized Bivalent Monoclonal	HER-2	Phase I
Bevacizumab (Avastin™)	Genentech Inc.	Humanized monoclonal	VEGF	Registered Phase II–III
IMC-1C11	ImClone	Chimeric Monoclonal	VEGFR2	Phase I
CDP-860	Celltech Group/ Zymogenetics	Humanized Monoclonal	PDGFRβ	Phase I–II

cancer progression has spurred intense efforts into identifying inhibitors directed against EGFR/Her2, as a consequence of which there are a large number of compounds currently in clinical trials. These include gefitinib (6, Fig. 3; iressa™, AstraZeneca PLC), erlotinib (7, Fig. 3; tarceva™, OSI Pharmaceuticals Inc./Genentech Inc./Roche Holdings AG), EKB-569 (8, Fig. 3; Howard Hughes Medical Institute/Wyeth Research), canertinib (9, Fig. 3; CI-1033, Pfizer Inc.), lapatinib (10, Fig. 3; GW-572016, GlaxoSmithKline PLC), and mubritinib (11, Fig. 3; TAK165, Takeda Inc.) (51). In addition, there are at least five antibodies directed against EGFR family members being tested in the clinic; see Table 3. Trastuzumab (Herceptin™, Genentech Inc./Roche AG), a recombinant monoclonal antibody directed against the ErbB2, received FDA approval after it was shown to improve the survival and delay disease progression in trials of patients with metastatic breast tumors that overexpress Her-2 (52). Cetuximab/ Erbitux/C225 (ImClone Systems Inc./Merck KgaA/Bristol-Myers Squibb) is a humanized antibody that recognizes the EGFR (53). The results from a phase II clinical study led to an approval by the FDA for the use of C225 in combination with irinotecan in the treatment of patients with EGFR-expressing, metastatic colorectal cancer who are refractory or intolerant to chemotherapy. ABX-EGF (Abgenix Inc./Amgen Inc.), in contrast, represents a fully humanized mAb, designed to minimize host immune responses directed against the therapeutic antibody. Directed against the EGFR, ABX-EGF inhibited the growth of xenografts and has displayed some activity in phase II clinical trials with renal carcinoma patients (54–56). Medarex has developed two bispecific antibodies comprising two humanized Fab domains, one of which recognizes EGFR (MDX-447) or Her2 (MDX-H210), the other specific for CD64, thereby directing cytotoxic effector cells to target cells overexpressing EGFR or Her2 respectively. MDX-447 has entered phase II clinical trials for the treatment of head and neck cancer that overexpress EGFR (51,57,58).

6. SMALL-MOLECULE INHIBITORS OF EGFR AND Her2

The two EGFR inhibitors that have progressed furthest in clinical trials are gefitinib (Iressa™, an EGFR inhibitor) and erlotinib (Tarceva™, a dual EGFR/HER2 inhibitor). Preclinical studies demonstrated activity against a wide spectrum of cell lines and both antitumor and antiangiogenic effects were observed in implanted tumors (59–62). Activity hints were observed in several tumor types expected to express high levels of EGFR, including colorectal cancer, NSCLC, and head and neck squamous cell carcinoma (HNSCC), in phase I clinical trials (63–65). Analysis of EGFR phosphorylation and activation of downstream signaling pathways in a surrogate tissue, demonstrated that inhibition of the target could be achieved at doses below the maximally tolerated dose (66,67). Single-agent phase II trials of both gefitinib and erlotinib have been carried out in patients with chemotherapy-refractory NSCLC. Similar response rates, approx 10%, were observed for both inhibitors, with higher response rates being observed in women and patients with adenocarcinoma and lower response rates in smokers (68). A particularly high response rate in a Japanese study led to approval in 2002 for the treatment of NSCLC patients with gefitinib; this was followed by approval by the American and Australian regulatory authorities for the treatment of NSCLC patients previously treated with chemotherapy. The most severe drug-related adverse event, a potentially fatal interstitial lung disease observed in less than 1% of patients, is currently under investigation. Phase III trials for both drugs as single agents are currently under way.

The success of combination therapy with trastuzumab and C225 suggests that small-molecule inhibitors directed against the EGFR or Her2 could also enhance the cytotoxic effects of conventional chemotherapeutic drugs. In agreement with this, additive or synergistic inhibition of the growth of animal tumor models was observed on combination of either gefitinib or erlotinib with irradiation or chemotherapeutic drugs (60). However, separate trials on NSCLC patients, which combined gefitinib or erlotinib with chemotherapy, failed to show any improvement in response, when compared to treatment with chemotherapy alone (62,69). Although various mechanistic reasons have been cited for the failure to observe a response in these trials, one of the central problems is the lack of a clear definition of which patients would respond to anti-EGFR therapy. Posttreatment studies failed to correlate expression of target and response to the drug, suggesting additional factors need to be considered. If these factors are not taken into consideration, clinical trials run the risk of being underpowered as subpopulations of patients that do benefit from treatment are overlooked.

Alternative strategies have been adopted in designing inhibitors directed against the EGFR family. EKB569 and canetinib are two irreversible inhibitors that are currently in clinical trials (70–72). Inhibiting all EGFR family members, canetinib displayed some activity, partial response and stable disease, in phase I trials and is now in trials of patients with refractory ovarian cancer. In addition to common EGFR inhibitor toxicities, such as rash and diarrhea, hypersensitivity and thrombocytopenia were observed in patients receiving canetinib (73,74). Presently, it is unclear if these effects are because of inhibition of all EGFR family members or an off-target effect.

The plethora of ligands and the ability of EGFR family members to both trans- and autophosphorylate obscures the role of an individual receptor in tumor growth. Recent studies have demonstrated that compounds displaying specificity against one kinase in vitro may also block the activity of heterodimerisation partners in the cellular context (74). As these compounds progress through the clinic, the relative advantages of inhibiting one, two, or all EGFR family members against possible disadvantages, e.g., toxicity and off-target effects, will become apparent.

7. VASCULAR ENDOTHELIAL GROWTH FACTOR RECEPTOR

Angiogenesis is the process by which capillaries sprout from established blood vessels. The vascular endothelial growth factor (VEGF) and its cognate receptors, VEGFRs, play an essential role in this process (75). As tumors grow beyond a certain size, simple diffusion of nutrients and oxygen becomes insufficient, necessitating the *de novo* establishment of a blood supply (76,77). Inhibition of tumor angiogenesis, by blocking the action of VEGF, would therefore be predicted to starve the tumors to death. It has been suggested that VEGF inhibitors would be most effective in a minimal disease state; however the ability of VEGF, also known as vascular permeability factor, to regulate vessel permeability suggests that its inhibition could also decrease the ascitic fluid formation and edema observed in tumors with established vascularization. Confirmation

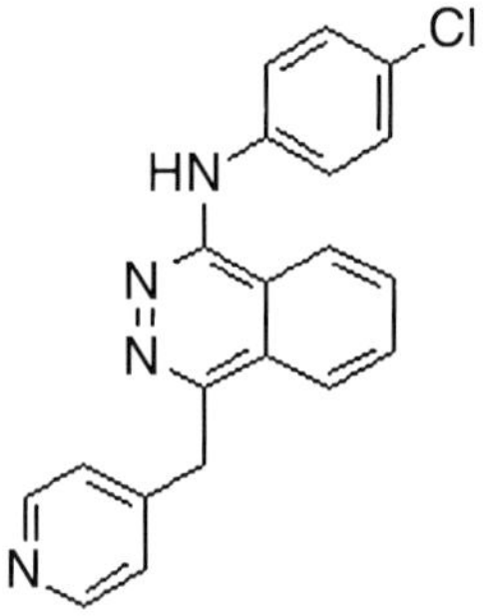

Vatalanib
12

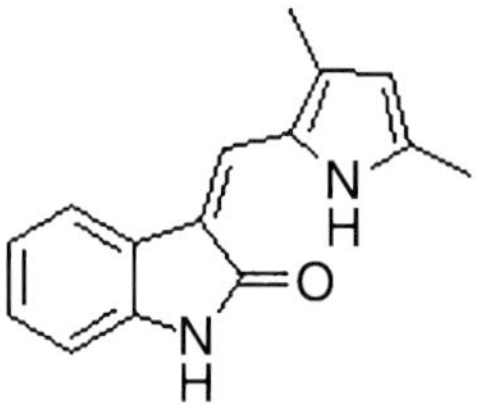

Semaxanib
13

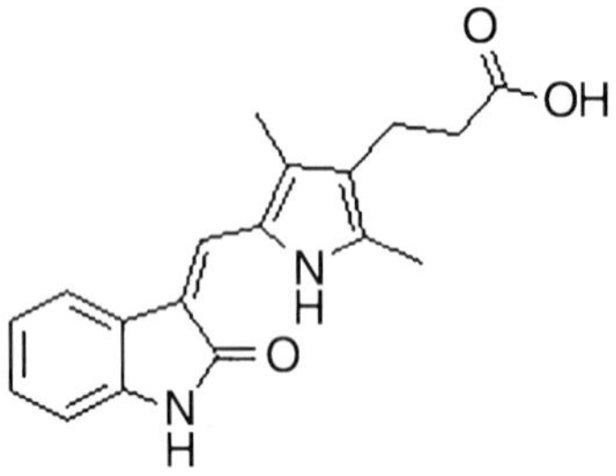

SU-6668
14

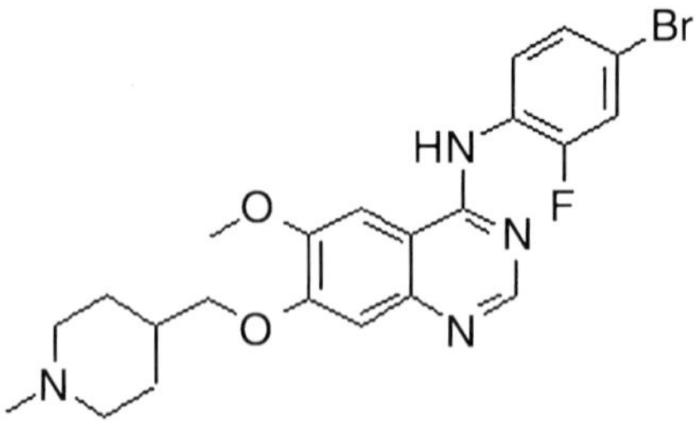

ZD-6474
15

Fig. 4. Representative examples of kinase inhibitors of the VEGFR system.

of this hypothesis has been provided by dynamic, contrast-enhanced molecular resonance imaging (DCE-MRI) studies of tumors treated with two VEGFR inhibitors vatalanib (12, Fig. 4; PTK787/ZK222584, Novartis Pharma AG/Schering AG) or bevacizumab (Avastin™; Genentech Inc./Roche AG).

Bevacizumab, a monoclonal antibody that inhibits interaction of VEGF with its cognate receptors, prolonged time to progression of disease in patients with metastatic renal cancer *(78)*, metastatic breast cancer *(79)*, and NSCLC in phase I and II clinical trials *(80)*. It has recently been approved by the FDA as a first-line treatment, in combination with 5-FU/leucovorin/CPT-11 therapy (Salz regimen), for treatment of patients with metastatic colorectal cancer *(81,82)*. Antibodies directed against the VEGFR2 have also been developed. The chimeric MAb IMC-1C11 (ImClone Systems Inc.) inhibited the proliferation of leukemia cells and prolonged the survival of xenotransplanted mice in preclinical studies. Induction of stable disease in some patients with colorectal cancer was observed in phase I clinical trials *(83)*.

In an alternative strategy, angiozyme, a stabilized ribozyme that targets the pre-mRNA of the VEGF-R1 has been tested in phase I trials in a population of healthy volunteers. No toxicities were observed, leading to trials in which angiozyme has been administered, in combination with the Salz regimen, in patients with metastatic colorectal cancer. Preliminary reports were promising, as ribozyme-treated patients, who displayed a reduction in VEGF-R1 levels, fared better in terms of stable disease and progression-free survival than those treated with chemotherapy alone *(84,85)*. Although this drug should be extremely specific, the expense of manufacturing and the inconvenience of the necessary daily subcutaneous injections render small-molecular-mass inhibitors of the VEGFR an attractive alternative.

One of the initial, and probably the most extensively studied KDR inhibitor, semaxanib (13, Fig. 4; SU5416, SUGEN Inc.), demonstrated promising anti-angiogenic and anti-tumor effects in animal models, however poor solubility and the lack of any survival advantage when administered in combination with chemotherapy led to the drug being discontinued *(86,87)*. SU-6668 (14, Fig. 4; SUGEN Inc.), which possesses a similar selectivity profile to SU5416 but greater solubility, also entered clinical trials, but was discontinued for the treatment of solid tumors as adverse events, including severe pain and serositis, necessitated reducing the dosing regimen below that predicted to be required to achieve pharmacologically active levels *(88)*. More recently, SUGEN Inc./Pfizer have introduced SU11248 (4; Fig. 2), based on an indolinone backbone, that is orally bioavailable and has a wider spectrum of activity than either SU5416 or SU-6668 *(89)*. Regression of wide-spectrum tumor xenografts was accompanied by long-term inhibition of both VEGFR and PDGFR phosphorylation in tumor tissue. Anti-tumor activity was accompanied by a reduction in microvessel density, inhibition of vascular permeability, and significant apoptosis. Phase I trials against patients with solid tumors and AML (*see* Section 3) indicated the drug was tolerated and displayed sufficient efficacy to support additional trials *(90)*.

Vatalanib (PTK787/ZK222584) (Novartis Pharma AG/Schering AG) is a phthalazine derivative displaying relative specificity for VEGFR-1, -2, and -3, but also inhibiting PDGFR-β, c-kit, and CSF-1R at higher concentrations. Preclinical studies demonstrated that vatalanib inhibited vascularization of ovarian and lung carcinoma as well as affecting the formation of ascites and the volume of the pleural effusate *(91–93)*. In phase I clinical studies, the compound was well tolerated and delayed disease progression. Similar evidence of efficacy was observed in combination with cytotoxic agents in patients with colorectal carcinoma and glioblastoma *(94)*. Treatment was associated with elevation of VEGF serum levels in patients. Tumor hypoxia, as a consequence of anti-angiogenesis therapy, would lead to stabilization of the transcription factor HIF-1α and thereby its target genes, which include VEGF.

The 4-anilino-quinazoline scaffold, utilized for the design of EGFR inhibitors, has provided a series of selective VEGFR2 inhibitors, such as ZD6474 (15, Fig. 4; AstraZeneca PLC) *(95)*. ZD6474 displayed promising anti-angiogenic and anti-tumor effects in preclinical animal studies and an acceptable safety profile in phase I studies leading to its promotion to phase II trials for NSCLC, SCLC, and myeloma *(96,97)*.

8. NONCANCER VEGF-TARGETED THERAPIES

Abnormal growth of blood vessels occurs in pathological conditions other than cancer. Age-related macular degeneration and diabetic retinopathy, the most significant causes of blindness in the developed world, are characterized by the growth of leaky, fragile blood vessels *(98)*. Current therapy to shrink these vessels, including laser surgery and photodynamic therapy, provides only temporary results. VEGF has been demonstrated to be a key regulator of ocular neovascularization and current trials are addressing the effects of anti-VEGF therapy in patients with age-related macular degeneration. Pegaptanib (Pfizer/Eyetech), an anti-VEGF pegylated aptamer, and Lucentis/Ranibizumab (Genetech Inc./Roche AG), an anti-VEGF humanized antibody, have both shown promising results in the clinic *(99)*. Neovascularization also accompanies chronic inflammation in conditions such as psoriasis, rheumatoid arthritis, and atherosclerosis *(100)*. Neovastat (Aeterna Laboratories), a naturally occurring antiangiogenic product that inhibits both VEGF signaling and matrix metalloproteinase activation, has shown clinical activity in phase I/II trials of psoriasis patients *(101)*. Recently, vatalanib has been demonstrated to have activity in several preclinical models of rheumatoid arthritis suggesting inhibition of VEGF signaling may also have beneficial effects in patients with this condition *(102)*.

9. NOVEL TYROSINE KINASE TARGETS

As described in the introduction, more than 20 different kinases are currently under evaluation as potential targets for the treatment of malignant disease. Subsequent to the advances in treatment of solid tumors with inhibitors directed against VEGFRs, other kinases involved in angiogenic signaling are being considered, including Tie-2 and fibroblast growth factor receptors (FGFRs). A large amount of data supports the involvement of FGFR family members in cancer *(103)*. Overexpression and amplification of FGFR1 has been observed in solid tumors and activating point mutations as well as translocations have been documented in multiple myeloma (MM) and bladder cancer. The activity of some of the VEGFR inhibitors against the FGFR has led to clinical trials assessing their effects in the context of cancers driven by the latter receptor.

Recently, a series of papers have been published describing the action of small-mol-wt inhibitors on three other novel tyrosine kinase targets, insulin-like growth factor I receptor (IGF-IR), c-Met, and JAK family members.

10. IGF-IR

IGF-IR activation stimulates proliferation and protection from apoptosis *(104,105)*. Overexpression is transforming, whereas genetic ablation of IGF-IR protects fibroblasts from transformation by most oncogenes *(106)*. Although activating mutations have not been documented, overexpression of both IGF-IR and its ligands occurs in several tumor types and is associated with a poorer prognosis *(107–109)*. Ablation of IGF-IR function by antisense treatment, inhibitory antibodies, or dominant negative mutants blocked growth, metastasis, and invasion of implanted tumors, providing a proof of concept that IGF-IR represents a valid cancer target *(110–114)*. The most preclinically advanced of these agents is A12 (Imclone Systems Inc.), a fully human antibody that prevents IGF-IR interacting with its ligand. Administration of A12 was shown to inhibit the growth of breast, renal, and pancreatic implanted tumors *(115)*.

The IGF-IR and insulin receptor are closely related, being 100% homologous within the ATP binding site. Selectivity over the insulin receptor is a critical requirement for any IGF-IR inhibitor to avoid disturbing glucose homeostasis. Two recent papers detail the identification of selective IGF-IR inhibitors *(116,117)*. NVP-AEW541 (Novartis Pharma AG) was obtained following a structure-based design approach. Although equipotent inhibition of the InsR and IGF-IR kinase domains was observed in vitro, NVP-AEW541 was 27-fold more potent toward the IGF-IR in cell-based assays. It is proposed that differences in the conformation of the full length and purified kinase domains underlie this specificity. Inhibition of IGF-IR phosphorylation by NVP-AEW541 induced apoptosis and prevented growth of an IGF-IR-transformed implanted fibrosarcoma. Long-term treatment with the compound was reported to not alter glucose/insulin homeostasis. A closely related compound, NVP-ADW742 (16, Fig. 5), inhibited growth of a wide spectrum of tumor cell lines, all of which express IGF-IR, although no direct correlation between levels of receptor expression and sensitivity to the drug was recorded. Inhibition of both freshly isolated and established MM cell lines was observed and treatment prolonged the survival of mice implanted with an MM cell line. Inhibition of the receptor resulted in sensitization to chemotherapeutic agents, indicating that such an approach could also be adopted in the clinic. IGF-IR is widely expressed in normal tissues and identification of those tumors that would respond best to its inhibition is an essential prerequisite for any clinical trials. A common genetic expression profile, observed on treatment with several different IGF-IR-blocking agents, may provide a set of markers that could be used for this purpose.

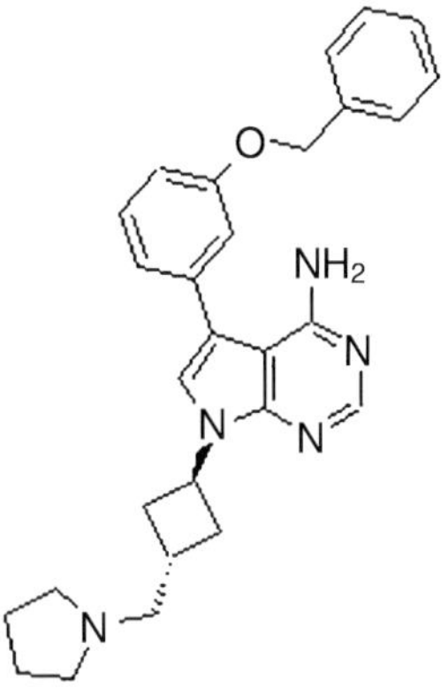

NVP-ADW742
16

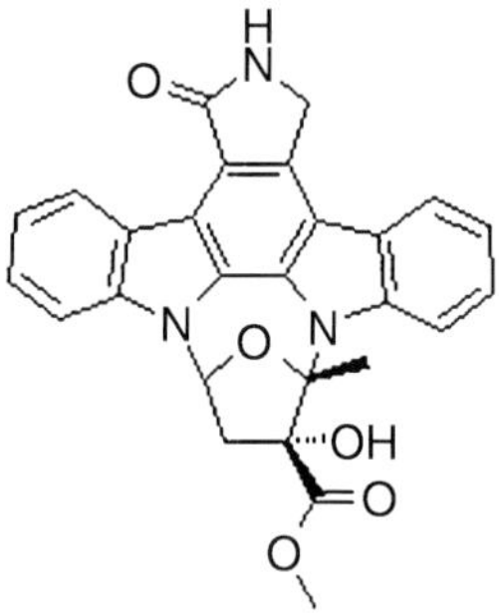

K-252a
17

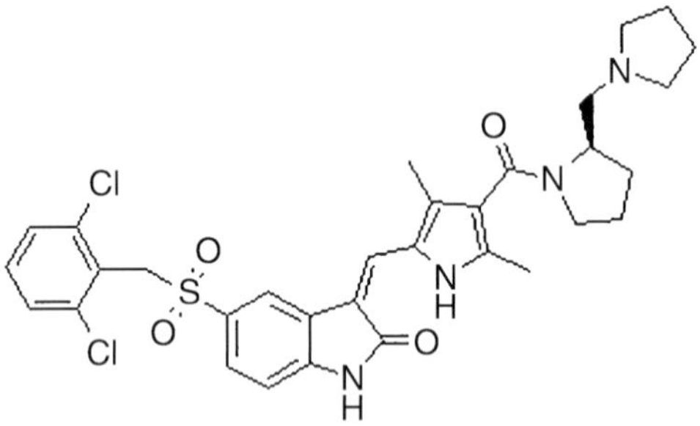

PHA-665752
18

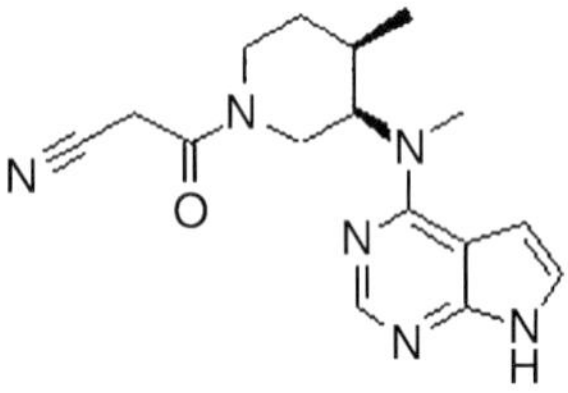

CP-690.440
19

Fig. 5. Kinase inhibitors of IGF-IR (16), c-Met (17-18), and JAK-3 (19).

11. c-MET

c-Met is considered to be an attractive target for cancer therapy owing to its role in migration, angiogenesis, epithelial-mesenchymal transition, invasion, and metastasis *(118,119)*. Aberrant expression of both c-Met and its cognate ligand hepatocyte growth factor/scatter factor (HGF/SF), is observed in many tumors, and may result in unregulated activation of the receptor *(120)*. Amplification and mutation of c-Met has been observed in familial and sporadic renal carcinoma *(121)*, as well as in metastases of HNSCC *(122)* and gastric carcinoma *(123)*. Overexpression of c-Met is associated with more aggressive, invasive disease and a poorer patient prognosis *(124)*. Experimental proof of concept for a role of c-Met in cancer progression has been provided by the use of c-Met-targeting antisense ribozymes *(125–128)*, dominant negative receptor mutants *(129)*, and naturally occurring inhibitory mutants of HGF/SF *(130,131)*.

Recently, K252a (17, Fig. 5), a relatively nonspecific staurosporine derivative, has been demonstrated to inhibit c-Met phosphorylation in cellular assays *(132)*. Subsequently, several publications have presented preclinical data on compounds, based on the pyrrole indolinone backbone, that have activity against c-Met *(133–135)*. Of these, the most advanced, PHA-665752 (18, Fig. 5), was shown to potently inhibit c-Met in cell-based autophosphorylation assays. PHA-665752 was 21-fold more potent against c-Met than the closely related receptor c-Ron and failed to inhibit most other kinases, with the exception of Flk-1. Despite the role of c-Met in normal tissue morphogenesis and development, PHA-665752 was reported to be well tolerated in mice. A dose-dependent inhibition of c-Met phosphorylation in implanted GTL-16 gastric carcinomas was accompanied by a block in tumor growth.

12. JAK FAMILY MEMBERS

Janus or just another kinase (JAK) is an intracellular kinase that plays a fundamental role in cytokine signaling *(136,137)*. Four human homologs exist— JAK-1, JAK-2, JAK-3, and TYK-2. Once activated, JAKs phosphorylate transcription factors of the STAT (signal transducers and activators of transcription). JAK-1 or JAK-2 are essential for developmental viability, whereas JAK-3 mutants display defects in T-cell and natural-killer (NK) cell development *(138,139)*. JAK-3 mutation in human results in a severe combined immunodeficiency (SCID) *(140–142)*. JAK-3 is activated on stimulation with pro-inflammatory cytokines and mutation of the common receptor subunit required for signaling by these cytokines also results in an SCID phenotype *(143)*. These data, in combination with the more restricted expression of JAK-3, have prompted the development of JAK-3 inhibitors as immunosuppressants for use in transplant patients. Several inhibitors that specifically target JAK-3 have been developed. Recently it has been reported that CP-690,440 (19, Fig. 5; Pfizer Inc.) was able to prevent allograft rejection in a mouse heart and monkey kidney transplantation models. Comparing favorably with clinically used immunosuppressants, CP-690,440 specifically inhibited JAK-3 activation by the pro-inflammatory cytokine IL-2, but did not alter T-cell receptor signaling or fibroblast proliferation *(144)*. WHI-P131/Janex-1(Hughes Institute) was also demonstrated to block JAK-3 activation in vivo and showed activity in a graft-vs-host-disease murine transplantation model *(145)*.

A role in cancer progression has also been identified for JAKs. Translocations involving JAK-2 have been identified in human leukemia. These mutations preserve the catalytically active JAK domain, resulting in its constitutive activation. Tel-JAK2 fusion proteins have been documented in acute lymphoblastoid leukemia and several cases of myeloid malignancy *(146)*. In addition, constitutive JAK activation is observed in leukemias containing other kinase fusion proteins such as Bcr-Abl and NPM-Alk *(147,148)*. Promoter methylation of negative regulators of JAKs, such as SOCS, has also been documented in both liquid and solid tumors in which JAK activation is observed *(149,150)*. Several JAK inhibitors, with varying specificities among family members, have been tested against preclinical cancer models. AG-490, a fairly nonspecific benzylidene derivative, has been shown to be able to inhibit the proliferation of leukemic cells containing all three of the kinase fusion proteins described above. Activity of AG-490, as well as Janex-1, has also been documented in murine pancreatic, HNSCC, and hepatocellular tumor models *(150–155)*. Although reported to be generally well tolerated, the essential roles of JAK-1 and JAK-2 in haematopoietic homeostasis demand thorough analysis of the safety profiles of such compounds before they enter the clinic.

13. CLINICAL QUESTIONS

13.1. Target Selection

The mixed experience with tyrosine kinase inhibitors as targeted therapies has demonstrated that a clear understanding of the role a potential target plays in disease progression is equally as important as the efficient optimization of a compounds potency, toxicity and pharmacokinetic profile. In the case of a single activating mutation, which is both necessary and sufficient to induce the disease state, as is the case for Bcr-Abl in chronic phase CML, treatment with a compound that specifically inhibits the kinase, imatinib, is therapeutically successful. The presence of activating mutations has been used as a rationale in oncology to determine potential kinase and disease targets. Genetic screening of families has linked germline mutations in kinases to a predisposition to cancer, e.g., the association of familial papillary renal cell carcinoma with mutations in c-Met or multiple endocrine neoplasia type 2 in families with c-Ret mutations *(156)*. The presence of receptor mutations in sporadic versions of the same cancer further reinforces the idea that they have a fundamental role in progression of the disease. Correlation of a particular mutation with the metastasis/aggressiveness of a type of cancer has been used to further validate a candidate kinase, e.g., EGFRvIII mutants. Currently, however, single kinase-activating point mutations or translocations have not been characterized in the majority of solid tumors, although a genomics-sequencing approach that has identified an unexpectedly high frequency of c-raf mutations in melanoma suggests that other cryptic mutations remain to be discovered *(157)*. Overexpression of the receptor tyrosine kinase, or its ligand, has therefore also been used as a rationale to design inhibitors, e.g., gefitinib and erlotinib. It is notable that whereas single-agent trials with these inhibitors have demonstrated remarkable activity against different tumors, the percentage of patients that responded was commonly low, 10% or less, despite activity against a wide spectrum of tumors in preclinical studies. The underlying reasons for this discrepancy are complex but are probably because of inappropriate selection of a patient population owing to a lack of understanding of the biology of the target rather than a reflection of the quality of the target *per se*. Unlike the correlation between response to imatinib and expression of Bcr-Abl in CML or activated c-kit in GIST, no convincing relationship between expression of the EGFR and activity of the drugs was observed. This suggests that additional factors, other than solely expression levels, lead to the tumor becoming dependent on signals emanating from the receptor. The advent of validated antibodies that specifically recognize phosphorylated, and therefore activated, receptors has reinforced this point. A study of HER-2 overexpressing breast carcinomas demonstrated wide variations (between 12 and 58%) in the number of tumors that contained phosphorylated receptor *(158,159)*. Even the detection of phosphorylated receptor may not be sufficient, as cellular assays have demonstrated that modulation of downstream signaling pathways can also affect sensitivity to EGFR inhibitors *(160)*. The use of phosphoproteomic and gene expression profiling analysis of well-annotated tumor banks, such as the one being established from NSCLC patients in a large Canadian trial of gefitinib, will hopefully provide some indications regarding patient selection for anti-EGFR therapy.

13.2. Emergence of Resistance Mutations

Successful therapeutic inhibition of any tyrosine kinase activity runs the risk of the emergence of mutations that circumvent the action of the inhibitor. Several nonexclusive mechanisms can be envisaged by which this could be achieved: (a) amplification of the target, (b) mutation of the target to prevent action of the inhibitor, (c) activation of a complementary pathway that bypasses requirement for the target, and (d) upregulation of drug efflux pumps to lower the intracellular concentrations of the inhibitor. Analysis of CML patients, who have relapsed on treatment with imatinib, has allowed the characterization of a series of point mutations that, when experimentally recapitulated, result in resistance to the drug *(161)*. The isolation of an analogous mutation in FipIL1-PDGFR from an HES patient, who relapsed after treatment with imatinib, indicates the appearance of resistance mutations is not a peculiarity of Bcr-Abl or CML patients *(18)*. Identical mutations were isolated from chronic-phase CML patients in remission after treatment with imatinib, but who eventually underwent relapse, and in two untreated blast-phase CML patients, who subsequently failed to respond to imatinib, suggesting that these mutations could explain primary resistance to the drug *(162,163)*. Structural studies indicated that Abl binds to imatinib in an unusual inactive conformation where the "activation loop," which regulates kinase activity, adopts a self-inhibitory conformation by imitating bound substrate *(164,165)*. These studies also demonstrated that imatinib extended much further into the catalytic site, making contact with residues that account for the specificity profile of the inhibitor. Mapping the drug-resistant point mutations suggests that whereas some appear to mutate residues that make direct contact with imatinib, e.g., T315-I315, others convey resistance by sterically hindering the ability of the kinase to achieve an inactive conformation *(163)*. Experimental mutagenesis of the Abl kinase has identified a series of point mutations that confer imatinib resistance, including all of the mutations that have been detected in patients, indicating that it is possible to predict which mutations will arise *(166)*. A similar approach is currently under way with other kinase inhibitors. Currently, attention is being focused on design of inhibitors that can overcome kinases bearing these resistance mutations. Interestingly, subsequent studies indicated that the imatinib resistant mutant of FipIL1-PDGFR was more sensitive to midostaurin than "wild-type" FipIL1-PDGFR, suggesting that inhibitors that specifically target the active kinase conformation could be used to therapeutically treat such point mutations *(167)*.

14. CONCLUSIONS AND PERSPECTIVES

Advances have been made in both identifying the role of tyrosine kinases in disease and characterizing inhibitors that can block their activity in a therapeutically relevant manner.

It is clear that the multiple steps that underlie the development of a disease such as cancer will result in increasing subfractionation of patient populations based on the presence of mutations in either tumor suppressors or oncogenes. Such stratification, in addition to allowing tailor-made therapies, will both reduce the numbers of patients that would benefit from administration of a particular kinase inhibitor and require the administration of cocktails of drugs directed against multiple targets. Both events will inevitably be associated with an increase in the cost of patient health care. The costs can only be justified by ensuring that a compound is administered in an appropriate manner. This requires a complete understanding of the pharmacokinetic, toxicity, specificity, and potency of a kinase inhibitor. Appropriate preclinical disease models and detailed analysis of the reasons underlying the failures, or poor responses, of certain inhibitors in clinical trials should be used to aid the development of more successful ones in the future.

REFERENCES

1. Blume-Jensen P, Hunter T. Oncogenic kinase signalling [Review]. *Nature* 2001; 411: 355–365.
2. Noble ME, Endicott JA, Johnson LN. Protein kinase inhibitors: insights into drug design from structure. *Science* 2004; 303:1800–1805.
3. Krystal–Imatinib mesylate (STI571) for myeloid malignancies other than CML. *Leuk Res* 2004; 28(Suppl 1):53–59.
4. Druker BJ, Sawyers CL, Kantarjian HR, et al. Activity of a specific inhibitor of the BCR-ABL tyrosine kinase in the blast crisis of chronic myeloid leukemia and acute lymphoblastic leukemia with the Philadelphia chromosome. *N Engl J Med* 2001; 344:1038–1042.

5. Heinrich MC, Griffith DJ, Druker BJ, et al. Inhibition of c-kit receptor tyrosine kinase activity by STI 571, a selective tyrosine kinase inhibitor. *Blood* 2000; 96:925–932.

6. Daley GQ, Van Etten RA, Baltimore D. Induction of chronic myelogenous leukemia in mice by the P210bcr/abl gene of the Philadelphia chromosome. *Science* 1990; 247:824–830.

7. Savage DG, and Antman KH. Imatinib mesylate; a new oral targeted therapy. *N Engl J Med* 2002; 346:683–693.

8. Griffin JD. Resistance to targeted therapy in leukaemia. *Lancet* 2002; 359:458–459.

9. Gorre ME, Mohammed M, Ellwood K, et al. Clinical resistance to STI-571 cancer therapy caused by BCR-ABL gene mutation or amplification. *Science* 2001; 293:876–880.

10. Gorre ME, Sawyers CL. Molecular mechanisms of resistance to STI571 in chronic myeloid leukemia. *Curr Opin Hematol* 2002; 9:303–307.

11. Demetri GD, von Mehren M, Blanke CD, et al. Efficacy and safety of imatinib mesylate in advanced gastrointestinal stromal tumors. *N Engl J Med* 2002; 347:472–480.

12. van Oosterom AJ, Judson I, Verweij J, et al. Safety and efficacy of imatinib (STI571) in metastatic gastrointestinal stromal tumours: a phase I study. *Lancet* 2001; 358:1421–1423.

13. Heinrich MC, Corless CL, Demetri GD, et al. Kinase mutations and imatinib response in patients with metastatic gastrointestinal stromal tumor. [Report]. *J Clin Oncol* 2003; 21:4342–4349.

14. Heinrich MC, Corless CL, Duensing A, et al. PDGFRA-activating mutations in gastrointestinal stromal tumors. *Science* 2003; 299:708–710.

15. Rubin BP, Schuetze SM, Eary JF, et al. Molecular targeting of platelet-derived growth factor B by imatinib mesylate in a patient with metastatic dermatofibrosarcoma protuberans [*see* comment]. *J Clin Oncol* 2002; 20:3586–3591.

16. Sawyers CL. Imatinib GIST keeps finding new indications: successful treatment of dermatofibrosarcoma protuberans by targeted inhibition of the platelet-derived growth factor receptor [comment]. *J Clin Oncol* 2002; 20:3568–3569.

17. Apperley JF, Gardembas M, Melo JV, et al. Response to imatinib mesylate in patients with chronic myeloproliferative diseases with rearrangements of the platelet-derived growth factor receptor beta. *N Engl J Med* 2002; 347:481–487.

18. Cools J, DeAngelo DJ, Gotlib J, et al. A tyrosine kinase created by fusion of the PDGFRA and FIP1L1 genes as a therapeutic target of imatinib in idiopathic hypereosinophilic syndrome [*see* comment]. *N Engl J Med* 2003; 348:1201–1214.

19. Kemmer K, Corless CL, Fletcher JA, et al. KIT mutations are common in testicular seminomas. *Am J Pathol* 2004; 164:305–313.

20. Pardanani A, Elliott M, Reeder T. Imatinib for systemic mast-cell disease. *Lancet* 2003; 362:535–536.

21. Tefferi A, Mesa RA, Gray LA, et al. Phase 2 trial of imatinib mesylate in myelofibrosis with myeloid metaplasia. *Blood* 2002; 99:3854–3856.

22. Pietras K, Ostman A, Sjoquist M, et al. Inhibition of platelet-derived growth factor receptors reduces interstitial hypertension and increases transcapillary transport in tumors. *Cancer Res* 1902; 61:2929–2934.

23. Myllarniemi M, Frosen J, Calderon Ramirez LG, et al. Selective tyrosine kinase inhibitor for the platelet-derived growth factor receptor in vitro inhibits smooth muscle cell proliferation after reinjury of arterial intima in vivo. *Cardiovasc Drugs Ther* 1999; 13: 159–168.

24. Savikko J, Taskinen E, Von Willebrand E. Chronic allograft nephropathy is prevented by inhibition of platelet-derived growth factor receptor: tyrosine kinase inhibitors as a potential therapy. *Transplantation* 2003; 75:1147–1153.

25. Lassila M, Allen TJ, Cao, Z, et al. Imatinib attenuates diabetes-associated atherosclerosis. *Arterioscler Thromb Vasc Biol* 2004; 24:935–942.

26. Sihvola RK, Tikkanen JM, Krebs R, et al. Platelet-derived growth factor receptor inhibition reduces allograft arteriosclerosis of heart and aorta in cholesterol-fed rabbits. *Transplantation* 2003; 75:334–339.

27. Ahuja H, Bar-Eli M, Advani SH, Benchimol S, Cline MJ. Alterations in the p53 gene and the clonal evolution of the blast crisis of chronic myelocytic leukemia. *Proc Nat Acad Sci US Am* 1989; 86:6783–6787.

28. Foti A, Ahuja HG, Allen SL, et al. Correlation between molecular and clinical events in the evolution of chronic myelocytic leukemia to blast crisis. *Blood* 1991; 77:2441–2444.

29. Sill H, Goldman JM, Cross NC. Homozygous deletions of the p16 tumor-suppressor gene are associated with lymphoid transformation of chronic myeloid leukemia. *Blood* 1995; 85:2013–2016.

30. Kiyoi H, Naoe T, Nakano Y, et al. Prognostic implication of FLT3 and N-RAS gene mutations in acute myeloid leukemia. *Blood* 1999; 93:3074–3080.

31. Nakao M, Yokota S, Iwai T, et al. Internal tandem duplication of the flt3 gene found in acute myeloid leukemia. *Leukemia* 1996; 10:1911–1918.

32. Gilliland DG, Griffin JD. The roles of FLT3 in hematopoiesis and leukemia [Review]. *Blood* 2002; 100:1532–1542.

33. Carnicer MJ, Nomdedeu JF, Lasa A, et al. FLT3 mutations are associated with other molecular lesions in AML. *Leuk Res* 2004; 28:19–23.

34. Sohal J, Phan VT, Chan PV. A model of APL with FLT3 mutation is responsive to retinoic acid and a receptor tyrosine kinase inhibitor, SU11657. *Blood* 2003; 101:3188–3197.

35. Levis M, Allebach J, Tse KF, et al. A FLT3-targeted tyrosine kinase inhibitor is cytotoxic to leukemia cells in vitro and in vivo. *Blood* 2002; 99:3885–3891.

36. Fabbro D, Ruetz S, Bodis S, et al. PKC412—a protein kinase inhibitor with a broad therapeutic potential. *Anticancer Drug Des* 2000; 15:17–28.

37. Weisberg E, Boulton C, Kelly LM, et al. Inhibition of mutant FLT3 receptors in leukemia cells by the small molecule tyrosine kinase inhibitors PKC412. *Cancer Cell* 2002; 1:433–443.

38. Abrams TJ, Lee LB, Murray LJ, Mendel DB, Cherrington JM. Inhibition of Kit-positive SCLC growth by SU11248, a novel tyrosine kinase inhibitors. *93rd Annual Meeting of the AACR Symposium on Molecular Targets and Cancer Therapeutics* 2002.

39. Mendel DB, Laird AD, Xin X, et al. In vivo antitumor and mechanism of action studies of SU11248, a potent and selective inhibito of the VEGF and PDGF receptors. *93rd Annual Meeting of the AACR Symposium on Molecular Targets and Cancer Therapeutics* 2002; 93:Abstr. 5349.

40. Kelly LM, Yu JC, Boulton CL, et al. CT53518, a novel selective FLT3 antagonist for the treatment of acute myelogenous leukemia (AML). *Cancer Cell* 2002; 1:421–432.

41. O'Farrell AM, Abrams TJ, Yuen HA, et al. SU11248 is a novel FLT3 tyrosine kinase inhibitor with potent activity in vitro and in vivo. *Blood* 2003; 101:3597–3605.

42. Smith BD, Levis M, Beran M, et al. Single agent CEP-701, a novel FLT3 inhibitor, shows initial response in patients with refractory acute myeloid leukemia. *Proc Am Soc Clin Oncol* 2003; 194.

43. Smith BD, Levis M, Beran M, et al. Single agent CEP-701, a novel FLT3 inhibitor, shows biologic and clinical activity in patients with relapsed or refractory acute myeloid leukemia. *Blood* 2004; 2003–2011.

44. Mendelsohn J, Baselga J. The EGF receptor family as targets for cancer therapy. *Oncogene* 2000; 19:6550–6565.

45. Velu TJ. Structure, function and transforming potential of the epidermal growth factor receptor [Review]. *Mol Cell Endocrinol* 1990; 70:205–216.

46. Yarden Y, Slimkowski MX. Untangling the erbB signalling network. *Nat Rev Mol Cell Biol* 2001; 2:127–137.

47. Yarden Y. The EGFR family and its lignad in human cancer signalling mechanisms and therapeutic opportunities. *Eur J Cancer* 2001; 37:S3–S8.

48. Salomon DS, Brandt R, Ciardiello F, Normanno N. Epidermal growth factor–related peptides and their receptors in human malignancies. *Crit Rev Oncol Hematol* 1995; 19:183–232.

49. Garcia IE, Adams, GP, Sundareshan P, et al. Expression of mutated epidermal growth factor receptor by non-small cell lung carcinomas. *Cancer Res* 1993; 53:3217–3220.

50. Libermann TA, Nusbaum HR, Razon N, Amplification, enhanced expression and possible rearrangement of EGF receptor gene in primary human brain tumours of glial origin. *Nature* 1985; 313:144–147.

51. Sridhar SS, Seymour L, Shepherd FA. Inhibitors of epidermal-growth-factor receptors: a review of clinical research with a focus on non-small-cell lung cancer [Review]. *Lancet Oncol* 2003; 4:397–406.

52. Hirsch FR, Langer CJ. The role of HER2/neu expression and trastuzumab in non-small cell lung cancer [Review]. *Semin Oncol* 2004; 31:Suppl 82.

53. Baselga J, Pfister D, Cooper MR, et al. Phase I studies of anti-epidermal growth factor receptor chimeric antibody C225 alone and in combination with cisplatin. *J Clin Oncol* 2000; 18:904–914.

54. Foon KA, Yang XD, Weiner LM, et al. Preclinical and clinical evaluations of ABX-EGF, a fully human anti-epidermal growth factor receptor antibody [Review]. *Int J Radiat Oncol Biol Phys* 2004;58:984–990.

55. Lynch DH, Yang XD. Therapeutic potential of ABX-EGF: a fully human anti-epidermal growth factor receptor monoclonal antibody for cancer treatment. *Semin Oncol* 2002; 29:47–50.

56. Yang XD, Jia XC, Corvalan JR, Wang P, Davis CG. Development of ABX-EGF, a fully human anti-EGF receptor monoclonal antibody, for cancer therapy. *Crit Rev Oncol Hematol* 2001; 38:17–23.

57. Repp R, van Ojik HH, Valerius T, et al. Phase I clinical trial of the bispecific antibody MDX-H210 (anti-FcgammaRI × anti-HER-2/neu) in combination with Filgrastim (G-CSF) for treatment of advanced breast cancer. *Br J Cancer* 2003; 89:2234–2243.

58. Wallace PK, Romet-Lemonne JL, Chokri M, Kasper LH, Fanger MW, Fadul CE. Production of macrophage-activated killer cells for targeting of glioblastoma cells with bispecific antibody to FcgammaRI and the epidermal growth factor receptor. *Cancer Immunol Immunother* 2000; 49:493–503.

59. Dancey J, Sausville EA. Issues and progress with protein kinase inhibitors for cancer treatment [Review]. *Nat Rev Drug Discov* 2003; 2:296–313.

60. Dancey J. Epidermal growth factor receptor inhibitors in clinical development [Review]. *Int J Radiat Oncol Biol Phys* 2004; 58:1003–1007.

61. Dancey J, Freidlin B. Targeting epidermal growth factor receptor—are we missing the mark? [Review]. *Lancet* 2003; 362:62–64.

62. Herbst RS, Giaccone G, Schiller JH, et al. Gefitinib in combination with paclitaxel and carboplatin in advanced non-small-cell lung cancer: a phase III trial—INTACT 2 [*see* comment]. *J Clin Oncol* 2004;22:785–794.

63. Rich JN, Reardon DA, Peery T, et al. Phase II trial of gefitinib in recurrent glioblastoma. *J Clin Oncol* 2004; 22:133–142.

64. Sanders. Investigations into the mechanism for suramin as an inhibitor of cAMP-dependent protein kinase.1996; BIOL-035.

65. Soulieres D, Senzer NN, Vokes EE, Hidalgo M, Agarwala SS, Siu LL. Multicenter phase II study of erlotinib, an oral epidermal growth factor receptor tyrosine kinase inhibitor, in patients with recurrent or metastatic squamous cell cancer of the head and neck. *J Clin Oncol* 2004; 22:77–85.

66. Baselga J, Rischin D, Ranson M, et al. Phase I safety, pharmacokinetic, and pharmacody-namic trial of ZD1839, a selective oral epidermal growth factor receptor tyrosine kinase inhibitor, in patients with five selected solid tumor types [*see* comment]. *J Clin Oncol* 2002; 20:4292–4302.

67. Cortes-Funes H, Soto Parra H. Extensive experience of disease control with gefitinib and the role of prognostic markers. *Br J Cancer* 2003; 89:Suppl 8.

68. Santoro A, Cavina R, Latteri F, et al. Activity of a specific inhibitor, gefitinib (Iressa, ZD1839), of epidermal growth factor receptor in refractory non-small-cell lung cancer. *Ann Oncol* 2004; 15(1):33–37.

69. Phase III trials of Tarceva(TM) plus chemotherapy in first-line non-small cell lung cancer do not meet primary efficacy endpoint. Genentech press release; 2003.

70. Allen LF, Lenehan PF, Eiseman IA, Elliott WL, Fry DW. Potential benefits of the irre-versible pan-erbB inhibitor, CI-1033, in the treatment of breast cancer. *Semin Oncol* 2002; 3:11–21.

71. Allen LF, Eiseman IA, Fry DW, Lenehan PF. CI-1033, an irreversible pan-erbB receptor inhibitor and its potential application for the treatment of breast cancer. *Semin Oncol* 2003; 5:65–78.

72. Wissner A, Overbeek E, Reich MF, et al. Synthesis and structure–activity relationships of 6,7-disubstituted 4-anilinoquinoline-3-carbonitriles. The design of an orally activa, irre-versible inhibitor of the tyrosine kinase activity of the epidermal growth factor receptor (EGFR) and the human epidermal growth factor receptor-2 (HER-2). *J Med Chem* 2003; 46:49–63.

73. Shin D, Nemunaitis J, Zinner RG, et al. A phase I clinical and biomarker study of CI-1033, a novel pan-ErbB tyrosine inhibitor in patients with solid tumors. *Proc Am Soc Clin Oncol* 2001; 20:Abst 324.

74. Anido J, Matar P, Albanell J, et al. ZD1839, a specific epidermal growth factor receptor (EGFR) tyrosine kinase inhibitor, induces the formation of inactive EGFR/HER2 and EGFR/HER3 heterodimers and prevents heregulin signaling in HER2-overexpressing breast cancer cells. *Clin Cancer Res* 2003; 9:1274–1283.

75. Ferrara N, Gerber H-P, LeCouter J. The biology of VEGF and its receptors [Review]. *Nat Med* 2003; 9:669–676.

76. Bergers G, Benjamin LE. Tumorigenesis and the angiogenic switch [Review]. *Nat Rev Cancer* 2003; 3:401–410.

77. Folkman J. Anti-angiogenesis: new concept for therapy of solid tumors. *Ann Surg* 1972; 175:409–419.

78. Yang JC, Haworth L, Sherry RM, et al. A randomized trial of bevacizumab, an anti-vascu-lar endothelial growth factor antibody, for metastatic renal cancer [*see* comment] *N Engl J Med* 2003; 349:427–434.

79. Cobleigh MA, Langmuir VK, Sledge GW, et al. A phase I/II dose-escalation trial of beva-cizumab in previously treated metastatic breast cancer. *Sem Oncol* 2003; 30: Suppl 24.

80. Kabbinavar F, Hurwitz HI, Fehrenbacher L, et al. Phase II, randomized trial comparing bevacizumab plus fluorouracil (FU)/leucovorin (LV) with FU/LV alone in patients with metastatic colorectal cancer [*see* comment]. *J Clin Oncol* 2003; 21:60–65.

81. FDA approves avastin, a targeted therapy for first-line metastatic colorectal cancer patients. Genentech press release; 2004.

82. Willett CG, Boucher Y, di Tomaso E, et al. Direct evidence that the VEGF-specific antibody bevacizumab has antivascular effects in human rectal cancer. *Nat Med* 2004; 10:145–147.

83. Posey JA, Ng TC, Yang B, et al. A phase I study of anti-kinase insert domain-containing receptor antibody, IMC-1C11, in patients with liver metastases from colorectal carcinoma. *Clin Cancer Res* 2003; 9:1323–1332.

84. Manley PW, Bold G, Fendrich G, et al. Advances in the structural biology, design and clinical development of VEGF-R kinase inhibitors for the treatment of angiogenesis. *Biochim Biophys Acta* 2004; 1697(1–2):17–27.

85. Sandberg JA, Parker VP, Blanchard KS, et al. Pharmacokinetics and tolerability of an antiangiogenic ribozyme (ANGIOZYME) in healthy volunteers. *J Clin Pharmacol* 2000; 40:1462–1469.

86. Kuenen BC, Tabernero J, Baselga J, et al. Efficacy and toxicity of the angiogenesis inhibitor SU5416 as a single agent in patients with advanced renal cell carcinoma, melanoma, and soft tissue sarcoma. *Clin Cancer Res* 2003; 9:1648–1655.

87. Mendel DB, Laird AD, Smolich BD, et al. Development of SU5416, a selective small molecule inhibitor of VEGFR receptor tyrosine kinase activity, as an anti-angiogenesis agent. *Anticancer Drug Des* 2000; 15:29–41.

88. Fabbro D, Manley PW. Su-6668.SUGEN. *Curr Opin Investig Drugs* 2001; 2:1142–1148.

89. Mendel DB, Laird AD, Xin X, et al. In vivo antitumor activity of SU11248, a novel tyrosine kinase inhibitor targeting vascular endothelial growth factor and platelet-derived growth factor receptors: determination of a pharmacokinetic/pharmacodynamic relationship. *Clin Cancer Res* 2003; 9:327–337.

90. O'Farrell AM, Abrams TJ, Yuen HA, et al. M.SU11248 is a novel FLT3 tyrosine kinase inhibitor with potent activity in vitro and in vivo. *Blood* 2003; 101:3597–3605.

91. Wedge SR, Ogilvie DJ, Dukes M, et al. VEGF receptor tyrosine kinase inhibitors as potential anti-tumor agents. *Proc Am Assoc Cancer Res* 2000; 41:3610.

92. Wood JM, Bold G, Buchdunger E, et al. PTK787/ZK 222584, a novel and potent inhibitor of vascular endothelial growth factor receptor tyrosine kinases, impairs vascular endothelial growth factor-induce responses and tumor growth after oral administration. *Cancer Res* 2000; 60:2178–2189.

93. Wood JM, Bold G, Buchdunger E, et al. PTK787/ZK 222584, a novel and potent inhibitor of vascular endothelial growth factor receptor tyrosine kinases, impairs vascular endothelial growth factor-induced responses and tumor growth after oral administration. *Cancer Res* 2000; 60:2178–2189.

94. Morgan B, Thomas AL, Drevs JN, et al. Dynamic contrast-enhanced magnetic resonance imaging as a biomarker for the pharmacological response of PTK787/ZK 222584, an inhibitor of the vascular endothelial growth factor receptor tyrosine kinases, in patients with advanced colorectal cancer and liver metastases: results from two phase I studies. *J Clin Oncol* 2003; 21:3955–3964.

95. Ciardiello F, Caputo R, Damiano V, et al. Antitumor effects of ZD6474, a small molecule vascular endothelial growth factor receptor tyrosine kinase inhibitor, with additional activity against epidermal growth factor receptor tyrosine kinase [*see* comment]. *Clin Cancer Res* 2003; 9:1546–1556.

96. Hurwitz HI, Eckhardt SG, Holden SN, et al. A phase I study of ZD6474, an oral VEGF receptor tyrosine kinase inhibitor, in patients with solid tumors. Proceedings of the 12th AACR-NCI-EORTC International Conference, Molecular Targets and Cancer Therapeutics, Discovery, Biology and Clinical Applications 2001; 7:5.

97. Hurwitz HI, Eckhardt SG, Holden SN, et al. Phase I pharmacokinetic and biological study of the angiogenesis inhibitor ZD6474, in patients with solid tumors. *Proc Am Soc Clin Oncol* 2001; 20:396.

98. Hampton T. Scientists take aim at angiogenesis to treat degenerative eye diseases. *JAMA* 2004; 291:1309–1310.

99. Vinores SA. Technology evaluation: pegaptanib, Eyetech/Pfizer.*Curr Opin Mol Ther* 2003; 5:673–679.

100. Ferrara N. Vascular endothelial growth factor. *Trends Cardiovasc Med* 1993; 3:244–250.

101. Sauder DN, DeKoven J, Champagne P, Croteau D, Dupont E. Neovastat (AE-941), an inhibitor of angiogenesis: randomized phase I/II clinical trial results in patients with plaque psoriasis. *J Am Acad Dermatol* 2002; 47:535–541.
102. Grosios K, Wood J, Esser R, Raychaudhuri A, Dawson J. Angiogenesis inhibition by the novel VEGF receptor tyrosine kinase inhibitor, PTK787/ZK222584, causes significant anti-arthritic effects in models of rheumatoid arthritis. *Inflamm Res* 2004; 53(4):133–142.
103. Jeffers M, LaRochelle WJ, Lichenstein HS. Fibroblast growth factors in cancer: therapeutic possibilities. *Expert Opin Ther Targets* 2002; 6(4):469–482.
104. Baserga R, Hongo A, Rubini M, Prisco M, Valentinis B. The IGF-I receptor in cell growth, transformation and apoptosis. *Biochim Biophys Acta* 1997; 1332:F105–F126.
105. Baserga R. The contradictions of the insulin-like growth factor 1 receptor [Review]. *Oncogene* 2000; 19:5574–5581.
106. Werner H, Leroith D. The role of the insulin-like growth factor system in human cancer [Review]. *Adv Cancer Res* 1996; 68:183–223.
107. Furstenberger G, Senn HJ. Insulin-like growth factors and cancer [Review]. *Lancet Oncol* 2002; 3:298–302.
108. Grimberg A, Cohen P. Role of insulin-like growth factors and their binding proteins in growth control and carcinogenesis [Review]. *J Cell Physiol* 2000; 183:1–9.
109. Yu H, Rohan T. Role of the insulin-like growth factor family in cancer development and progression. *J Natl Cancer Inst* 2000; 92:1472–1489.
110. Kalebic T, Blakesley V, Slade C, Plasschaert S, Leroith D, Helman LJ. Expression of a kinase-deficient IGF-I-R suppresses tumorigenicity of rhabdomyosarcoma cells constitutively expressing a wild type IGF-I-R. *Int J Cancer* 1998; 76:223–227.
111. Reiss K, D'Ambrosio C, Tu X, Tu C, Baserga R. Inhibition of tumor growth by a dominant negative mutant of the insulin-like growth factor I receptor with a bystander effect. *Clin Cancer Res* 1998; 4:2647–2655.
112. Scotlandi K, Benini S, Nanni P, et al. Blockage of insulin-like growth factor-I receptor inhibits the growth of Ewing's sarcoma in athymic mice. *Cancer Res* 1998; 58:4127–4131.
113. Scotlandi K, Avnet S, Benini S, et al. Expression of an IGF-I receptor dominant negative mutant induces apoptosis, inhibits tumorigenesis and enhances chemosensitivity in Ewing's sarcoma cells. *Int J Cancer* 2002; 101:11–16.
114. Scotlandi K, Maini C, Manara MC, et al. Effectiveness of insulin-like growth factor I receptor antisense strategy against Ewing's sarcoma cells. *Cancer Gene Ther* 2002; 9:296–307.
115. Burtrum D, Zhu Z, Lu D, et al. A fully human monoclonal antibody to the insulin-like growth factor I receptor blocks ligand-dependent signaling and inhibits human tumor growth in vivo. *Cancer Res* 2003; 63:8912–8921.
116. Garcia-Echeverria C, Pearson MA, Marti A, et al. In vivo antitumor activity of NVP-AEW541—A novel, potent, and selective inhibitor of the IGF-IR kinase. *Cancer Cell* 2004; 5(3):231–239.
117. Mitsiades CS, Mitsiades NS, McMullan CJ, et al. Inhibition of the insulin-like growth factor receptor-1 tyrosine kinase activity as a therapeutic strategy for multiple myeloma, other hematologic malignancies, and solid tumors. *Cancer Cell* 2004; 5(3):221–230.
118. Birchmeier C, Birchmeier W, Gherardi E, Vande Woude GF. Met, metastasis, motility and more [Review]. *Nat Rev Mol Cell Biol* 2003; 4:915–925.
119. Birchmeier W. Signaling by HGF/SF and the receptor tyrosine kinase Met. *Eur J Biochem* 2003; 270:Suppl 1:8–9.
120. Longati P, Comoglio PM, Bardelli A. Recepor tyrosine kinases as therapeutic targets: the model of the Met oncogene. *Curr Drug Targets* 2001; 2:41–55.
121. Schmidt L, Duh FM, Chen F, et al. Germline and somatic mutations in the tyrosine kinase domain of the Met proto-oncogene in papillary renal carcinomas. *Nat Gen* 1997; 16:68–73.

122. Lorenzato A, Olivero M, Patane S, et al. Novel somatic mutations of the MET oncogene in human carcinoma metastases activating cell motility and invasion. *Cancer Res* 2002; 62:7025–7030.

123. Lee JH, Han SU, Cho H, et al. A novel germ line juxtamembrane Met mutation in human gastric cancer. *Oncogene* 2000; 19:4947–4953.

124. Yamashita J, Ogawa M, Yamashita S, et al. Immunoreactive hepatocyte growth factor is a strong and independent predictor of recurrence and survival in human breast cancer. *Cancer Res* 1994; 54:1630–1633.

125. Abounader R, Ranganathan S, Lal B, et al. Reversion of human glioblastoma malignancy by U1 small nuclear RNA/ribozyme targeting of scatter factor/hepatocyte growth factor and c-met expression. *J Natl Cancer Inst* 1999; 91:1548–1556.

126. Herynk MH, Stoeltzing O, Reinmuth N, et al. Down-regulation of c-met inhibits growth in the liver of human colorectal carcinoma cells. *Cancer Res* 2003; 63:2990–2996.

127. Jiang WG, Grimshaw D, Lane J, et al. A hammerhead ribozyme suppresses expression of hepatocyte growth factor/scatter factor receptor c-MET and reduces migration and invasiveness of breast cancer cells. *Clin Cancer Res* 2001; 7:2555–2562.

128. Jiang WG, Grimshaw D, Martin TA, et al. Reduction of stromal fibroblast-induced mammary tumor growth, by retroviral ribozyme transgenes to hepatocyte growth factor/scatter factor and its receptor, c-MET. *Clin Cancer Res* 2003; 9:4274–4281.

129. Furge KA, Kiewlich D, Le P, et al. Suppression of Ras-mediated tumorigenicity and metastasis through inhibition of the Met receptor tyrosine kinase. *Proc Natl Acad Sci USA* 2001; 98:10722–10727.

130. Maemondo M, Narumi K, Saijo Y, et al. Targeting angiogenesis and HGF function using an adenoviral vector expressing the HGF antagonist NK4 for cancer therapy. *Mol Ther* 2002; 5:177–185.

131. Matsumoto, and Nakamura T. Suppression of tumor malignancy by NK4/malignostatin: a new cancer therapy by inhibition of tumor invasion-metastasis and angiogenesis. *Saishin Igaku* 2000; 55:1960–1968.

132. Morotti A, Mila S, Accornero P, Tagliabue E, Ponzetto C. K252a inhibits the oncogenic properties of Met, the HGF receptor. *Oncogene* 2002; 21:4885–4893.

133. Christensen JG, Schreck R, Burrows J, et al. A selective small molecule inhibitor of c-Met kinase inhibits c-Met-dependent phenotypes in vitro and exhibits cytoreductive antitumor activity in vivo. *Cancer Res* 2003; 63:7345–7355.

134. Sattler M, Pride YB, Ma P, et al. A novel small molecule Met inhibitor induces apoptosis in cells transformed by the oncogenic Tpr-Met Tyrosine kinase. *Cancer Res* 2003; 63:5462–5469.

135. Wang X, Le P, Liang C, et al. Potent and selective inhibitors of the Met [hepatocyte growth factor/scatter factor (HGF/SF) receptor] tyrosine kinase block HGF/SF-induced tumor cell growth and invasion. *Mol Cancer Ther* 2003; 2(11):1085–1092.

136. Liu WM, Stimson LA, Joel SP. The in vitro activity of the tyrosine kinase inhibitor STI571 in BCR-ABL positive chronic myeloid leukaemia cells: synergistic interactions with anti-leukaemic agents. *Br J Cancer* 2002; 86:1472–1478.

137. Shuai K, Liu B. Regulation of JAK-STAT signalling in the immune system [Review]. *Nat Rev Immunol* 2003; 3:900–911.

138. Igaz A, Toth S, Falus A. Biological and clinical significance of the JAK-STAT pathway; lessons from knockout mice [Review]. *Inflamm Res* 2001; 50:435–441.

139. Ihle JN, Stravapodis D, Parganas E, et al. The roles of Jaks and Stats in cytokine signaling. [Review] [49 refs]. *Cancer J Sci Am* 1998; 4:Suppl 91.

140. Macchi P, Villa A, Giliani S, et al. J. Mutations of Jak-3 gene in patients with autosomal severe combined immune deficiency (SCID). *Nature* 1995; 377:65–68.

141. Notarangelo LD, Mella P, Jones A, et al. Mutations in severe combined immune deficiency (SCID) due to JAK3 deficiency [Review]. *Hum Mutat* 2001; 18:255–263.
142. Russell SM, Tayebi N, Nakajima H, et al. Mutation of Jak3 in a patient with SCID: essential role of Jak3 in lymphoid development. *Science* 1995; 270:797–800.
143. Tsuge I, Matsuoka H, Abe T, Kamachi Y, Torii S. Interleukin-2 receptor gamma-chain mutations in severe combined immunodeficiency with B-lymphocytes. *Eur J Pediatr* 1996; 155:1018–1024.
144. Changelian PS, Flanagan ME, Ball DJ, et al. Prevention of organ allograft rejection by a specific Janus kinase 3 inhibitor. *Science* 2003; 302:875–878.
145. Cetkovic-Cvrlje M, Dragt AL, Vassilev A, Liu XP, Uckun FM. Targeting JAK3 with JANEX-1 for prevention of autoimmune type 1 diabetes in NOD mice. *Clin Immunol* 2003; 106:213–225.
146. Lacronique V, Boureux A, Valle VD, et al. A TEL-JAK2 fusion protein with constitutive kinase activity in human leukemia. *Science* 1997; 278:1309–1312.
147. Ruchatz H, Coluccia AM, Stano P, Marchesi E, Gambacorti-Passerini C. Constitutive activation of Jak2 contributes to proliferation and resistance to apoptosis in NPM/ALK-transformed cells. *Exp Hematol* 2003; 31:309–315.
148. Sun X, Layton JE, Elefanty A, Lieschke GJ. Comparison of effects of the tyrosine kinase inhibitors AG957, AG490, and STI571 on BCR-ABL-expressing cells, demonstrating synergy between AG490 and STI571. *Blood* 2001; 97:2008–2015.
149. Fukushima N, Sato N, Sahin F, Su GH, Hruban RH, Goggins M. Aberrant methylation of suppressor of cytokine signalling-1 (SOCS-1) gene in pancreatic ductal neoplasms. *Br J Cancer* 2003; 89:338–343.
150. Yoshikawa H, Matsubara K, Qian GS, et al. SOCS-1, a negative regulator of the JAK/STAT pathway, is silenced by methylation in human hepatocellular carcinoma and shows growth-suppression activity [*see* comment]. *Nat Gen* 2001; 28:29–35.
151. Burke WM, Jin X, Lin HJ, et al. Inhibition of constitutively active Stat3 suppresses growth of human ovarian and breast cancer cells. *Oncogene* 2001; 20:7925–7934.
152. Li L, Shaw PE. Autocrine-mediated activation of STAT3 correlates with cell proliferation in breast carcinoma lines. *J Biol Chem* 2002; 277:17397–17405.
153. Sriuranpong V, Park JI, Amornphimoltham P, et al. Epidermal growth factor receptor-independent constitutive activation of STAT3 in head and neck squamous cell carcinoma is mediated by the autocrine/paracrine stimulation of the interleukin 6/gp130 cytokine system. *Cancer Res* 2003; 63:2948–2956.
154. Toyonaga T, Nakano K, Nagano M, et al. Blockade of constitutively activated Janus kinase/signal transducer and activator of transcription-3 pathway inhibits growth of human pancreatic cancer. *Cancer Lett* 2003; 201:107–116.
155. Uckun FM, Sudbeck EA, Mao C, et al. Structure-based design of novel anticancer agents [Review]. *Curr Cancer Drug Targets* 2001; 1:59–71.
156. Hunt JL. Molecular mutations in thyroid carcinogenesis [Review]. *Am J Clin Pathol* 2002; 118:Suppl 27.
157. Davies H, Bignell GR, Cox C, et al. Mutations of the BRAF gene in human cancer [*see* comment]. *Nature* 2002; 417:949–954.
158. DiGiovanna MP, Chu P, Davison TL, et al. active signaling by HER-2/neu in a subpopulation of HER-2/neu-overexpressing ductal carcinoma in situ: clinicopathological correlates. *Cancer Res* 2002; 62:6667–6673.
159. Thor AD, Liu S, Edgerton S, et al. Activation (tyrosine phosphorylation) of ErbB-2 (HER-2/neu): a study of incidence and correlation with outcome in breast cancer. *J Clin Oncol* 2000; 18:3230–3239.
160. She QB, Solit D, Basso A, Moasser MM. Resistance to Gefitinib in PTEN-null HER-overexpressing tumor cells can be overcome through restoration of PTEN function or

pharmacologic modulation of constitutive phosphatidylinositol 3'-Kinase/Akt pathway signaling. *Clin Cancer Res* 2003; 9:4340–4346.

161. Sawyers CL. Opportunities and challenges in the development of kinase inhibitor therapy for cancer [Review]. *Genes Dev* 2003; 17:2998–3010.

162. Shah NP, Nicoll JM, Nagar B, et al. L.Multiple BCR-ABL kinase domain mutations confer polyclonal resistance to the tyrosine kinase inhibitor imatinib (STI571) in chronic phase and blast crisis chronic myeloid leukemia [comment]. *Cancer Cell* 2002; 2:117–125.

163. Shah NP, Sawyers CL. Mechanisms of resistance to STI571 in Philadelphia chromosome-associated leukemias [Review]. *Oncogene* 2003; 22:7389–7395.

164. Nagar B, Hantschel O, Young MA, et al. Structural basis for the autoinhibition of c-Abl tyrosine kinase [*see* comment]. *Cell* 2003; 112:859–871.

165. Schindler T, Bornmann W, Pellicena P, et al. Structural mechanism for STI-571 inhibition of Abelson tyrosine kinase. *Science* 2000; 289:1938–1942.

166. Azam M, Latek RR, Daley GQ. Mechanisms of autoinhibition and STI-571/imatinib resistance revealed by mutagenesis of BCR-ABL [*see* comment]. *Cell* 2003; 112:831–843.

167. Cools J, Stover EH, Boulton CL, et al. PKC412 overcomes resistance to imatinib in a murine model of FIP1L1-PDGFRalpha-induced myeloproliferative disease [*see* comment]. *Cancer Cell* 2003; 3:459–469.

2 Inhibitors of Signaling Interfaces

Targeting Src Homology 2 Domains in Drug Discovery

Carlos García-Echeverría, *PhD*

CONTENTS

INTRODUCTION
ANTAGONISTS OF THE PP60[C-SRC] SH2 DOMAIN
ANTAGONISTS OF THE LCK SH2 DOMAIN
ANTAGONISTS OF THE SH2 DOMAIN OF GRB2
CONCLUSIONS
REFERENCES

1. INTRODUCTION

The Src homology 2 (SH2) domain is a noncatalytic module of approx 100 amino acids that is mainly involved in directing interactions with cellular substrates *(1,2)*. The SH2 domain recognizes short phosphotyrosine-containing sequences in which specificity is conveyed by the residues immediately *C*-terminal to the phosphotyrosine amino acid *(3–5)*. The specific association of an SH2 domain with a phosphotyrosine-containing sequence of another protein precipitates a cascade of intracellular protein–protein interactions that result in signal propagation *(1,6,7)*. Antagonists of critical SH2-binding events can be reasonably result in the inactivation of undesirable signal transduction networks and can represent a targeted treatment of a broad range of human diseases (e.g., cancer, osteoporosis, disorders of the immune and cardiovascular systems) *(8)*.

Highly potent antagonists of specific SH2 domains have been identified largely by structure-based drug design efforts, but these compounds had several undesirable features *(9–12)*. In addition to the intrinsic problems associated with the peptidic nature of some of these molecules (e.g., degradation, rapid elimination from plasma, high first-pass metabolism, and very low oral bioavailability),

From: *Cancer Drug Discovery and Development:*
Protein Tyrosine Kinases: From Inhibitors to Useful Drugs
Edited by: D. Fabbro and F. McCormick © Humana Press Inc., Totowa, NJ

the phosphate group of phosphotyrosine, which is an essential element for binding to the SH2 domain, is metabolically unstable to phosphatases present in cells and further limits the ability of these compounds to reach efficacious concentrations inside the cell *(13)*. Attempts to address these issues have appeared in the literature for a variety of SH2 domains and, recently, significant advances have been accomplished in the identification of antagonist of the SH2 domains of Src, Lck, and Grb2 that showed activity in cell-based assays and animal models.

2. ANTAGONISTS OF THE PP60$^{C\text{-}SRC}$ SH2 DOMAIN

The nonreceptor tyrosine kinase pp60$^{c\text{-}src}$ contains a myristoylated *N*-terminal domain, a catalytic kinase region, an SH2 domain, an SH3, and a short *C*-terminal regulatory peptide sequence, often called the tail *(14–15)*. Overexpression or hyperactivation of this protein has been implicated in the development of human breast adenocarcinomas and colon carcinoma *(16–17)*. Furthermore, pp60$^{c\text{-}src}$, which is normally expressed at high levels in osteoclasts, has also been implicated in regulating osteoclast-mediated resorption of bone *(18–19)*. The kinase domain of Src can activate intracellular signaling networks by phosphorylating docking proteins that can then cause the relocation of SH2 domain-containing signaling proteins. The SH2 domain of Src has been postulated to play an important role in regulating these intracellular signaling events by acting as an adaptor that brings specific protein substrates into the signaling complex (e.g., truncation of the SH2 domain in Src affects its transforming abilities without affecting its kinase activity) *(20)*. Compounds that modulate pp60$^{c\text{-}src}$ regulated signal transduction pathways by blocking its SH2 domain offer potential value as antiproliferative agents or in the treatment of osteoporosis.

Extensive nuclear magnetic resonance (NMR) and X-ray crystallographic studies of the Src SH2 protein complexed with peptides containing the Tyr(PO$_3$H$_2$)-Glu-Glu-Ile motif *(4–5)* provided a detailed molecular map of the binding pockets of this therapeutic target *(21–24)*. Examination of the three-dimensional structures revealed the presence of two major binding pockets, one interacting with phosphotyrosine and the other with the side chain of isoleucine at the X$_{+3}$ position. The two glutamic acid residues do not make a strong interaction with the protein and mainly serve to orient the phosphotyrosine and isoleucine side chains at their respective binding pockets *(21,25)*. This structural information has been instrumental in the design of potent and selective Src SH2 antagonist and representative compounds have been selected to illustrate the approaches used by different groups.

Using a dipeptide framework *(26)*, compound **1**[1] (Fig. 1) was designed with the intent of replacing the three C-terminal residues in Ac-Tyr(PO$_3$H$_2$)-

---

[1]To normalize data from different experiments, results for individual antagonists were presented in this paper as a ratio, IC$_{50}$(test) = IC$_{50}$(standard).

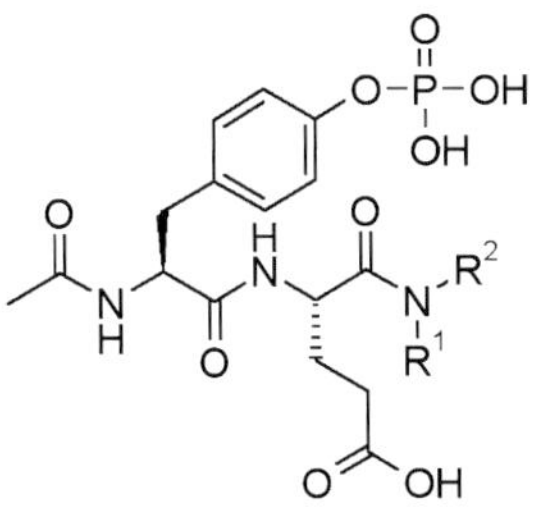

R¹= R²= n-C₅H₁₁ 1
R¹= CH₃, R₂= n-C₇H₁₄OH 2
R¹= CH₃, R₂= n-C₈H₁₆OH 3
R¹= CH₃, R₂= n-C₆H₁₂OH 4

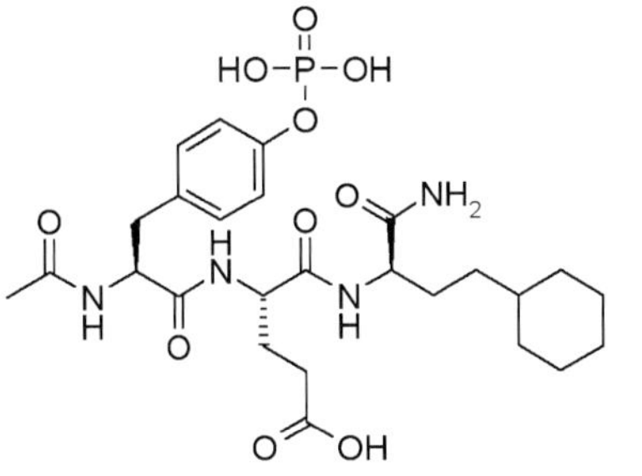

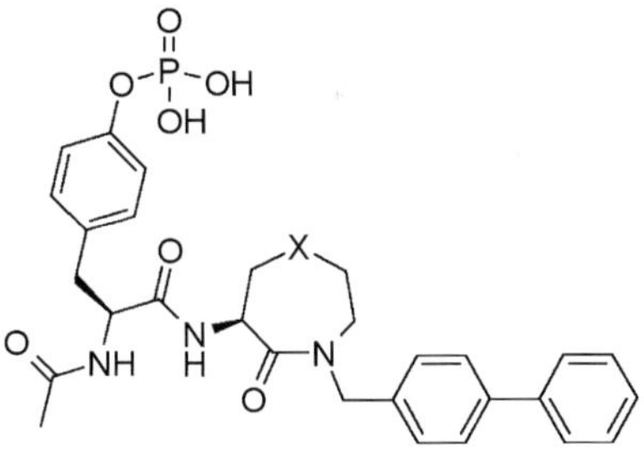

X= C, RU 81843 9
X= S 10

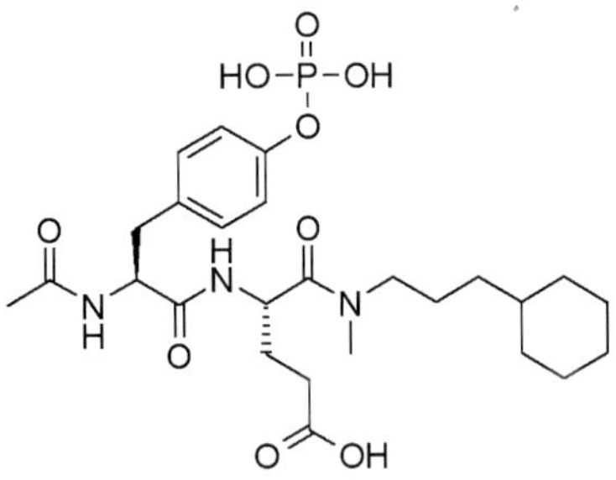

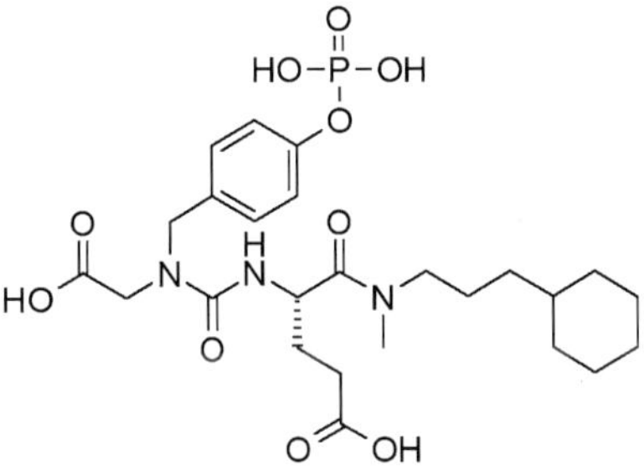

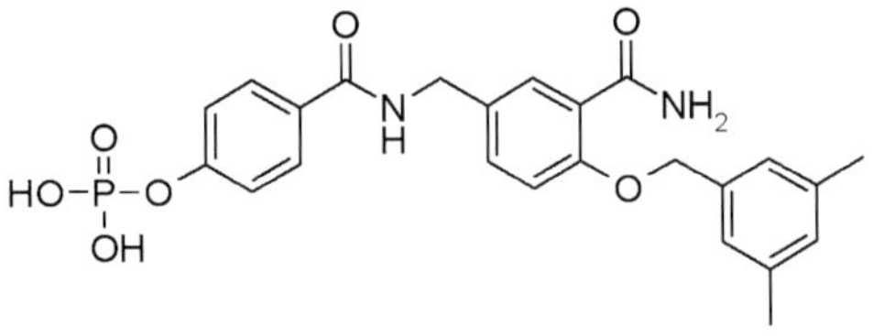

Fig. 1. Antagonists of the Src SH2 domain.

Glu-Glu-Ile-Glu-OH with a fragment that would access the hydrophobic binding pocket at the X₊₃ position without compromising interactions within the phosphotyrosine binding pocket (26,27). Following this approach, the C-terminal heptanol (compound **2**, IC₅₀(**2**)/IC₅₀(standard = 3.6; Fig. 1) and octanol

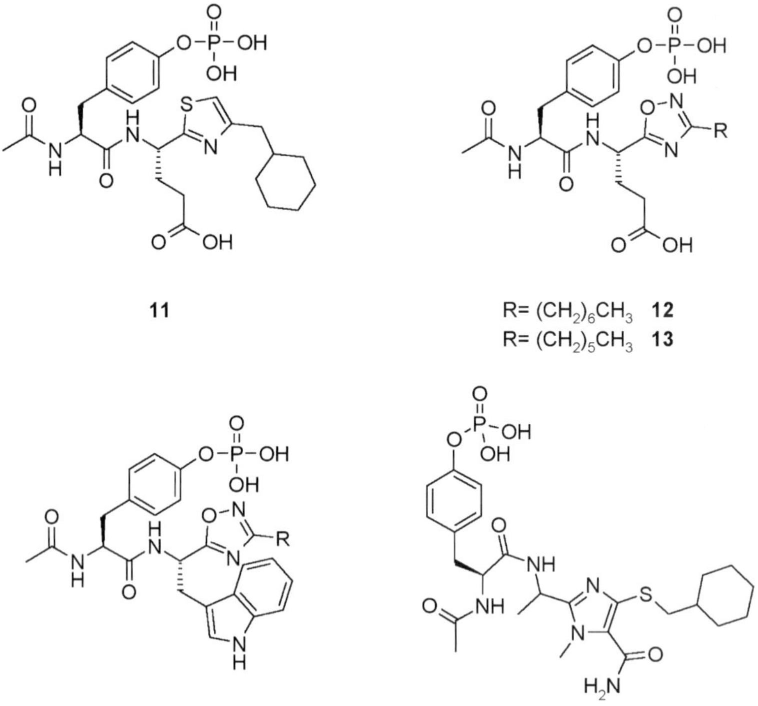

11

R= $(CH_2)_6CH_3$ **12**
R= $(CH_2)_5CH_3$ **13**

R= $(CH_2)_5CH_3$ **14**

15

Fig. 2. Antagonists of the Src SH2 domain based on the use of heterocyclic scaffolds.

(compound **3**, IC_{50}(**3**)/IC_{50}[standard] = 2.2; Fig. 2) analog were synthesized to form a hydrogen-bonding interaction with amino acid residues lining the X_{+3} pocket. This interaction was confirmed by solving the X-ray structure of the Src SH2 domain co-crystallized with compound **4** (IC_{50}(**4**)/IC_{50}[standard] = 6.4; Fig. 1). Efforts to enhance binding affinity by increasing intramolecular association and filling or shielding the phosphotyrosine binding site with a set of different N-terminal groups were unsuccessful.

A further step in the dipeptidation of a previously reported antagonist (compound **5**, Fig. 1; IC_{50} = 8.5 μ*M*) *(28)* was obtained by removing the C-terminal carboxamide. This moiety was previously thought to be involved in an intramolecular hydrogen bond with the oxygen of the backbone carbonyl group of phosphotyrosine. Compounds with binding affinities in the low micromolar range were obtained by changing the rotational freedom and hydrophobicity of the C-terminus (e.g., compound **6**, Fig. 1; IC_{50} = 0.79 μ*M*). Modeling studies with these new ligands provided the basis for the design of nonpeptide Src antagonists containing urea linkage (compound **7**, Fig. 1; IC_{50} = 7.0 μ*M*) *(29)*

The X-ray crystal structure of compound **7** complexed with the Src SH2 domain revealed a conformation and binding interactions quite different from the initial predictions: (1) the orientation of the phenyl phosphate ring is nearly orthogonal to that exhibited by phosphotyrosine-containing compounds in other X-ray structures; (2) a *cis*-amide bond between glutamic acid and the C-terminal groups; and (3) a lack of interaction between the side chain of Arg^{12} and both the N-terminal carboxyl moiety and the phosphate group of the ligand. These factors can account for the low binding affinity observed for these urea-containing compounds, but, more important, they illustrate some of the difficulties associated with the prediction of ligand binding by computational methods. This issue was also encountered in another *de novo* series of nonpeptide antagonists *(30)*. In this case, the X-ray structure of compound **8** (Fig. 1; IC_{50} = 6.6 μM) with the Src SH2 domain revealed that the phenylphosphate group is capable of a binding mode at the phosphotyrosine site substantially different from that observed for the phosphotyrosine side chain in peptides bound to the Src SH2 domain.

In a modular approach to identify nonpeptidic Src SH2 antagonists, different scaffolds have been utilized to mimic the Glu–Glu dipeptide. Structure–activity relationship (SAR) studies with caprolactam/thioazepinone derivatives led to the identification of potent Src SH2 inhibitors (compounds **9** and **10**, IC_{50} = 9 nM and IC_{50} = 87 nM, respectively; Fig. 1) *(31,32)*. The X-ray structures of some of these compounds complexed with the Src SH2 domain revealed that these templates deliver their substituents into the phosphotyrosine and X_{+3} pockets.

Optimization of 2,4- and 2,5-substituted thiazole, and 1,2,4-oxadiazole derivatives resulted in Src SH2 inhibitors (e.g., compounds **11**, IC_{50} = 26 μM; **12**, IC_{50} = 7 μM; **13**, IC_{50} = 8 μM; and **14**, IC_{50} = 16 μM; Fig. 2) that showed binding affinities similar to the reference tetrapeptide *(33,34)*. In a very similar approach, tetrasubstituted imidazoles (e.g., compound **15**, IC_{50} = 8.6 μM; Fig. 2) were used to mimic the interactions made by the corresponding amino acid side chains of the peptide antagonist, but these compounds showed weaker activity than the reference peptide *(35)*.

Recently, phosphate, phosphonate, or phosphonic acid derivatives, which were prepared by parallel synthesis, were screened (BIAcore technology) to identify potential phosphotyrosine mimetics that bind to the Src SH2 domain. Napthyl, tetrahydroquinoline, and amido phosphonate derivatives with binding affinities in the mM range were discovered *(36)*. Parallel to this strategy, small fragments with low binding affinity for Src SH2 were screened using crystal soaking *(37)*. Structure determination of more than 20 of these compounds bound into the phosphotyrosine pocket of Src SH2 domain allowed a selection of the best fragments to incorporate into antagonist platforms. Malonate-type

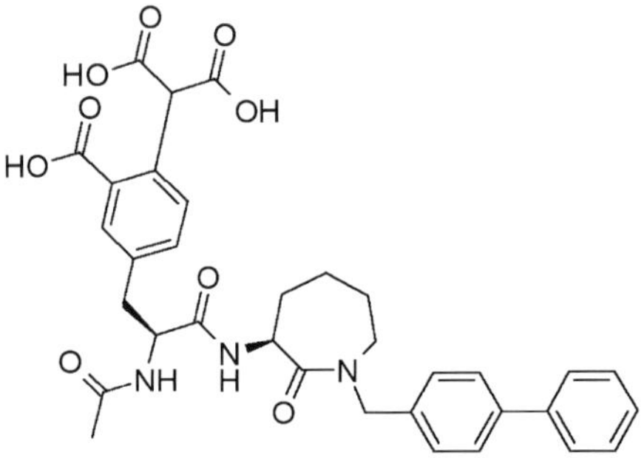

16

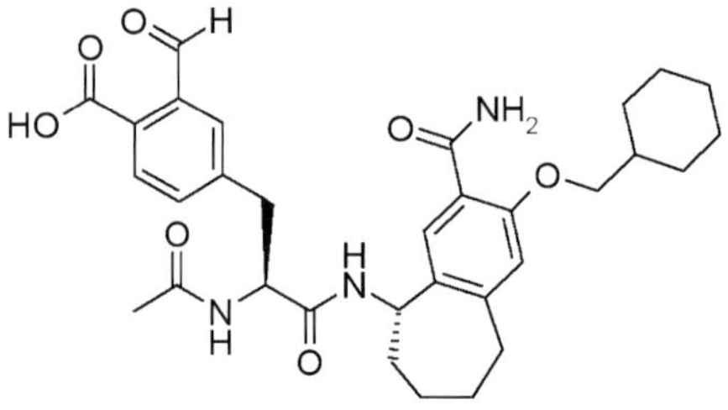

17

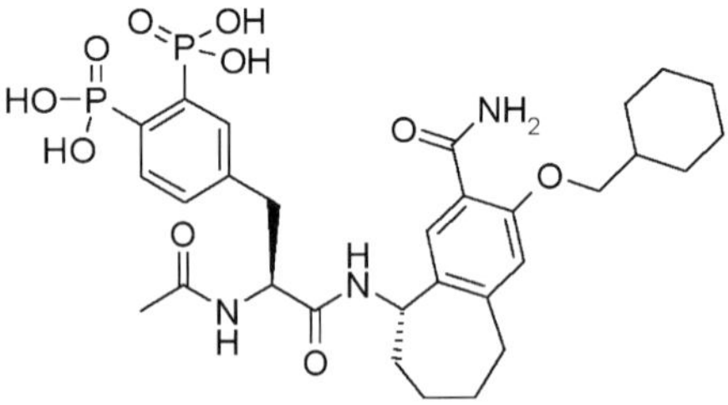

18

Fig. 3. Antagonists of the Src SH2 domain containing phosphotyrosine mimetics.

inhibitors (compound **16**, Fig. 3), which combine a previously published antagonist scaffold *(31,32)* and a phosphotyrosine mimetic identified by this X-ray
technique, showed potent in vitro activity (IC_{50} = 5 n*M*) and good stability in
rat and human plasma.

In addition to the preceding screening approaches, extensive synthetic efforts
have been devoted to the identification and preparation of phosphotyrosine
mimics with improved stability and cellular permeability *(38–44)*. Replacement
of phosphotyrosine by some of these nonhydrolyzable analogs has advanced
the discovery of compounds for cellular and in vivo studies *(40–42)*. Some of
these antagonist are briefly described in the following.

Compound **17** (AP22161, Fig. 3) binds in vitro with high affinity to the Src
SH2 domain (IC_{50} = 0.24 µ*M*) and inhibits Src cellular activity *(41)*. The formyl
group was introduced to form a hemithioacetal with the side chain of cysteine-
188 when the compound is bound to the protein *(45–47)*. NMR experiments
and in vitro studies with wild-type Src and mutant cysteine-188 SH2 domains
demonstrated that compounds containing the 3-formyl-4-carboxy-substituted
phenylalanine building block target cysteine-188 and that this binding is

reversible *(41)*. Cysteine-188 is unique for the Src SH2 domain providing selectivity against other members of the Src family of tyrosine kinases (e.g., IC_{50} = 29.4 μM and 421.9 μM for the SH2 domains of Yes and ZAP 70, respectively). Compound **17** showed partial morphological reversion of the transformed phenotype of cSrcY527F cells at 100 μM and inhibited osteoclast-mediated resorption of dentine with an IC_{50} value of 42.9 μM. A better in vivo inhibitory profile was obtained with compounds containing 3',4'-diphosphonophenylalanine. This phosphotyrosine mimetic exhibits bone-targeting properties that confer osteoclast selectivity, hence reducing undesired effects in other cell types. Compound **18** (AP22408, Fig. 3) inhibited Src SH2 binding with an IC_{50} of 0.30 μM *(42,48)* and blocks rabbit osteoclast-mediated resorption of dentine slices (IC_{50} = 1.6 μM). Furthermore, this compound showed a statistically significant (p = 0.0379) antiresorptive activity in an in vivo thyroparathyroidectomized model of parathyroid hormone-induced bone resorption when administered intravenously (50 mg/kg, twice daily). Although these type of derivatives allowed to validate Src as a therapeutic target for the treatment of osteoporosis in preclinical models, they did not progress into clinical trials.

3. ANTAGONISTS OF THE LCK SH2 DOMAIN

The Src family tyrosine kinase p56[lck] (Lck: lymphoid T-cell tyrosine kinase) is predominantly expressed in T-lymphocytes and natural killers *(49–51)*. p56[lck] is absolutely required in the early phase of T-cell antigen receptor (TCR) activation and plays a critical role in T-cell-mediated immune responses *(52)*. Like all Src homologs, it comprises a unique amino-terminal region, followed by Src-homology domains SH3 and SH2, and a tyrosine kinase domain *(53)*. Lck is localized to the plasma membrane through myristolytation and palmitylation, and is associated through its unique amino-terminal segment with the cytoplasmic tails of the T-cell co-receptor glycoprotein CD4 or CD8 *(54,55)*. Upon kinase activation *(56)*, p56[lck] phosphorylates specific tyrosine residues of the ξ-chain of TCR within a motif termed immunoglobulin receptor family tyrosine-based activation motives (ITAMs). The phosphorylated residues recruit a second cytoplasmic protein tyrosine kinase called ZAP-70 to promote T-cell activation *(57,58)*. The SH2 domain of Lck regulates the kinase activity of the protein *(59)* and may mediate protein–protein interactions with ZAP-70 and/or the ξ subunit of the T-cell receptor *(60)*. Several lines of evidence have established a critical role for p56[lck] in antigen-induced T-cell responses *(61–63)*. Consequently, there has been much effort in developing kinase inhibitors and SH2 antagonists for p56[lck] with a view for therapeutic use in a number of diseases such as multiple sclerosis, asthma, inflammatory bowel disease, rheumatoid arthritis, and T-cell-based leukemias and lymphomas *(64,65)*.

The SH2 domain of Lck exhibits a marked preference for the sequence Tyr(PO_3H_2)-Glu-Glu-Ile *(5)* and this motif served as the starting point for the design of antagonist for this protein. This effort was guided by the early report of an X-ray structure of Lck SH2 in complex with a phosphopeptide (EPQpYEEIPYL) *(66)*. Analysis of this structure revealed the presence of a large hydrophobic pocket for the residue at the X_{+3} position (isoleucine) and the lack of grooves or pockets for the X_{+1} and X_{+2} (glutamic acids) residues. The interactions made by the side chains of phosphotyrosine and isoleucine are complemented by a network of hydrogen bonds to the peptide backbone. As for the design of Src SH2 antagonists, the identification of an uncharged replacement for the doubly charged Glu–Glu sequence has been an additional challenge for this target. Even though relatively limited work has been reported to date on antagonist of Lck SH2 domain, significant progress has been made over the past few years.

SAR studies and computational chemistry were initially utilized to reduce the overall charge of Ac-Tyr(PO_3H_2)-Glu-Glu-Ile-OH (K_d = 0.1 μM) while maintaining Lck SH2 inhibitory activity *(67)*. The most potent compound was obtained by replacing the C-terminal dipeptide Glu-Ile-OH with (S)-1-(4-isopropylphenlyl)ethylamine and the glutamic acid at the X_{+1} position with leucine (compound **19**, K_d = 0.2 μM; Fig. 4). This last modification was guided by early SAR studies that indicated that a wide range of amino acids with uncharged side chains were well tolerated at X_{+1} *(68)*. In an attempt to identify a replacement of the phosphotyrosine residue and using the skeleton of compound **20** (K_d = 0.18 μM; Fig. 4), a series of compounds that contain monocharged, nonhydrolyzable phosphate groups were synthezised (e.g., difluoroacetic acid, (R/S)-hydroxyacetic, sulfonic acid, oxamic acid) *(69)*. The drop in activity observed for these derivatives (75- to 920-fold decrease in potency relative to compound **20**) was partially compensated by incorporating lipophilic substituents, particularly naphthylacetyl groups, at the N-terminus (e.g., compound **21**, K_d = 5.0 μM; Fig. 4). Recently, this type of derivatives was elaborated to an Lck SH2 antagonist (compound **22**, K_d = 1 μM; Fig. 4) that shows relatively good cell permeability *(70)*. Compound **22** is the result of a design effort to rigidify the backbone of the molecule (introduction of the pyridone ring), enforce a favorable conformation for binding (incorporation of a methyl group at the 4-position of the pyridine ring), and increase cell permeability by enhancing desolvation (dimethyl substitution adjacent to the carboxylate group). The preceding properties allowed the effect of Lck SH2 inhibition in a cellular setting, to be demonstrated for the first time. Calcium mobilization is a very early event in T-cell activation and an antagonist of Lck SH2 should inhibit receptor-mediated increase in cytosolic calcium. Inhibition of intracellular calcium concentration was observed (ED_{50} = 10 μM) when compound **22** was tested in Jurkat T-cells activated with mouse antihuman CD3 antibody. Additional

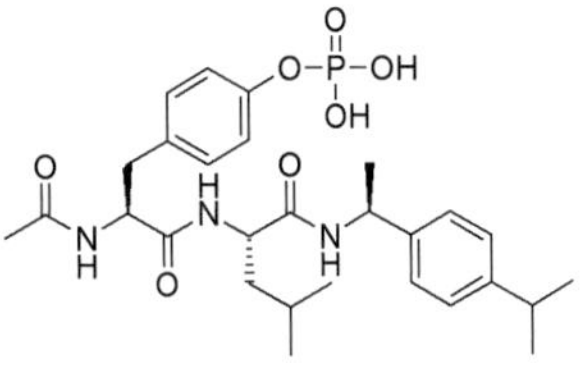

Fig. 4. Antagonists of the Lck SH2 domain.

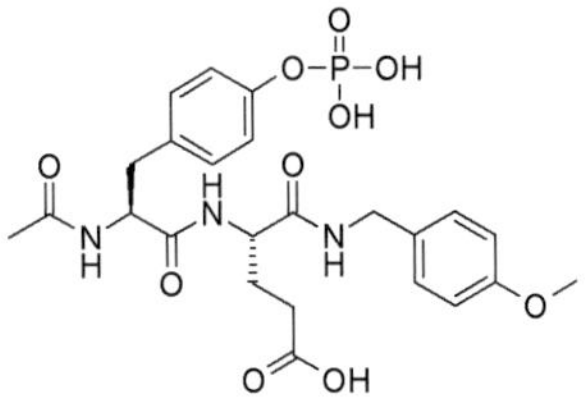

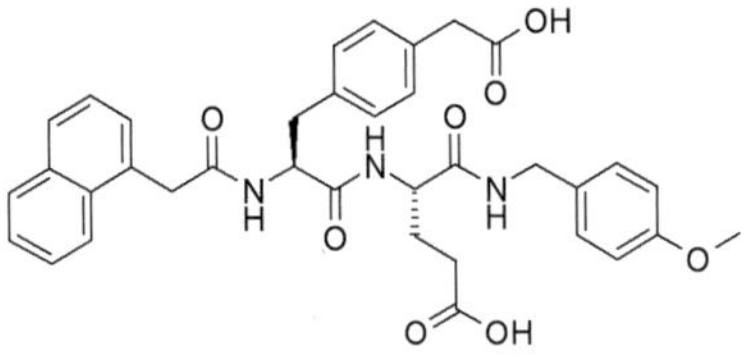

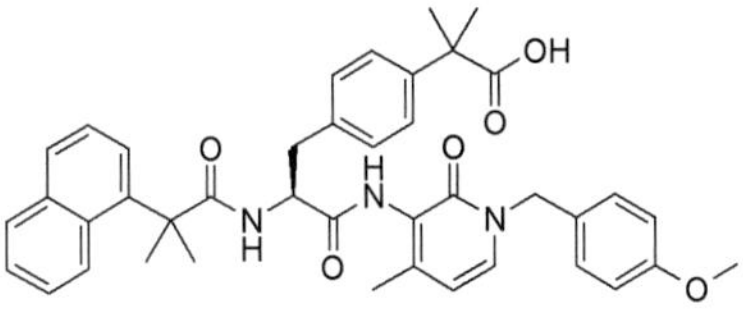

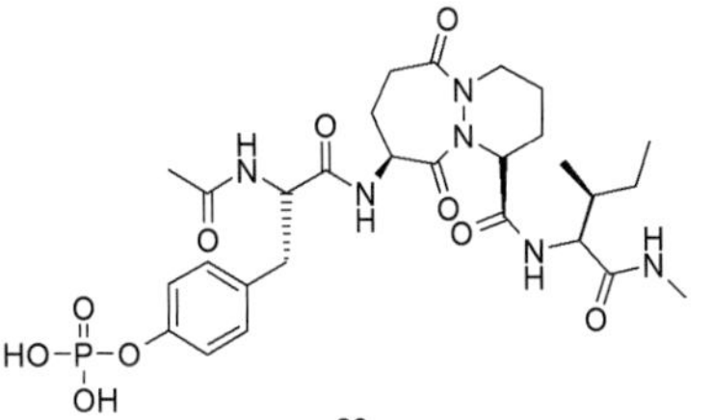

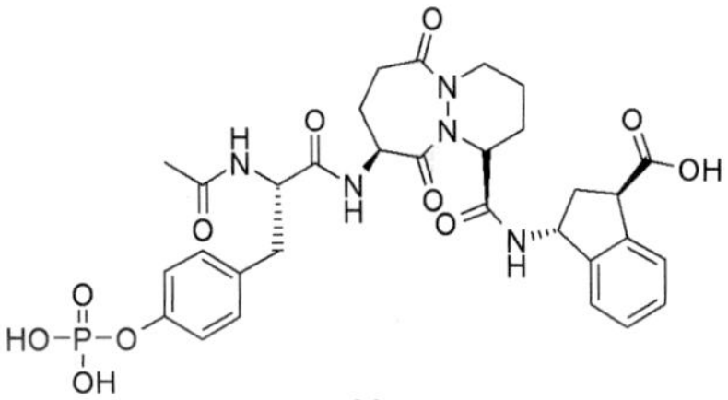

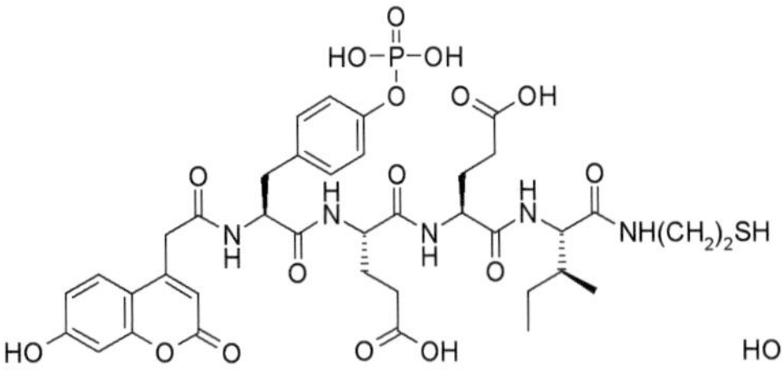

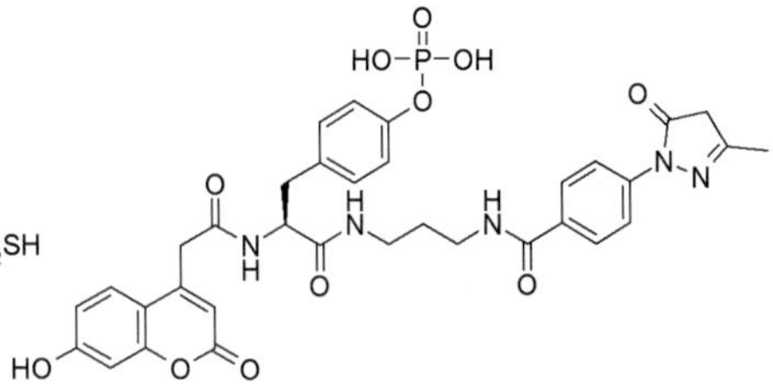

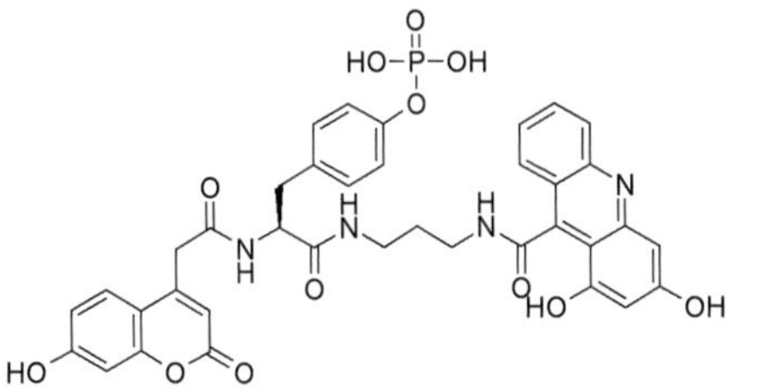

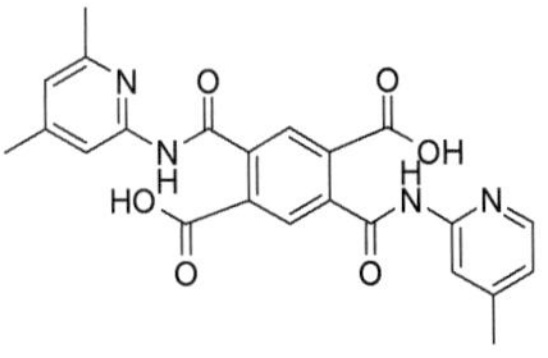

experiments show that this effect was enantiomer and TCR-signaling dependent, confirming that this antagonist function via the intended mechanism.

Recently, a new class of structure-based Lck SH2 antagonists incorporating 9-aminopyridazinodiazepine as a Glu–Glu mimetic has met with success *(71)*. After confirming that the diazepine moiety was a suitable Glu–Glu replacement, the X-ray structure of compound **23** (IC_{50} = 1.6 µ*M*; Fig. 4) bound to Lck SH2 was utilized to improve the potency of this type of derivatives by optimizing the interactions with the X_{+3} pocket. The best result was obtained when the *C*-terminal isoleucine methylamide was replaced with (1R,3R)-3-amino-indan-1-carboxylic acid (compound **24**, IC_{50} = 0.03 µ*M*). This moiety was introduced to favor a hydrogen bond between the C-terminal carboxylic acid and the side chain of an arginine (Arg-67) on the rim of the X_{+3} pocket.

Synthetic combinatorial libraries have also been exploited to identify Lck SH2 antagonists *(72)*. A focused library was prepared by coupling 900 carboxylic acids to the N-terminus of Tyr(PO_3H_2)-Glu-Glu-Ile (73). Despite previous observations on the lack of SH2 selectivity or affinity for modifications introduced at the N-terminus of phosphotyrosine in most of the SH2 consensus sequences *(74)*, highly potent and selective phosphopeptides were identified by screening this library. For example, compound **25** (Fig. 4) exhibits a nearly five fold preference for the Lck SH2 domain vs that of Fyn (K_d^{Lck} = 35 n*M* vs K_d^{Fyn} = 150 n*M*) and good general SH2 selectivity (e.g., $K_d^{PLC\gamma l}$ = 4.9 µ*M*, K_d^{p85} = 9.3 µ*M* and K_d^{Grb2} = 11.3 µ*M*). In a continuation of this strategy *(72)*, two separate libraries were constructed to identify a Glu–Glu surrogate (84-member library) and a replacement of isoleucine (900-member library). This approach ultimately afforded compounds **26** (IC_{50} = 1.4 µ*M*, K_d = 2.9 µ*M*; Fig. 4) and **27** (IC_{50} = 2.4 µ*M*; Fig. 4), which exhibit binding affinities for p56^{lck} SH2 that are comparable to the activity observed for Ac-Tyr(PO_3H_2)-Glu-Glu-Ile-NH_2 (IC_{50} = 0.66 µ*M*; K_d = 1.3 µ*M*).

As an alternative to conventional screening assays, an NMR-based screening approach called SAR by NMR™ *(75,76)* was used to identify novel phosphotyrosine mimetics that bind to Lck SH2 *(77)*. This NMR method allows building blocks that bind to a target protein with low affinity (m*M* to µ*M* range) to be reliably identified. A number of phosphonates, multiply charged aromatic acids, and phthalamate analogs with binding affinities in the millimolar were identified (e.g., compound **28**, IC_{50} = 0.06 m*M*; Fig. 4), but no Lck SH2 antagonists containing these molecules have so far been disclosed.

4. ANTAGONISTS OF THE SH2 DOMAIN OF Grb2

Growth factor receptor-bound protein 2 (Grb2) is an adapter protein composed of two SH3 domains flanking a single SH2 domain *(78,79)*. This protein plays a key role in the Ras signal-transduction pathway by linking a variety of tyrosine kinase receptors (e.g., EGFR, erbB-2, c-MET), receptor-associated proteins (e.g. Syp, IRS-1), and oncogenic proteins (e.g. BCR-Abl) to the mitogen-activated protein (MAP) kinase cascade *(80,81)*. In unstimulated cells, Grb2 is located in the cytosol in complex with the guanine nucleotide exchange factor for Ras, son of sevenless (Sos), through SH3-mediated protein–protein interactions. Upon activation of receptor tyrosine kinases by growth factors, the Grb2–Sos complex translocates to the plasma membrane, and converts the inactive Ras·GDP to active Ras·GTP *(82–86)*. Activated Ras triggers the MAP kinase cascade that is essential for cell growth and differentiation. The interaction between Grb2 and the activated kinases or the phosphorylated receptor-associated proteins is mediated by the SH2 domain of the signaling protein that recognizes specific phosphotyrosine sequences. Agents that specifically disrupt the protein–protein interactions mediated by Grb2 SH2 could potentially shut down the Ras signaling pathway and present an intervention point for the treatment of hyperproliferative diseases *(8,87–89)*.

Phosphotyrosyl peptide libraries have shown that the consensus sequence for Grb2 SH2 is $Tyr(PO_3H_2)$-X_{+1}-Asn-X_{+3} and the residue that determines specificity is asparagine *(4,5,90–92)*. Numerous X-ray and NMR structures of the Grb2 SH2 domain and complexed thereof with phosphotyrosyl peptides containing the preceding consensus sequence or peptidomimetic inhibitors have been determined (for representative examples, *see* refs. *93* and *94)*. The folding of the ligand-bound SH2 domain of Grb2 shows a general pattern consisting of a central antiparallel β-sheet flanked by two α-helices, and the phosphopeptide ligand binds in a type I β-turn conformation centered around the X_{+1} and X_{+2} residues. The folded conformation of the ligand, the exclusive selectivity of Grb2 SH2 for asparagine at the X_{+2} position, and specific protein–ligand interactions identified in structural studies have been extensively exploited in the design and identification of potent and selective Grb2 SH2 antagonists *(12)*. Representative examples have been selected to illustrate this structure-based design approach.

Starting from the tripeptide Ac-Tyr-(PO_3H_2)-Ile-Asn-NH_2, which is the minimal recognition motif retaining μ*M* affinity for the Grb2 SH2 domain (IC_{50} = 8.64 μ*M*), the N-terminal acetyl group was replaced with 3-aminobenzyloxycarbonyl to form a stacking interaction with the side chain of an arginine residue (ArgαA2) and a hydrogen bond interaction with the phosphate group of phosphotyrosine *(95)*. This single replacement resulted in a 133-fold increase in binding activity (IC_{50} = 0.065 μ*M*), and was the first breakthrough in the identification of potent antagonists of the Grb2-SH2 domain. A further improvement in activity was obtained when isoleucine was replaced by 1-aminocyclohexanecarboxylic acid (compound **29**, IC_{50} = 1 n*M*; Fig. 5) *(96)*. This α,α-disubstituted cyclic α-amino acids was selected by molecular modeling to induce a local right-handed 3_{10} helical conformation and to establish multiple van der Waals interactions with the amino acids forming the Grb2 SH2 X_{+1}

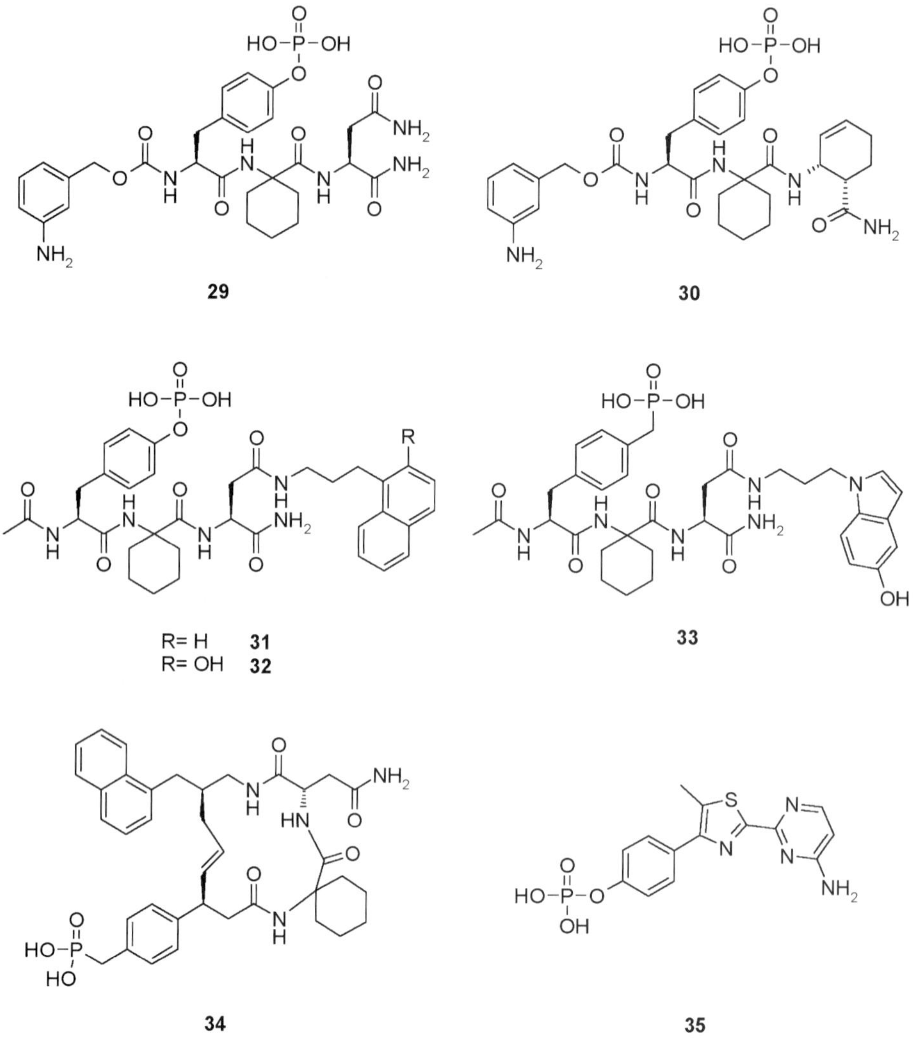

Fig. 5. Antagonists of the Grb2 SH2 domain.

binding pocket. This molecular prediction was confirmed later by X-ray crystallography *(97)*. Incorporation of Ac_6c had a positive impact not only in the Grb2 SH2 binding affinity of the modified antagonist, but also in its specificity profile against other SH2 domains. Thus, compound **29** shows 240- and 1500-fold preferential binding to Grb2 SH2 over p85 *N*-terminal SH2 and Lck SH2, respectively, and at least 1200-fold selectivity to Grb2 SH2 over Shp2 SH2 in competitive binding phosphopeptide assays *(96)*.

The core sequence of compound **29** has been extensively utilized as a "molecular platform" to identify asparagine mimetics, increase the number of interactions with the protein, or introduce conformational constraints. An overview of some of these modifications is covered in the following.

Experiments with degenerate phosphopeptide libraries (4,5,91,92) and structural studies (94) have identified and confirmed the exclusive selectivity of Grb2 SH2 for asparagine at the X_{+2} position. Asparagine occupies the $i + 2$ position of a type-I β-turn and its side chain forms hydrogen bond interactions with the backbone carbonyl groups of Lys βD6 and Leu βE4, and the NH of Lys βD6. Interactive molecular modeling showed that (1S,2R)-cyclic β-amino acids with different ring sizes were able to preserve the orientation of the carboxamide group of asparagine without clashing with the side chain of Trp EF1 (97,98). Replacement of asparagine by (1S,2R)-2-amino-cyclohex-3-ene carboxylic acid (Achec) resulted in compound 30 (Fig. 5) which has a binding affinity almost identical to that observed for compound 29 ($IC_{50} = 1.6$ nM vs $IC_{50} = 1$ nM). The X-ray crystal structure of Grb2 SH2 bound to compound 30 revealed the expected (1S,2R) configuration for Achec and showed that this β-amino acid perfectly mimics the intermolecular hydrogen bond interactions of the side chain of asparagine (97).

Several reports have described the design of inhibitors containing C-terminal substituted carboxamides (99,100). These modifications were introduced in the parent compound to create additional van der Waals contacts with an extended hydrophobic region on the surface of Grb2 SH2 (100). Noteworthy in this series is the high potency achieved with the naphtyl (e.g., compounds 31 and 32, $IC_{50} = 47$ nM and 11 nM, respectively; Fig. 5) and 3-indol-1-yl-propyl (e.g., compound 33, $IC_{50} = 0.3$ nM; Fig. 5) derivatives. Macrocyclic variants of these derivatives have also been reported (e.g., compound 34, $IC_{50} = 0.02$ µM) (101,102). The conformational constraints imposed by the cyclization were expected to increase binding affinity by favoring a β-bend conformation and reducing entropy penalties. Recently, trans- and cis-cyclopropanes have also been used to enforce locally extended and reversed turn peptide conformations in Grb2 SH2 antagonists (103).

Structure-based design efforts have also advanced nonpeptidic Grb2 SH2 antagonists. Following a minimal pharmacophore strategy, a rigid thiazole spacer was used to link two moieties that mimic the main pharmacophores of the natural ligand, the phenylphosphate of the phosphotyrosine residue, and the β-carboxamide of asparagine. The Grb2–SH2 binding affinity of the mimetic (compound 35, $IC_{50} = 26$ µM; Fig. 5) was found to be in the same range as that measured for two reference phosphopeptides (Ac-Tyr(PO$_3$H$_2$)-Xxx-Asn-NH$_2$, $IC_{50} = 67$ µM, Xxx = Gly; $IC_{50} = 4$ µM, Xxx = Val) (104).

Until very recently, the investigation of intracellular signal transduction pathways triggered by the interaction of Grb2 SH2 with phosphoproteins was compromised by the paucity of reported antagonists of Grb2 SH2 that are effective in cell-based assays. Phosphotyrosine mimetics (12,101,105–108), prodrug

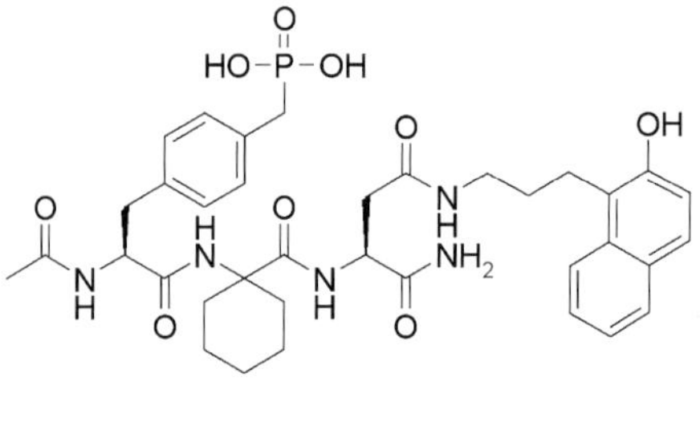

Fig. 6. Representative examples of Grb2 SH2 antagonists active in cell-based assays.

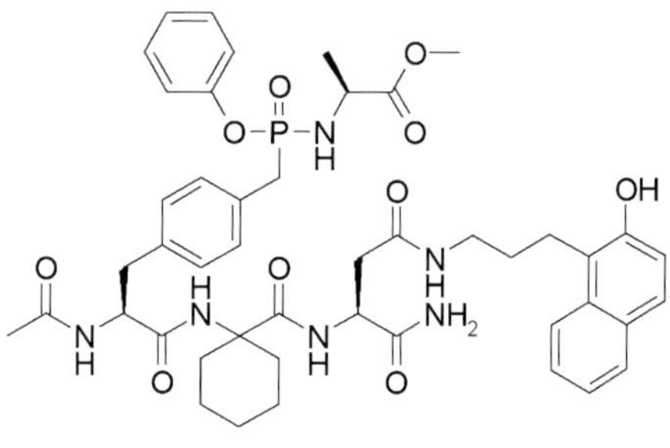

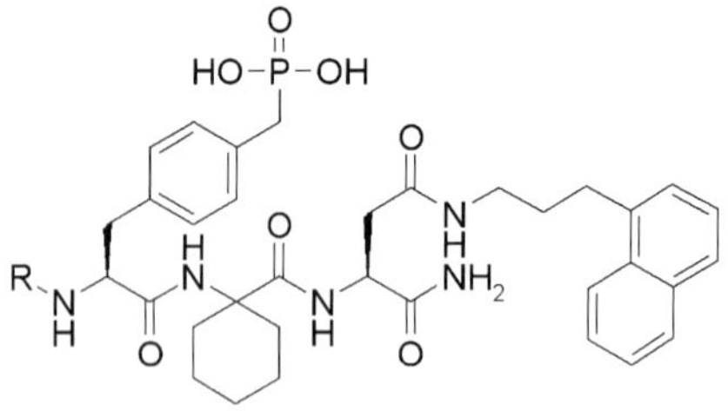

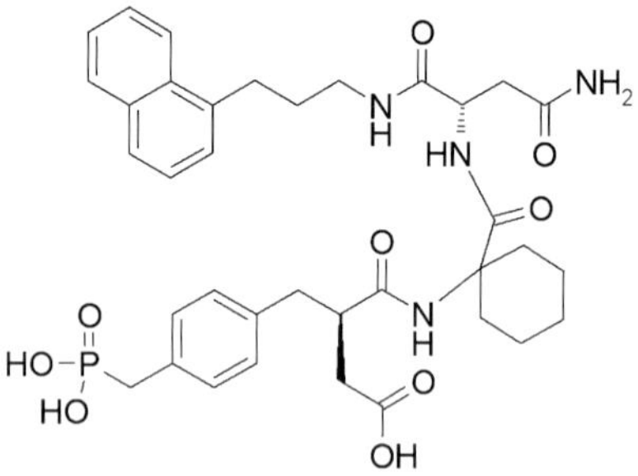

systems *(109)*, and cell-permeable vectors *(110)* have been used to improve the cellular permeability and intracellular stability of Grb2 SH2 antagonist *(109,111–113)*. These approaches have allowed the identification of tool compounds to validate the Grb2 target in cellular settings. Thus, compound **36** (Fig. 6) blocks EGFR- and Shc-Grb2 protein–protein interactions in human mammary carcinoma MDA-MB-468 cell and inhibits the anchorage-independent growth of these cells *(112)*. Grb2 SH2 inhibition by compound **37** (Fig. 6), a prodrug derivative of compound **36** with improved cell permeability, induced expression of the cell cycle inhibitors p21[Wafl/Cipl/CAPl] and p27[Kipl], reverse transformation *(112)*, and inhibit hepatocyte growth factor– induced motility *(113)*. All together these findings provide experimental evidence that targeting Grb2 SH2 is sufficient to block normal Ras function and mitogenesis, and to reverse transformation in a cellular setting. Inhibition of Grb2 SH2 binding in MDA-MB-453 cell-based systems has also been reported for analog of compound **36** containing several phosphotyrosine mimetics and N-terminal capping groups *(113,116)*. These derivatives achieved intracellular inhibition of Grb2 SH2 binding at micromolar to submicromolar concentrations (e.g., compound **38**, $IC_{50} = 0.5$ μ*M*; Fig. 6). Of particular note in this study is the usefulness of the

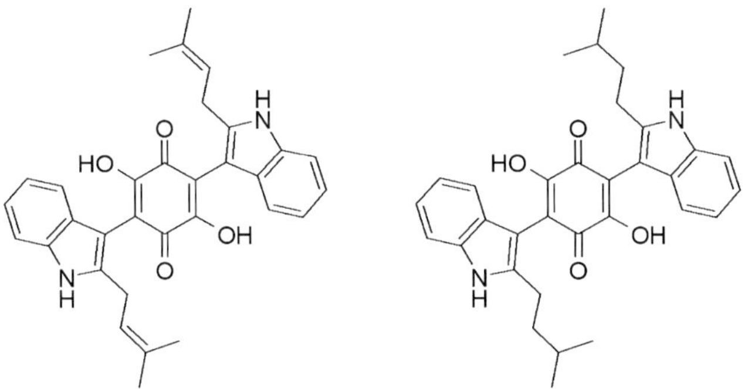

41 **42**

Fig. 7. Dihydroxyquinoline derivatives that inhibit Grb2 SH2-b mediated protein–protein interactions.

N-terminal oxalyl group to enhance the binding potency of both phosphorous and non-phosphorous-containing phosphotyrosine mimetics. Molecular modeling dynamics simulations of compound **38** in the phosphotyrosine-binding pocket of the Grb2 SH2 domain identified new interactions between the positively charged arginine-67 residue and elements of the N^α-oxalyl group. These interactions can account for the positive binding effect observed for the oxalyl-containing phosphotyrosine mimetics. The potential binding enhancements afforded by the oxalyl group have been further explored using other acidic N^α-terminal substituents *(116)*. N^α-oxalyl can be replaced by N^α-malonyl without affecting the ex-cellular Grb2 binding inhibition activity of the antagonist. Consistent with this result, compound **39** was equipotent to compound **38** in inhibiting MDA-MB-453 cell growth. Recently, antimitogenic activity ($IC_{50} = 8$ µ*M*) has also been reported for a Grb2 SH2 antagonist containing a new phosphotyrosyl mimetic *(105)*. Compound **40** (Fig. 6) exhibits potent in vitro Grb2 SH2 inhibitory activity ($IC_{50} = 8$ n*M*) and is able to inhibit intracellular association of Grb2 protein with phosphorylated p185[erbB-2] at concentrations equivalent to its antimitogenic activity. Similar results have also been obtained with derivatives of compound **34** *(101)*.

In addition to the previous synthetic compounds, several natural products and derivatives thereof have also been described to interfere with the protein–protein interactions mediated by Grb2. A series of naturally occurring *bis*(indolyl)dihydroxyquinolines have been shown to inhibit the binding of Grb2 to the tyrosine-phosphorylated form of the EGFR tyrosine kinase (e.g., compound **41**, Asterriquinone E, $IC_{50} = 2.9$ µ*M*; Fig. 7) *(117)*. Promising inhibitory properties have also been observed with a synthetic analog of the above natural product (e.g., compound **42**, $IC_{50} = 1.2$ µ*M*; Fig. 7) *(118)*. Other natural products (e.g., actinomycin D, C_2 and VII, lutein, 8-O-metylslerotiori-namine, sclerotiorin, and isochromophilone iv) have also been reported to inhibit the Shc/Grb2 protein–protein interaction in a dose-dependent manner *(119–124)*. Although the published results suggest that the preceding natural products could be nonphosphorylated Grb2 antagonists, so far no evidence has been given to confirm that these compounds interact with the pockets that exist on the phosphotyrosine-binding surface of Grb2 SH2.

5. CONCLUSIONS

Although some of the SH2 antagonists reviewed and discussed herein have been valuable tool compounds for improving our understanding of several intracellular signaling pathways, the identification of potential therapeutic agents for the treatment of diseases associated with Src, Lck, or Grb2 molecular interactions has met with limited success. Structure-based design approaches have been used to reduce the size, charge, and peptide character of SH2 antagonists while increasing their potency, selectivity, and ability to penetrate and reach the intracellular target. These medicinal chemistry efforts have led to true mechanism-based agents that have validated the concept of SH2 domain inhibition in cellular settings and animal models, but, unfortunately, none of these antagonists has so far entered clinical trials.

REFERENCES

1. Koch CA, Anderson D, Moran MF, Ellis C, Pawson T. SH2 and SH3 domains: elements that control interactions of cytoplasmic signaling proteins. *Science* 1991; 252:668–674.
2. Sadowski I, Stone JC, Pawson T. A noncatalytic domain conserved among cytoplasmic protein tyrosine kinases modifies the kinase function and transforming activity of Fujinami sarcoma virus p130$^{gag-fps}$. *Mol Cell Biol* 1986; 6:4396.
3. Ladbury JE, Arold S. Searching for specificity in SH domains. *Chem Biol* 2000; 7:R3–R8.
4. Songyang Z, Shoelson SE, McGlade J, et al. Specific motifs recognized by the SH2 domain of Csk, 3BP2, fps/fes, Grb-2, HCP, SHC, Syk, and Vav. *Mol Cell Biol* 1994; 14:2777–2785.
5. Songyang Z, Shoelson SE, Chaudhuri M, et al. SH2 domains recognize specific phospho-peptide sequences. *Cell* 1993; 72:767–778.
6. Pawson T, Gish GD, Nash P. SH2 domains, interaction modules and cellular wiring. *Trends Cell Biol* 2001; 11:504–511.
7. Pawson T. Protein modules and signalling networks. *Nature* 1995; 373:573–580.
8. Botfield MC, Green J. SH2 and SH3 domains: choreographers of multiple signaling pathways. *Annu Rep Med Chem* 1995; 30:227–237.
9. García-Echeverría C. Antagonist of Src homology 2 (SH2) domains of Grb2, Src, Lck and ZAP-70. *Curr Med Chem* 2001; 8:1715–1730.
10. Shakespeare WC. SH2 domain inhibition: a problem solved? *Curr Opin Chem Biol* 2001; 5:409–415.
11. Vu CB. Recent advances in the design and synthesis of SH2 inhibitors of Src, Grb2, and ZAP-70. *Curr Med Chem* 2000; 7:1081–1100.

12. Fretz H, Furet P, García-Echeverría C, Rahuel J, Schoepfer J. Structure based design of compounds inhibiting Grb2 SH2 mediated protein–protein interactions. *Curr Pharm Des* 2000; 6:1777–1796.

13. Gilmer T, Rodriguez M, Jordan S, et al. Peptide inhibitors of src SH3-SH2-phosphoprotein interactions. *J Biol Chem* 1994; 269:31711–31719.

14. Resh MD. Myristylation and palmitylation of Src family members: the fats of the matter. *Cell* 1994; 76:411–413.

15. Superti-Furga G. Regulation of the Src protein tyrosine kinase. *FEBS Lett* 1995; 369:62–66.

16. Boerner RJ, Kassel DB, Barker SC, Ellis B, DeLacy P, Knight WB. Correlation of the phosphorylation states of pp60c-src with tyrosine kinase activity: the intramolecular pY530-SH2 complex retains significant activity if Y419 is phosphorylated. *Biochemsitry* 1996; 35:9519–9525.

17. Luttrell DK, Lee A, lansing TJ, et al. Involvement of pp60[src] with the major signaling pathways in human breast cancer. *Proc Nat Acad Sci USA* 1994; 91:83–87.

18. Boyce BF, Yoneda T, Lowe C, Soriano P, Mundy GR. Requirement of pp60[csrc] expression for osteoclasts to form ruffled borders and resorb bone in mice. *J Clin Invest* 1992; 90:1622–1627.

19. Soriano P, Montgomery C, Geske R, Bradley A. Targeted disruption of the c-src protooncogene leads to osteopetrosis in mice. *Cell* 1991; 64:693–702.

20. Courtneidge SA, Fumagalli S, Koegl M, Superti-Furga G, Twamley-Stein GM. The Src family of protein tyrosine kinases: regulation and functions. *Dev Suppl* 1993;57–64.

21. Davidson JP, Lubman O, Rose T, Waksman G, Martin SF. Calorimetric and structural studies of 1,2,3-trisubstituted cyclopropanes as conformational constrained peptide inhibitors of Src SH2 domain binding. *J Am Chem Soc* 2002; 124:205–215.

22. Xu W, Harrison SC, Eck MJ. Three dimensional structure of the tyrosine kinase c-Src. *Nature* 1997; 385:595–602.

23. Waksman G, Shoelson SE, Pant N, Cowburn D, Kuriyan J. Binding of a high affinity phosphotyrosyl peptide the the Src SH2 domain: Crystal structures of the complexed and peptide-free forms. *Cell* 1993; 72: 779–790.

24. Waksman G, Kominos D, Robertson SC, et al. Crystal structure of the phosphotyrosine recognition domain SH2 of v-src complexed with tyrosine-phosphorylated peptides. *Nature* 1992; 358:646–653.

25. Verkhivker GM, Bouzida D, Gehlhaar DK, et al. Hierarchy of simulation models in predicting structures and energetics of the Src SH2 domains binding to tyrosyl phosphopeptides. *J Med Chem* 2002; 45:72–89.

26. Charifson PS, Shewchuk LM, Rocque W, et al. Peptide ligands of pp60c-src SH2 domains: a thermodynamic and structural study. *Biochemistry* 1997; 36:6283–6293.

27. Pacofsky GJ, Lackey K, Alligood KJ, et al. Potent dipeptide inhibitors of the pp60c-src SH2 domain. *J Med Chem* 1998; 41:1894–1908.

28. Plummer MS, Prasad JV, Para KS, et al. Hydrophobic D-amino acids in the design of peptide ligands for the pp60[src] SH2 domain. *Drug Des Discov* 1996; 13:75–81.

29. Plummer MS, Holland DR, Shahripour A, et al. Design, synthesis, and cocrystal structure of a nonpeptide Src SH2 domain ligand. *J Med Chem* 1997; 40:3719–3725.

30. Lunney EA, Para KS, Rubin JR, et al. Structure based design of novel series of nonpeptide ligands that bind to the pp60[src] SH2 domain. *J Am Chem Soc* 1997; 119:12471–12476.

31. Deprez P, Baholet I, Burlet S, et al. Discovery of highly potent Src SH2 binders: structure-activity studies and X-ray structures. *Bioorg Med Chem Lett* 2002; 12:1291–1294.

32. Lesuisse D, Deprez P, Albert E, et al. Discovery of thioazepinone ligands for the Src SH2: from non-specific to specific binding. *Bioorg Med Chem Lett* 2001; 11:2127–2131.

33. Buchanan JL, Vu ChB, Merry TJ, et al. Structure activity relationships of a novel class of Src SH2 inhibitors. *Bioorg Med Chem Lett* 1999; 9:2359–2364.
34. Buchanan JL, Bohacek RS, Luke GP, et al. Structure based design and synthesis of a novel class of Src SH2 inhibitors. *Bioorg Med Chem Lett* 1999; 9:2353–2358.
35. Deprez P, Mandine E, Vermond A, Lesuisse D. Imidazole-based ligands of the Src SH$_2$ protein. *Bioorg Med Chem Lett* 2002; 12:1287–1289.
36. Deprez P, Mandine E, Gofflo D, Meunier S, Lesuisse D. Small ligands interacting with the phosphotyrosine binding pocket of the Src SH$_2$ protein. *Bioorg Med Chem Lett* 2002; 12:1295–1298.
37. Lesuisse D, Lange G, Deprez P, et al. SAR and X-ray. A new approach combining fragment-based screening and rational drug design: application to the discovery of nanomolar inhibitors of Src SH2. *J Med Chem* 2002; 45:2379–2387.
38. Kawahata N, Yang MG, Luke GP, et al. A novel phosphotyrosine mimetic 4'-carboxymethyloxy-3'-phosphonophenylalanine (Cpp): exploitation in the design of nonpeptide inhibitors of pp60[Src] SH2 domain. *Bioorg Med Chem Lett* 2001; 11:2319–2323.
39. Sundaramoorthi R, Siedem C, Vu CB, et al. Selective inhibition of Src SH2 by a novel thiol-targeting tricarbonyl-modified inhibitor and mechanistic analysis by ^{1}H/^{13}C NMR spectroscopy. *Bioorg Med Chem Lett* 2001; 11:1665–1669.
40. Bohacek RS, Dalgarno DC, Hatada M, et al. X-ray structure of citrate bound to Src SH2 leads to a high-affinity, bone-targeted Src SH2 inhibitor. *J Med Chem* 2001; 44:660–663.
41. Violette SM, Shakespeare WC, Barlett C, et al. A Src SH2 selective binding compound inhibits osteoclast mediated resorption. *Chem Biol* 2000; 7:225–235.
42. Shakespeare W, Yang M, Bohacek R, et al. Structure-based design of an osteoclast selective, nonpeptide Src homology 2 inhibitor with in vivo antiresorptive activity. *Proc Nat Acad Sci USA* 2000; 97:9373–9378.
43. Rickles RJ, Henry PA, Guan W, et al. A novel mechanism-based mammalian cell assay for the identification of SH2-domain-specific protein–protein inhibitors. *Chem Biol* 1998; 5:529–538.
44. Stankovic CJ, Surendran N, Lunney EA, et al. The role of 4-phosphonodifluoromethyl- and 4-phosphono-phenylalanine in the selectivity and cellular uptake of SH2 domain ligands. *Bioorg Med Chem Lett* 1997; 7:1909–1914.
45. Shakespeare WC, Bohacek RS, Narula SS, et al. An efficient synthesis of a 4'phosphonodifluoromethyl3'formylphenylalanine containing Src SH2 ligand. *Bioorg Med Chem Lett* 1999; 9:3109–3112.
46. Alligood KJ, Charifson PS, Crosby R, et al. The formation of a covalent complex between a dipeptide ligand and the Src SH2 domain. *Bioorg Med Chem Lett* 1998; 8:1189–1194.
47. Machida K, Mayer BJ. The SH2 domain: versatile Signaling module and pharmaceutical target. *Biochem Biophys Act.* 2003; 1747:4–25.
48. Violette SM, Guan W, Barlett C, et al. Bone-targeted Src SH2 inhibitors block Src cellular activity and osteoclast-mediated resorption. *Bone* 2001; 28:54–64.
49. Chan AC, Desia DM, Weiss A. The role of protein-tyrosine kinases and protein-tyrosine phosphatases in T-cell antigen receptor signal-transduction. *Ann Rev Immunol* 1994; 12:555–592.
50. Weiss A, Littman D. Signal transduction by lymphocyte antigen receptors. *Cell* 1994; 76:263–266.
51. Cooper JA, Howell B. The when and how of Src regulation. *Cell* 1993; 73:1051–1054.
52. Perlmutter RM, Marth JD, Lewis DB, Peet R, Ziegler SF, Wilson CB. Structure and expression of lck transcripts in human lymphoid cells. *J Cell Biochem* 1988; 38:117–126.
53. Eck MJ, Atwell SK, Shoelson SE, Harrison SC. Structure of the regulatory domains of the Src-family tyrosine kinase Lck. *Nature* 1994; 368:764–769.

54. Turner JM, Brodsky MH, Irving BA, Levin SD, Perlmutter RM, Littman DR. Interaction of the unique N-terminal region of the tyrosine kinase p56lck with cytoplasmic domains of CD4 and CD8 is mediated by cysteine motifs. *Cell* 1990; 60:755–765.

55. Shaw AS, Amrein KE, Hammond C, Stern DF, Sefton BM, Rose JK. The lck tyrosine protein kinase interacts with the cytoplasmic tail of CD4 glycoprotein through its unique amino terminal domain. *Cell* 1989; 59:627–636.

56. Veillette A, Fournel M. The CD4 associated tzrosine protein kinase p56lck is positively regulated through its site of autophosphorylation. *Oncogene* 1990; 5:1455–1462.

57. Denny MF, Patai B, Straus DB. Differential T-cell antigen receptor signaling mediated by the Src family kinases Lck and Fyn. *Mol Cell Biol* 2000; 20:1426–1435.

58. Iwashima M, Irving BA, van Oers NSC, Chan AC, Weiss A. Sequential interactions of the TCR with two distinct cytoplasmic tyrosine. *Science* 1994; 263:1136–1139.

59. Couture C, Songyang Z, Jascur T, et al. Regulation of the Lck SH2 domain by tyrosine phosphorylation. *J Biol Chem* 1996; 271:24880–24884.

60. Strauss DB, Chan AC, Patai B, Weiss A. SH2 domain function is essential for the role of the Lck tyrosine kinase in T cell receptor signal transduction. *J Biol Chem* 1996; 271:9976–9981.

61. Sundvold V, Torgersen KM, Post NH, et al. Cutting edge: T cell-specific adapter protein inhibits T cell activation by modulating lck activity. *J Immunol* 2000; 165:2927–2931.

62. van Oers NSC, Lowinkropf B, Finlay D, Connolly K, Weiss A. αβ T cell development is abolished in mice lacking both Lck and Fyn protein tyrosine kinases. *Immunity* 1996; 5:429–436.

63. Straus DB, Weiss A. Genetic evidence for the involvement of the Lck tyrosine kinase in signal transduction through the T-cell antigen receptor. *Cell* 1992; 70:585–593.

64. Dowden J, Ward SG. Inhibitors of p56[lck]: assessing their potential as tools as manipulating T-lymphocyte activation. *Expert Opin Ther Patents* 2001; 11:295–306.

65. Broadbridge RJ, Sharma RP. The Src homology-2 domains (SH2 domains) of the protein tyrosine kinase p56[lck]: structure, mechanism and drug design. *Curr Drug Targets* 2000; 1:365–386.

66. Eck MJ, Shoelson SE, Harrison SC. Recognition of a high-affinity phosphotyrosyl peptide by the Src homology-2 domain of p56lck. *Nature* 1993; 362:87–91.

67. Llinas-Brunet M, Beaulieu PL, Cameron DR, et al. Phosphotyrosine-containing dipeptides as high-affinity ligands for the p56[lck] SH2 domain. *J Med Chem* 1999; 42:722–729.

68. Morelock MM, Ingraham RH, Betageri R, Jakes S. Determination of receptor-ligand kinetics and equilibrium binding constants using surface plasmon resonance: application to the Lck SH2 domain. *J Med Chem* 1995; 38:1309–1318.

69. Beaulieu PL, Cameron DR, Ferland JM, et al. Ligands for the tyrosine kinase p56lck SH2 domain: discovery of potent dipeptide derivatives with monocharged, nonhydrolyzable phosphate replacements. *J Med Chem* 1999; 42:1757–1766.

70. Proudfoot JR, Betageri R, Cardozo M, et al. Nonpeptidic, monocharged, cell permeable ligands for the p56lck SH2 domain. *J Med Chem* 2001; 44:2421–2431.

71. Hobbs CJ, Bit RA, Cansfield AD, et al. Structure-based design of peptidomimetic antagonists of p56[lck] SH2 domain. *Bioorg Med Chem Lett* 2002; 12:1365–1369.

72. Lee TR, Lawrence DS. Acquisition of highaffinity, SH2 targeted ligands via a spatially focused library. *J Med Chem* 1999; 42:784–787.

73. Lee TR, Lawrence DS. SH2-directed ligands for the Lck tyrosine kinase. *J Med Chem* 2000; 43:1173–1179.

74. Kuriyan J, Cowburn D. Modular peptide recognition domains in eukaryotic signaling. *Annu Rev Biophys Biomol Struct* 1997; 26:259–288.

75. Shuker SB, Hajduk PJ, Meadows RP, Fesik SW. Discovering high-affinity ligands for proteins: SAR by NMR. *Science* 1996; 274:1531–1534.

76. Hajduk PJ, Meadows RP, Fesik SW. Drug design: discovering high-affinity ligands for proteins. *Science* 1997; 278:497–499.

77. Hajduk PJ, Zhou MM, Fesik SW. NMR-based discovery of phosphotyrosine mimetics that bind to the Lck SH2 domain. *Bioorg Med Chem Lett* 1999; 9:2403–2406.

78. Lowenstein EJ, Daly RJ, Batzer AG, et al. The SH2 and SH3 domain-containing protein GRB2 links receptor tyrosine kinases to ras signaling. *Cell* 1992; 70:431–442.

79. Chardin P, Cussac D, Maignan S, Ducruix A. The Grb2 adaptor. *FEBS Lett* 1995; 369: 47–51.

80. Downward J. Control of ras activation. *Cancer Surv* 1996; 27:87–100.

81. Downward J. The GRB2/Sem-5 adaptor protein. *FEBS Lett* 1994; 338:113–117.

82. Buday L, Egan SE, Rodriguez V, Cantrell DA, Downward J. A complex of Grb2 adaptor protein, Sos exchange factor, and a 36-kDa membrane-bound tyrosine phosphoprotein is implicated in Ras activation in T cells. *J Biol Chem* 1994; 269:9019–9023.

83. Reif K, Buday L, Downward J, Cantrell DA. SH3 domains of the adapter molecule Grb2 complex with two proteins in T cells: the guanine nucleotide exchange protein Sos and a 75-kDa protein that is a substrate for T cell antigen receptor-activated tyrosine kinases. *J Biol Chem* 1994; 269:14081–14087.

84. Li N, Batzer A, Daly R, et al. Guanine-nucleotide-releasing factor hSos1 binds to Grb2 and links receptor tyrosine kinases to Ras signalling. *Nature* 1993; 363:85–88.

85. Gale NW, Kaplan S, Lowenstein EJ, Schlessinger J, Bar-Sagi D. Grb2 mediates the EGF-dependent activation of guanine nucleotide exchange on Ras. *Nature* 1993; 363:88–92.

86. Buday L, Downward J. Epidermal growth factor regulates p21ras through the formation of a complex of receptor, Grb2 adapter protein, and Sos nucleotide exchange factor. *Cell* 1993; 73:611–620.

87. Verbeek BS, Adriaansen-Slot SS, Rijksen G, Vroom TM. Grb2 overexpression in nuclei and cytoplasm of human breast cells: a histochemical and biochemical study of normal and neoplastic mammary tissue specimens. *J Pathol* 1997; 183:195–203.

88. Lowe PN, Skinner R. Regulation of Ras signal transduction in normal and transformed cells. *Cell Signalling* 1994; 6:109–123.

89. Dati C, Muraca R, Tazartes O, et al. CerbB2 and Ras expression levels in breast cancer are correlated and show a cooperative association with unfavorable clinical outcome. *Int J Cancer* 1991; 65:45–50.

90. Kessels HWHG, Ward AC, Schumacher TNM. Specificity and affinity motifs for Grb2 SH2-ligand interactions. *Proc Natl Acad Sci USA* 2002; 99:8524–8529.

91. Gram H, Schmitz R, Zuber JF, Baumann G. Identification of phosphopeptide ligands for the Src-homology 2 (SH2) domain of Grb2 by phage display. *Eur J Biochem* 1997; 246: 633–637.

92. Mueller K, Gombert FO, Manning U, et al. Rapid identification of phosphopeptide ligands for SH2 domains. Screening of peptide libraries by fluorescence-activated bead sorting. *J Biol Chem* 1996; 271:16500–16505.

93. Ogura K, Tsuchiya S, Terasawa H, et al. Solution structure of the SH2 domain of Grb2 complexed with the Shc-derived phosphotyrosine-containing peptide. *J Mol Biol* 1999; 289: 439–445.

94. Rahuel J, Gay B, Erdmann D, et al. Structural basis for specificity of Grb2SH2 revealed by a novel ligand binding mode. *Nat Struct Biol* 1996; 3:586–589.

95. Furet P, Gay B, Garcia-Echeverria C, et al. Discovery of 3-Aminobenzyloxycarbonyl as an N-Terminal group conferring high affinity to the minimal phosphopeptide sequence recognized by the Grb2-SH2 domain. *J Med Chem* 1997; 40:3551–3556.

96. Garcia-Echeverria C, Furet P, Gay B, et al. Potent antagonists of the SH2 domain of Grb2: optimization of the X+1 position of 3-Amino-Z-Tyr(PO$_3$H$_2$)-X$_{+1}$-Asn-NH$_2$. *J Med Chem* 1998; 41:1741–1744.

97. Furet P, Garcia-Echeverria C, Gay B, Schoepfer J, Zeller M, Rahuel J. Structure-based design, synthesis, and X-ray crystallography of a high-affinity antagonist of the Grb2-SH2 domain containing an asparagine mimetic. *J Med Chem* 1999; 42:2358–2363.

98. Furet P, Gay B, Schoepfer J, et al. Potent Grb2 SH2 antagonists containing asparagine mimetics. *Proc Am Pept Symp* 2000; 17:573–575.

99. Schoepfer J, Fretz H, Gay B, et al. Highly potent inhibitors of the Grb2-SH2 domain. *Bioorg Med Chem Lett* 1999; 9:221–226.

100. Furet P, Gay B, Caravatti G, et al. Structure-based design and synthesis of high affinity tripeptide ligands of the Grb2-SH2 domain. *J Med Chem* 1998; 41:3442–3449.

101. Wei C-Q, Gao Y, Lee K, et al. Macrocyclization in the design of Grb2 SH2 domain-binding ligands exhibiting high potency in whole-cell systems. *J Med Chem* 2003; 46:244–254.

102. Gao Y, Voigt J, Wu JX, Yang D, Burke TR, Jr. Macrocyclization in the design of a conformationally constrained Grb2 SH2 domain inhibitor. *Bioorg Med Chem Lett* 2001; 11:1889–1892.

103. Plake HR, Sundberg TB, Woodward AR, Martin SF. Design and synthesis of conformationally constrained, extended and reverse turn pseudopeptides as Grb2-SH2 domain antagonists. *Tetrahedron Lett* 2003; 44:1571–1574.

104. Caravatti G, Rahuel J, Gay B, Furet P. Structure-based design of a non-peptidic antagonist of the SH2 domain of Grb2. *Bioorg Med Chem Lett* 1999; 9:1973–1978.

105. Wei C-Q, Li B, Guo R, Yang D, Yang D, Burke TR, Jr. Development of a phosphatase-stable phosphotyrosyl mimetic suitably protected for the synthesis of high-affinity Grb2 SH2 domain-binding ligands. *Bioorg Med Chem Lett* 2002; 12:2781–2784.

106. Cody WL, Lin Z, Panek RL, Rose DW, Rubin JR. Progress in the development of inhibitors of SH2 domains. *Curr Pharm Des* 2000; 6:59–98.

107. Garcia-Echeverria C. Preparation of glyco-, phospho-, and sulfopeptides. In: Kates SA, Albericio F, eds. *Solid-Phase Synthesis: A Practical Guide.* New York Basel: Marcel Dekker, 2000:419–473.

108. Burke TR, Jr., Gao Y, Yao Z-J, Voigt J, Luo J, Yang D. Potent non phosphate containing Grb2 SH2 domain inhibitors. *Pept Sci* 2000;49–52.

109. Liu Q-W, Vidal M, Mathé C, Périgaud C, Garbay C. Inhibition of Ras-dependent mitogenic pathway by phosphopeptide prodrugs with antiproliferative properties. *Bioorg Med Chem Lett* 2000; 10:669–672.

110. Dunican DJ, Doherty P. Designing cell-permeant phosphopeptides to modulate intracellular signaling pathways. *Biopolymers* 2001; 60:45–60.

111. Gay B, Suarez S, Caravatti G, Furet P, Meyer Thomas, Schoepfer J. Selective Grb2 SH2 inhibitors as anti-ras therapy. *Int J Cancer* 1999; 83:235–241.

112. Gay B, Suarez S, Weber C, et al. Effect of potent and selective inhibitors of the Grb2 SH2 domain on cell motility. *J Biol Chem* 1999; 274: 23311–23315.

113. Yao ZJ, King CR, Cao T, et al. Potent inhibition of Grb2 SH2 domain binding by non-phosphate-containing ligands. *J Med Chem* 1999; 42:25–35.

114. Burke TR, Jr., Luo J, Yao ZJ, et al. Monocarboxylic-based phosphotyrosyl mimetics in the design of Grb2 SH2 domain inhibitors. *Bioorg Med Chem Lett* 1999; 9:347–352.

115. Ye B, Akamatsu M, Shoelson SE, et al. L-O-(2-Malonyl)tyrosine: a new phosphotyrosyl mimetic for the preparation of Src homology 2 domain inhibitory peptides. *J Med Chem* 1995; 38:4270–4270.

116. Burke TR, Jr., Yao Z-J, Gao Y, et al. N-terminal carboxyl and tetrazole-containing amides as adjuvants to Grb2 SH2 domain ligand binding. *Bioorg Med Chem Lett* 2001; 9:1439–1445.

117. Alvi KA, Pu H, Luche M, et al. Asterriquinones produced by Aspergillus candidus inhibit binding of the Grb-2 adapter to phosphorylated EGF receptor tyrosine kinase. *J Antibiot* 1999; 52:215–223.

118. Harris GD, Jr., Nguyen A, App H, Hirth P, McMahon G, Tang C. A One-pot, two-step synthesis of tetrahydro Asterriquinone E. *Org Lett* 1999; 1:431–433.
119. Nam J-Y, Son K-H, Kim H-K, et al. Sclerotiorin and isochromophilone IV: inhibitors of the Grb2-Shc interaction, isolated from *Penicillium multicolor* F1753. *J Microbiol Biotechnol* 2000; 10:544–546.
120. Kim H-K, Nam J-Y, Han MY, et al. Natural and synthetic analogues of Actinomycin D as Grb2-SH2 domain blockers. *Bioorg Med Chem Lett* 2000; 10:1455–1457.
121. Nam J-Y, Kim H-K, Son K-H, et al. 8-O-Methylsclerotiorinamine, antagonists of the Grb2-SH2 domain, isolated from *Penicillium multicolor*. *J Nat Prod* 2000; 63:1303–1305.
122. Lee E-S, Ahn B-T, Lee S-B, Kim H-K, Bok S-H, Jeong T-S. Isolation of Grb2-Shc domain binding inhibition component from Agastache rugosa. *Saengyak Hakoechi* 2000; 30:404–408.
123. Kim HK, Nam JY, Han MY, et al. Actinomycin D as a novel SH2 domain ligand inhibits Shc/Grb2 interaction in B104-1-1 (neu-transformed NIH3T3) and SAA (hEGFR-overexpressed NIH3T3) cells. *FEBS Lett* 1999; 453:174–178.
124. Nam JY, Kim HK, Son KH, et al. Actinomycin D, C2 and VII, inhibitors of GRB2-SHC interaction produced by Streptomyces. *Bioorg Med Chem Lett* 1998; 8:2001–2002.

3 PI3-Kinase Inhibition
A Target for Therapeutic Intervention

Peter M. Finan, PhD,
and Stephen G. Ward, PhD

CONTENTS

THE PI3-KINASE FAMILY OF ENZYMES
PHOSPHOINOSITIDE PHOSPHATASES
DOWNSTREAM EFFECTORS OF PI3-K SIGNALING
PI3-K INHIBITORS
GENE TARGETING OF PI3-K ISOFORMS
PI3-K INHIBITION: TARGETING MULTIPLE COMPONENTS
 OF TUMORIGENESIS
PI3-K AND CELL MIGRATION
INHIBITION OF PI3-K SIGNALING IN TUMOR CELLS
PI3-K INHIBITION: PROSPECTS FOR DRUG DEVELOPMENT
REFERENCES

1. THE PI3-KINASE FAMILY OF ENZYMES

The activation of cells by a wide variety of stimuli leads to rapid changes in 3-phosphorylated inositol lipids through the action of a family of enzymes known as phosphoinositide 3-kinases (PI3-Ks). PI3-K activation is central to the coordinated control of multiple cell signaling pathways leading to cell growth, cell proliferation, cell survival, and cell migration. The PI3-Ks have been classified into three groups according to their primary sequence and domain structure, mode of regulation, and substrate specificity in vitro (Fig. 1) (1). The class IA PI3-K subgroup consist of three catalytic subunits, p110α, β, and δ, which form heterodimers with one of five SH2 (Shc homology) domain-containing regulatory subunits, p85α, p85β, p55γ, p55α, and p50α. The class IA heterodimer can be recruited either directly to cell surface recep-

From: *Cancer Drug Discovery and Development:*
Protein Tyrosine Kinases: From Inhibitors to Useful Drugs
Edited by: D. Fabbro and F. McCormick © Humana Press Inc., Totowa, NJ

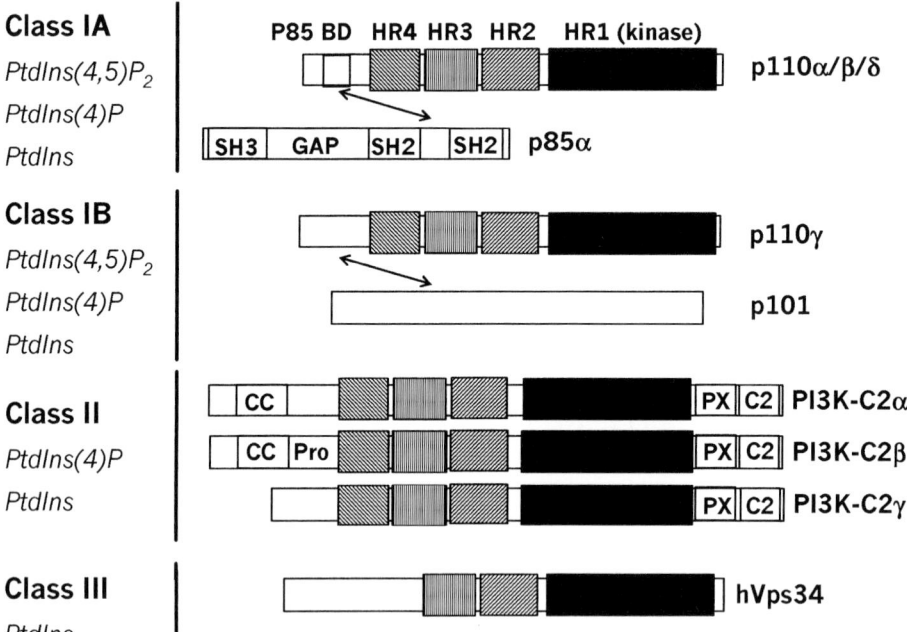

Fig. 1. The phosphoinositide (PI3-K) superfamily. Schematic representation of the modular structure of PI3-K catalytic and regulatory subunits (adapted from refs. *1, 106*). In vitro substrates for each class of PI3-K are shown in italics. The adaptor subunit p85α is shown. The domain structure of p85β is similar; p55α, p50α, and p55γ are shorter forms consisting of the SH2 domains and inter-SH2 domain alone. Class IA heterodimer formation is mediated by interactions between the N-terminal region of the catalytic subunit (P85BD) and the inter-SH2 domain of the adaptor subunit. Class IB heterodimer protein–protein interactions are less well defined and occur in multiple regions. HR, homology region; P85BD, P85 Binding domain; SH3, Src homology 3 domain; SH2, Src homology 2 domain; GAP, RhoGAP domain; CC, Coiled-coil region; Pro, Proline rich region; PX, Phox domain; C2, C2 domain.

tors, e.g., growth factor receptors, or indirectly by adaptor molecules such as Shc, growth factor receptor bound protein, (Grb2) or insulin receptor substrate (IRS)-1 *(2)*. P110δ was originally identified in leukocytes but is also expressed in other cell types including breast tissue and melanocytes *(3)*. The p110α and p110β isoforms are ubiquitous. Class IB consists of one member, a heterodimer of p110γ and a regulatory subunit termed p101, and is activated by G protein βγ subunits following the stimulation of G protein-coupled receptors (GPCRs). The expression of p110γ is predominantly in leukocytes but this isoform is also found in cardiac tissue. Both class IA and IB catalyze the formation of PtdIns

(3,4,5)P$_3$ in vitro and in addition to regulation by cell surface receptors, they can be activated directly by the small GTPase, Ras *(4)*. Class II consists of three isoforms, PI3-K-C2α, β, and γ, which phosphorylate PtdIns and PtdIns (4)P in vitro. PI3-K-C2α and β are widely expressed whereas PI3-K-C2γ is restricted to liver, breast, and prostate. More recently, a systematic immunohistochemical analysis has revealed the differential expression of PI3-K-C2α and C2β in smooth muscle and peripheral nerves respectively *(5)*. Class II PI3-Ks are predominantly associated with both plasma membrane and intracellular membrane compartments and their mechanism of regulation is not well understood. Recent evidence has indicated that clathrin functions as an adaptor for PI3-K-C2α, binding to its N-terminal region and stimulating its activity. In addition, the activation of Class II enzymes by growth factors *(6,7)* and the chemokine MCP-1 *(8)* has been reported. The class III PI3-K displays a high degree of homology to the yeast vacuolar sorting protein, VPS34, and utilizes PtdIns as a substrate. The mammalian enzyme controls the maturation of phagosomes *(9)* and also plays a role in vesicular transport *(10)*. Finally, a number of other kinases (mTOR, DNA-PK, ATM, SMG-1) are distantly related to the PI3-Ks; but as these enzymes do not phosphorylate inositol lipids, they will not be detailed further.

2. PHOSPHOINOSITIDE PHOSPHATASES

The cellular level of PtdIns (3,4,5)P$_3$ is tightly controlled by the regulation of PI3-K activation and by the action of multiple phosphoinositide phosphatases *(1)*. One of the most interesting of these is PTEN (also known MMAC1), which was originally described as a tumor suppressor mapping to 10q23 *(11)*. PTEN catalyzes the removal of phosphate from the 3′ position of the inositol ring and will convert PtdIns (3,4,5)P$_3$ to PtdIns (4,5)P$_2$. In contrast, Src homology α-containing inositol (SHIP) 5-phosphatases possess 5′ phosphatase activity and convert PtdIns (3,4,5)P$_3$ to PtdIns (3,4)P$_2$ *(12)*. PtdIns (3,4)P$_2$ can also act as a second messenger by recruiting pleckstrin homology (PH) domain-containing proteins to the cell membrane so, rather than eliminating signaling downstream of PI3-K activation, SHIP may act as a switch controlling different cellular events.

3. DOWNSTREAM EFFECTORS OF PI3-K SIGNALING

The varied cellular responses following PI3-K activation are controlled, in part, by proteins containing PH domains that bind directly to phosphoinositides including PtdIns (3,4,5)P$_3$ and relocate signaling complexes to the plasma membrane *(13)*. PH-domain-containing proteins include the serine/threonine kinases, phosphoinositide-dependent kinase 1 (PDK1), protein kinase B (PKB), and members of the protein kinase C family; and tyrosine kinases from the Tec family such as Btk, phospholipase Cγ, and guanosine diphosphate (GDP)-GTP (guanosine triphosphate) exchange factors for small GTPases, Rac, and ARF6 *(14)*. One of the most studied targets downstream of PI3-K activation is PKB (also known as Akt). This serine/threonine kinase is activated in part via phosphorylation by PDK-1, which itself is regulated by PI3-K *(15)*. PKB activation leads to the phosphorylation of a multitude of proteins affecting cell cycle progression, cell survival, and cell growth; these substrates are detailed elsewhere in this chapter.

4. PI3-K INHIBITORS

The dissection of PI 3–kinase-signaling pathways has been greatly aided by the availability of two pharmacological tools, wortmannin and LY294002 (Fig. 2). The fungal metabolite, wortmannin, inhibits PI3-K in the low-nanomolar range *(16)* and binds covalently to a conserved lysine residue in the adenosine triphosphate (ATP)–binding site of PI3-K *(17)*. LY294002 is a reversible, ATP-competitive inhibitor with an IC_{50} for recombinant PI3-K in the low-micromolar range *(18)*. The binding mode of wortmannin, LY294002, and other nonselective kinase inhibitors has been further refined through the elucidation of several X-ray crystal structures of p110γ-inhibitor complexes *(19)*. It is important to point out that neither inhibitor displays selectivity for different members of the PI3-K family with the exception of PI3-K-C2α, which is less sensitive to both inhibitors *(20)*. LY294002 also inhibits casein kinase 2 with similar potency to PI3-K and has been shown to block kV currents directly *(21)*. At higher concentrations, wortmannin inhibits PI3-K-related enzymes (e.g., mTOR, ATM), PI4-K β, and myosin light chain kinase. In addition, methylxanthines such as caffeine and theophylline (Structures 3 and 4, Fig. 2) inhibit p110δ although their activity is rather weak *(22)*. Recently, a number of patent specifications have been published that describe inhibitors of PI3-K, including compounds that exhibit some selectivity for individual isoforms. PIramed have described several imidazopyridine derivatives (e.g. compounds of general Structure 5, Fig. 2). These are claimed to exhibit excellent PI3-K inhibitory activities, especially against p110α, although no isoform selectivity data is provided. Kinacea have disclosed a series of morpholino-substituted compounds related to LY294002 that showed isoform selectivity. They describe quinolone and pyridopyrimidine compounds (Structures 6 and 7, Fig. 2) that are approx 100-fold more potent against p110α/β isoforms compared to the p110γ isoform. ICOS Corporation have recently described IC87114, which is a selective p110δ inhibitor containing a quinazoline core structure (Structure 8, Fig. 2). This compound has an IC_{50} of 0.5 μ*M* for p110δ inhibition and more than 50-fold selectivity over the other class I PI3-K isoforms *(23)*. Finally, a patent spec-

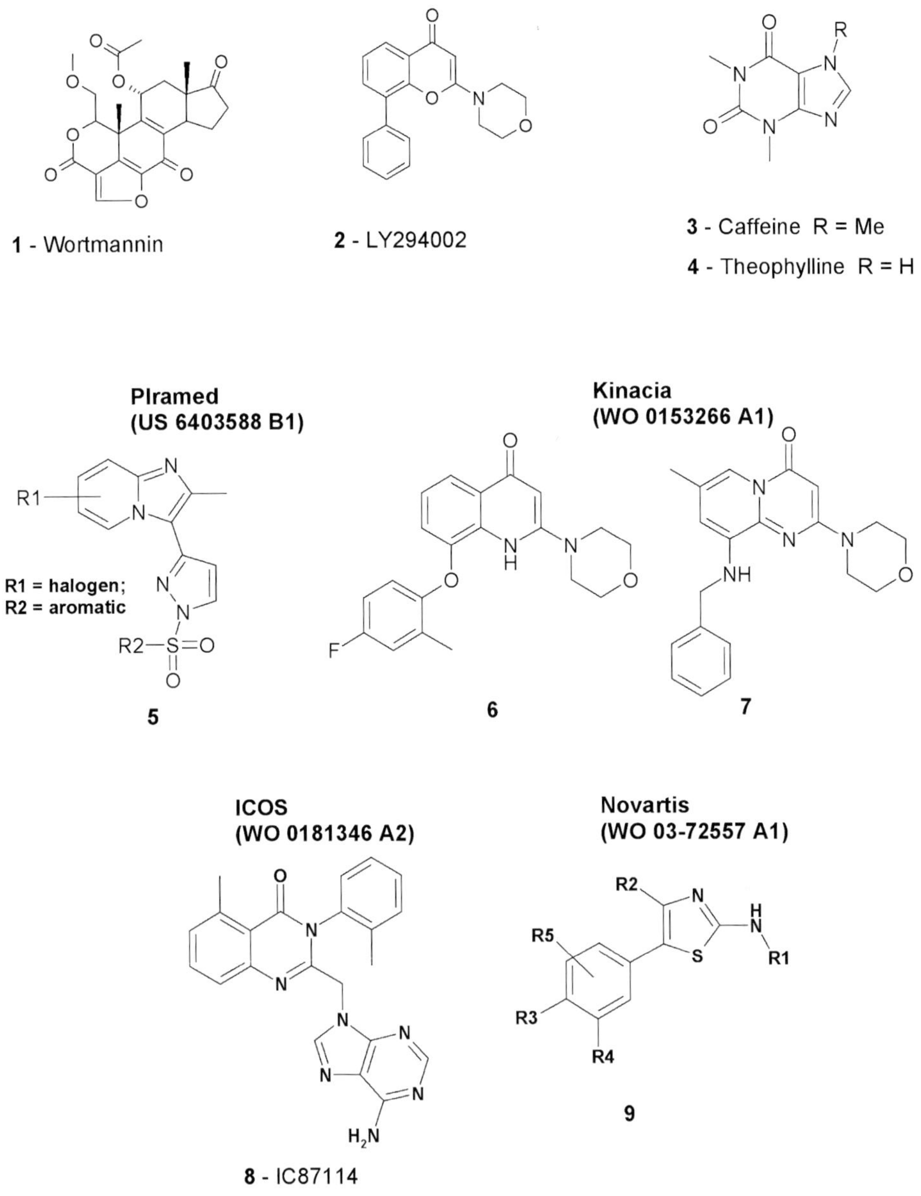

Fig. 2. PI3-K inhibitor structures. Chemical structures of known PI3-K inhibitors from the scientific and patent literature (patent numbers are in parentheses).

ification from Novartis describes 5-phenylthiazole derivatives as PI3-K inhibitors. The diverse chemical scaffolds reported in these patents represent promising lead compounds for the future development of PI3-K isoform selective inhibitors.

5. GENE TARGETING OF PI3-K ISOFORMS

With the limited availability of isoform selective inhibitors, much of our understanding of the roles of different PI3-Ks has come from gene targeting in mice, and this topic has been subject of some excellent reviews *(24,25)*. Targeting of the regulatory and catalytic subunits of PI3-K has generated much-needed insight into the roles of PI3-Ks in vivo but has been complicated by the level of redundancy between adaptor molecules and uncontrollable changes in protein expression that occur when one component of the heterodimer is removed. All four class I catalytic subunits have been inactivated by gene targeting in mice. P110α knockout mice die between days 9.5 and 10.5, and fibroblasts derived from these mice fail to proliferate *(26)*. There is a large increase in the expression of the adaptor molecules p85α and p55γ in these cells, and the authors do not rule the possibility that the phenotype may be partly due to the adaptor subunits acting as dominant negative proteins rather than a simple loss of function of p110α. Mice lacking p110β die by day 3.5 *(27)*. Interesting to note, both p110α and p110β heterozygous mice and double heterozygous mice are healthy and viable. This suggests a lack of redundancy between the isoforms. Further elucidation of the roles of p110α and p110β should be forthcoming when tissue-specific knockouts are produced.

Two different approaches have been adopted to address the functional role of p110δ: the elimination of the expression of p110δ *(28,29)* and the generation of "knock-in" mice that express a mutated form of p110δ *(30)*. These mice contain a point mutation (p110δ^{D910A}) within the catalytic domain that abolishes lipid kinase activity while maintaining normal levels of protein expression. All three p110δ-mutant mice display a block in B-cell differentiation at the bone marrow stage, reduced B-cell numbers in the spleens of adult mice, and a profound reduction in serum immunoglobulin production. The lack of responsiveness to anti-IgM correlates with reduced PI3-K-dependent signaling events in B-cells such as impaired calcium flux and severely abrogated PKB phosphorylation in both p110δ$^{-/-}$ and p110δ$^{D910A/D910A}$ cells. In addition, T-cells isolated from p110δ$^{D910A/D910A}$ mice show reduced proliferation in response to anti-CD3 but responded normally to anti-CD3 and anti-CD28 co-stimulation.

P110γ knockout mice are healthy and viable but display impaired neutrophil and macrophage chemotaxis in vitro and in vivo *(31–33)*. Further examination of neutrophil migration from these mice suggests that p110γ is essential for the direction of cell movement along a chemoattractant gradient *(34)*. In addition, mast cells derived from p110γ mice fail to degranulate fully in vitro in response to GPCR agonists such as adenosine *(35)*. Mice lacking p110γ fail to form edema after intradermal injection of adenosine or when challenged by passive systemic anaphylaxis. In platelets, p110γ mediates ADP-induced platelet aggregation. This protects p110γ-deficient mice from death caused by ADP-induced platelet-dependent thromboembolic vascular occlusion. However, bleeding time in the mice was unaffected *(36)*. Finally, p110γ negatively modulates cardiac contractility, supporting the potential of PI3-K γ inhibition in the treatment of heart failure *(37,38)*.

Two deletions of the p85α regulatory subunit have been reported. A deletion of the first exon of p85α interrupts the expression of the full-length protein but does not affect the p55α and p50α splice variants *(39)*. These animals develop a B-cell immunodeficiency and, surprisingly, increased insulin sensitivity and hypoglycemia. Deletion of all splice variants of p85α and leads to death of most of the animals within 4 d *(40)*. Interpretation of this phenotype is hampered owing to upregulation of p85β and a reduction in the levels of p110α and p110β, supporting a role for p85α in protecting the catalytic subunits from proteolysis. P85β deletion also results in mice with improved insulin sensitivity and hypoglycemia *(41)*.

6. PI3-K INHIBITION: TARGETING MULTIPLE COMPONENTS OF TUMORIGENESIS

Many oncogenic signaling pathways are controlled by PI3-K and phosphatidylinositol phosphate 3´-phosphate (PTEN), so it would seem likely that there would be potential benefit from developing inhibitors of the PI3-K/PKB-signaling cassette *(42–44)*. The first evidence for the role of PI3-K in cellular transformation came from studies showing the association of PI3-K activity with the oncogenic proteins polyoma middle-T antigen and Rous sarcoma pp60v-src *(45,46)*. Later, the analysis of the avian sarcoma virus genome identified an ortholog of p110α that was capable of transforming chicken embryo fibroblasts and inducing hemangiosarcomas in chickens *(47)*. A truncated form of the regulatory subunit p85α, termed p65, has been isolated from a murine thymic lymphoma cell line *(48)*. The p65/p110α complex is constitutively active, and transgenic mice expressing p65 in T-lymphocytes develop a lymphoproliferative disorder and renal autoimmune disease *(49)*. Another mutated form of p85α has been described in a Hodgkin's lymphoma-derived cell line *(50)*. Amplification of the *PIK3CA* gene has been observed in some ovarian, cervical, and head and neck squamous-cell carcinomas *(51–53)*. In addition, a high frequency of *PIK3CA* mutations have been identified in a variety of cancers including colon, breast, gastric, lung, and ovary *(53a–53c)*. These mutations cause amino acid substitutions in hotspots throughout the p110α protein, but regions in the helical domain and kinase domain are most frequently mutated. These mutants enhance lipid kinase activity and transform chicken embryonic fibroblasts with high efficiency *(53d)*. PKB amplification and increased PKB kinase activity have been widely reported in tumor tissues *(54)*. Somatic mutations of the lipid phosphatase PTEN are found in an increasing number of human cancers and can lead to partial or total absence of the protein or complete inactivation of the phosphatase activity *(55)*. In other tumors, the reduction of expression occurs through epigenetic silencing of PTEN *(56,57)*. Increased PI3-K activity or loss of PTEN function leads to aberrant control of the cell cycle, increased survival from apoptosis, increased cell growth, and promotion of a metastatic phenotype. PI3-K inhibitors may block each of these facets of tumorigenesis and will be discussed in more detail.

6.1. PI3-K and Control of the Cell Cycle

Cell cycle progression requires the coordination of multiple signaling pathways including the Ras/MAPK and PI3-K/PKB pathways. However, the role of PI3-K is both cell type and stimulus dependent. For example, microinjection of antibodies that block p110α activity inhibits DNA synthesis in response to epidermal growth factor (EGF) and PDGF but not to other growth factors *(58)*. The PI3-K pathway controls both the degradation and the subcellular localization of many well-characterized gatekeepers of the cell cycle and a number of these proteins are PKB substrates. The cyclin-dependent kinase p27^{Kip1} normally resides in the nucleus but is located in the cytoplasm in a variety of tumors. Recent studies have shown that p27^{Kip1} phosphorylation by PKB blocks entry into the nucleus and that p27^{Kip1} cytosolic localization correlates with poor patient survival *(59–61)*. GSK3 is phosphorylated and inactivated by PKB, leading to stabilization of cyclin D1 and c-Myc *(62,63)*. In addition, phosphorylation of the FOXO family of forkhead transcription factors by PKB retains them in the cytoplasm. This promotes cyclin D1 expression and represses the expression of the cell cycle inhibitors p27^{Kip1} and p130RB2 *(64)*.

6.2. PI3-K and Cell Survival

The PI3-K/PKB pathway plays a central role in the control of apoptosis or programmed cell death. The expression of dominant-negative PKB blocks cell survival *(65)* whereas expression of activated PI3-K protects cells from apoptosis upon withdrawal of growth factors *(66)*. These effects are mediated in part through the phosphorylation of the Bcl-2 family member BAD on serine 112 by PKB *(67)*. Phosphorylated BAD is no longer able to bind Bcl-2 and Bcl-X$_L$, releasing them to restore a cell survival signal. In addition, PKB phosphorylates and inactivates the pro-apoptotic protease, caspase 9 *(68)*, and blocks the transcription of the pro-apoptotic forkhead-regulated genes such as FasL *(69)*. Another substrate of PKB is the oncoprotein, Mdm2. Phosphorylation of Mdm2 promotes entry into the nucleus, leading to the degradation of the pro-apoptotic tumor suppressor, p53 *(70)*.

6.3. PI3-K and Cell Growth

The regulation of cell growth, i.e., cell size, is another key feature of tumorigenesis that is under the control of the PI3-K pathway. Central to the control of cell growth is the mammalian target of rapamycin (mTOR), which activates p70S6 kinase and inhibits elongation-initiation factor 4E binding protein (4E-BP1), culminating in the stimulation of protein synthesis *(71)*. Recently, the pathway linking PI3-K to p70S6 kinase has been further defined at the biochemical and genetic level following the elucidation of the function of two proteins, hamartin and tuberin, which are mutated in the genetic disorder tuberous sclerosis *(72)*. These proteins form a complex, and tuberin acts as a GTPase-activating protein that inhibits the activity of the small GTPase, Rheb. Phosphorylation of tuberin by PKB inhibits GTPase activity. In its GTP-bound state, Rheb activates mTOR via an undefined mechanism *(73)*. Bearing in mind that mTOR is upstream of p70S6kinase, rapamycin has emerged as a potential treatment for both tuberous sclerosis and cancer. An ester analog of rapamycin, CCI-779, inhibits growth of PTEN-deficient human myeloma cell lines *(74)* and reverses growth of PTEN-deficient prostate cancer cell xenografts *(75)*.

6.4. PI3-K and Angiogenesis

Tumor growth and survival depends on the formation of new blood vessels to supply the tumor with necessary nutrients. The PI3-K pathway is activated by a number of pro-angiogenic stimuli, suggesting that inhibition of PI3-K is a potential point of intervention in this process. One critical regulator of angiogenesis is vascular endothelial growth factor (VEGF). PI3-K inhibition leads to the blockade of a VEGF production following growth factor stimulation in a variety of cell types *(76–78)*. In addition, signaling downstream of VEGF itself is also mediated by PI3-K *(79)*. PI3-K and the ERK/MAPK pathway also regulate the expression of the transcription factor, hypoxia-inducible factor 1 (HIF-1) *(80,81)*. HIF-1 regulates the expression of a wide variety of genes central to angiogenesis, including VEGF and the VEGF receptor, and functions as a sensor of oxygen balance *(82)*. Interestingly, a recent report has also suggested that HIF-1 activates the *CXCR4* gene (*see* Section 7), providing a potential link between hypoxia and tumor cell migration and homing *(83)*. In summary, by sharing common pathways between tumor cell growth and angiogenesis, a tumor maintains a nutrient- rich environment and adequate oxygen supply.

7. PI3-K AND CELL MIGRATION

Activation of PI3-K and the production of PtdIns $(3,4,5)P_3$ plays a pivotal role in the remodeling of the actin cytoskeleton and the control of cell movement. Studies in leukocytes and model organisms such as *Dictyostelium amoebae* have shed considerable light on the PI3-K-dependent signaling pathways involved in cell movement and highlighted the level of conservation in these pathways *(84)*. Movement along a chemoattractant gradient (chemotaxis) requires the rearrangement of the actin cytoskeleton, leading to cell elongation and the formation of lamellipodia at the front edge of the cell. Recent studies in the neutrophilic cell line HL60 have shown recruitment of a fluorescently tagged PKB-PH domain to the edge of the cell exposed to the highest concentration of chemoattractant. This has led to the suggestion that PI3-K and the localization of PtdIns $(3,4,5)P_3$ act as a "molecular compass," guiding the direction of neutrophils toward the source of chemoattractant *(85)*. The role of PI3-K activation in the movement of nonleukocyte cell types has also been established; for example, the migration of breast cancer cells in response to EGF stimulation *(86)*. A comprehensive analysis of chemokine receptor expression in breast tumor cell lines has revealed an upregulation of the chemokine receptors CXCR4 and CCR7 *(87)*. Stimulation by stromal cell-derived factor (SDF)-1, the ligand for CXCR4, leads to PI3-K activation in a variety of cell types.

Considerable efforts have been made to identify the PI3-K isoforms involved in the control of cell migration in response to different stimuli. In neutrophils, data from gene targeting in mice would support a pivotal role for p110γ in chemokine-induced cell movement *(31–33)*. Breast cancer cell chemotaxis in response to EGF stimulation is inhibited by microinjection of neutralizing antibodies that block p110δ and to lesser extent by antibodies to p110β *(3)*. Furthermore, migration of leukemic T-cells to SDF-1 requires the activation of both Class IA and Class IB PI3-Ks *(88)*.

8. INHIBITION OF PI3-K SIGNALING IN TUMOR CELLS

The first-generation PI3-K inhibitors, wortmannin and LY294002, have been shown to block the proliferation of numerous cancer cell lines in vitro and have been tested in a variety of tumor models both as single agents and in combination with radiation and traditional chemotherapeutic drugs *(89)*. Wortmannin blocks tumor growth and PKB phosphorylation in a number of xenograft studies in mice with severe combined immunodeficiency (SCID) *(90,91)*. In a more recent study, systemic wortmannin administration (1 mg/kg intraperitoneally) significantly increased mean survival time in mice following xenotransplantation of squamous-cell carcinoma or adenocarcinoma *(92)*. It is worth noting that a number of side effects were observed in this study including more than 10% weight loss, bloody diarrhea, and reduced physical activity. Wortmannin has been shown to enhance the effects of radiation and cytotoxic agents *(93,94)*. These effects may be partially owing to the inhibition of the closely related kinases, ATM and DNA-PK, but at low concentrations of wortmannin may also occur via a direct effect on PI3-K and the enhancement of apoptosis in treated cells.

LY294002 significantly suppresses tumor volume in a human colon cancer xenograft model *(95)*. In a pancreatic tumor model, treatment with LY294002 resulted in a dose-dependent inhibition of tumor growth *(96)*. In addition, a suboptimal dose of LY294002 produced further inhibition of tumor growth when combined with a suboptimal dose of the nucleoside analog, gemcitabine. Similar synergistic effects of chemotherapeutic agents and LY294002 have been observed in vitro with nonsmall-cell lung carcinoma and breast cancer cell lines *(97,98)*. More recently, LY294002 was shown to have minimal anti-tumor effect in mice bearing bladder tumor cell xenografts. However, in combination with radiation therapy, LY294002 treatment resulted in a significant and synergistic reduction in clonogenicity and growth delay *(99)*.

A number of groups have used gene therapy approaches to demonstrate the relevance of PI3-K signaling in tumorigenesis. Adenoviral-mediated transfer of PTEN has been shown to inhibit proliferation and metastasis of orthotopic prostate tumor cells *(100)* and colorectal cancer cells *(101)*. Treatment of bladder cancer cell models with adenoviral PTEN results in complete regression of the tumor *(102)*. In addition, downregulation of VEGF and decreased microvessel density was observed. Inhibition of PI3-K gene expression by RNA interference has also provided further evidence for the role of PI3-K in cancer cell proliferation and tumor growth. Transient transfection of HeLa cells with siRNA directed against p110β blocks HeLa cell growth on matrigel in vitro and tumor growth when transplanted in vivo *(103)*. A stable, inducible shRNA system for studying PI3-K signaling has also been described *(104)*. In a prostate tumor model, expression of shRNA directed against p110β reduced the formation of lymph node metastases but resulted in no reduction in tumor size, suggesting that in this model p110β may regulate cell migration.

9. PI3-K INHIBITION: PROSPECTS FOR DRUG DEVELOPMENT

The prospects for the use of PI3-K inhibitors in the treatment of various diseases including cancer, other proliferative disorders, inflammation, and cardiovascular disease look promising *(2,105)*. However, a limited number of in vivo studies (including side-effect and toxicological assessment) have been reported with the two well-characterized inhibitors, wortmannin and LY294002. Consequently, a number of questions regarding the merits of PI3-K inhibition remain unanswered. The PI3-K family of enzymes control a wide variety of cellular processes, thereby questioning whether an acceptable therapeutic index can be achieved from targeting this pathway. This may be possible when targeting tumors with *PIK3CA* mutations that activate the PI3-K pathway. In the last few years it has become apparent that individual PI3-K isoforms regulate distinct cellular events; e.g., chemokine-stimulated neutrophil migration is controlled by PI3-Kγ, and B-cell receptor signaling is mediated by PI3-Kδ. Therefore, it is reassuring that isoform-selective inhibitors can be generated despite the similarities in the ATP binding sites of the Class I PI3-Ks.

Based on the evidence generated so far, the PI3-K/PKB/PTEN signaling cassette appears integral to a number of facets of tumorigenesis. The frequency with which PI3-K signaling is deregulated in cancer and the target patient populations needs to be further defined but there is now a great deal of evidence to support the use of PI3-K inhibitors alone or in combination with radiation or traditional chemotherapeutic agents. The outcome of drug discovery efforts focused toward the development of PI3-K inhibitors is eagerly awaited.

REFERENCES

1. Vanhaesebroeck B, Leevers SJ, Ahmadi K, et al. Synthesis and function of 3-phosphorylated inositol lipids. *Annu Rev Biochem* 2001; 70:535–602.
2. Wymann MP, Zvelebil M, Laffargue M. Phosphoinositide 3-kinase signalling—which way to target? *TIPS* 2003; 34:366–376.
3. Sawyer C, Sturge J, Bennett DC, et al. Regulation of breast cancer cell chemotaxis by the phosphoinositide 3-kinase p110δ. *Cancer Res* 2003; 63:1667–1675.
4. Rodriguez-Viciana P, Warne PH, Dhand R, et al. Phosphatidylinositol-3-OH kinase as a direct target of Ras. *Nature* 1994; 370:527–532.
5. El Sheikh SS, Domin J, Tomtitchong P, Abel P, Stamp G, Lalani E-N. Topographical expression of class IA and class II phosphoinositide 3-kinase enzymes in normal human tissues is consistent with a role in differentiation. *BMC Clin Pathol* 2003; 3:4.
6. Arcaro A, Zvelebil MJ, Wallasch C, Ullrich A, Waterfield MD, Domin J. Class II phosphoinositide 3-kinases are downstream targets of activated polypeptide growth factor receptors. *Mol Cell Biol* 2000; 20:3817–3830.
7. Ktori C, Shepard PR, O'Rourke L. TNF-a and leptin activate the a-isoform of class II phosphoinositide 3-kinase. *Biochem Biophys Res Commun* 2003; 306:139–143.
8. Turner SJ, Domin J, Waterfield MD, Ward SG, Westwick J. The CC chemokine monocyte chemotactic peptide-1 activates both the class I p85/p110 phosphatidylinositol 3-kinase and the class II PI3-K-C2alpha. *J Biol Chem* 1998; 273:25987–29595.
9. Viera OV, Botelho RJ, Rameh L, et al. Distinct roles of class I and class III phosphatidylinositol 3-kinases in phagosome formation and maturation. *J Cell Biol* 2001; 155:19–25.
10. Futter CE, Collinson LM, Backer JM, Hopkins CR. Human VPS34 is required for internal vesicle formation within multivesicular endosomes. *J Cell Biol* 2001; 155:1251–1263.
11. Sulis ML, Parsons R. PTEN: from pathology to biology. *Trends Cell Biol* 2003; 13:478–483.
12. Krystal G. Lipid phosphatases in the immune system. *Semin Immunol* 2000; 12:397–403.
13. Cozier GE, Carlton J, Bouyoucef D, Cullen PJ. Membrane targeting by pleckstrin homology domains. *Curr Top Microbiol Immunol* 2004; 282:49–88.
14. Cantley LC. The phosphoinositide 3-kinase pathway. *Science* 2002; 296:1655–1657.
15. Chan TO, Rittenhouse SE, Tsichlis PN. AKT/PKB and other D3 phosphoinositide-regulated kinases: kinase activation by phosphoinositide-dependent phosphorylation. *Annu Rev Biochem* 1999; 68:965–1014.
16. Arcaro A, Wymann MP. Wortmannin is a potent phosphatidylimositol 3-kinase inhibitor: the role of phosphatidylinositol 3,4,5-trisphosphate in neutrophil responses. *Biochem J* 1993; 269:297–301.
17. Wymann MP, Bulgarelli-Leva G, Zvelebil MJ, et al. Wortmannin inactivates phosphoinositide 3-kinase by covalent modification of Lys-802, a residue involved in the phosphate transfer reaction. *Mol Cell Biol* 1996; 16:1722–1733.

18. Vlahos CJ, Matter MF, Hui KY, Brown RF. A specific inhibitor of phosphatidylinositol 3-kinase, 2-(-4-morpholinyl)-8-phenyl-4H-1-benzopyran-4-one (LY294002). *J Biol Chem* 1994; 269:5241–5248.

19. Walker EH, Pacold ME, Perisic O, et al. Structural determinants of phosphoinositide 3-kinase inhibition by wortmannin, LY294002, quercetin, myricetin, and staurosporine. *Mol Cell* 2000; 6:909–919.

20. Domin J, Pages F, Volinia S, et al. Cloning of a human phosphoinositide 3-kinase with a C2 domain that displays reduced sensitivity to the inhibitor wortmannin. *Biochem J* 1997; 326:139–147.

21. El-Kholy W, MacDonald PE, Lin JH, et al. The phosphatidylinositol 3-kinase inhibitor LY294002 potently blocks Kv currents via a direct mechanism. *FASEB J* 2003; 17:720–722.

22. Foukas LC, Daniele N, Ktori C, Anderson KE, Jensen J, Shepherd P. Direct effects of caffeine and theophylline on p110δ and other phosphoinositide 3-kinases. *J Biol Chem* 2002; 277:37124–37130.

23. Sadhu C, Masinovsky B, Dick K, Sowell CG, Staunton DE. Essential role of phosphatinositide 3-kinase in neutrophil directional movement. *J Immunol* 2003; 170:2647–2654.

24. Foukas LC, Okkenhaug K. Gene-targeting reveals physiological roles and complaex regulation of the phosphoinositide 3-kinases. *Arch Biochem Biophys* 2003; 414:13–18.

25. Sasaki T, Suzuki A, Sasaki J, Penninger JM. Phosphoinositide 3-kinases in immunity: lessons from knockout mice. *J Biochem (Tokyo)* 2002; 131:495–501.

26. Bi L, Okabe I, Bernard DJ, Wynshaw-Boris A, Nussbaum RL. Proliferative defect and embryonic lethality in mice homozygous for a deletion in the p110alpha subunit of phosphoinositide 3-kinase. *J Biol Chem* 1999; 274:10963–10968.

27. Bi L, Okabe I, Bernard DJ, Nussbaum RL. Early embryonic lethality in mice deficient in the p110beta catalytic subunit of PI3-K. *Mamm Genome* 2002; 13:169–172.

28. Clayton E, Bardi G, Bell SE, et al. A crucial role for the p110delta subunit of phosphatidylinositol 3-kinase in B cell development and activation. *J Exp Med* 2002; 196:753–763.

29. Jou ST, Carpino N, Takahashi Y, et al. Essential, non-redundant role for the phosphoinositide 3-kinase p110delta in signalling by the B cell receptor complex. *Mol Cell Biol* 2002; 22:8580–8591.

30. Okkenhaug K, Bilanchio A, Farjot G, et al. Impaired B and T cell antigen receptor signalling in p110δ PI3-kinase mutant mice. *Science* 2002; 297:1031–1034.

31. Hirsch E, Katanaev VL, Garlanda C, et al. Central role for G protein-coupled phosphoinositide 3-kinase gamma in inflammation. *Science* 2000; 287:1049–1053.

32. Sasaki T, Irie-Sasaki J, Jones RG, et al. Function of PI3-Kγ in thymocyte development, T cell activation, and neutrophil migration. *Science* 2000; 287:1040–1046.

33. Li Z, Jiang H, Xie W, Zhang Z, Smrcka AV, Wu D. Roles of PLC-β2 and –β3 and PI3-Kγ in chemoattractant-mediated signal transduction. *Science* 2000; 287:1046–1049.

34. Hannigan M, Zhan L, Li Z, Ai Y, Wu D, Huang C-K. Neutrophils lacking phosphoinositide 3-kinase γ show loss of directionality during N-formyl-Met-Leu-Phe- induced chemotaxis. *Proc Natl Acad Sci USA* 2002; 99:3603–3608.

35. Laffargue M, Calvez R, Finan P, et al. Phosphoinositide 3-kinase gamma is an essential amplifier of mast cell function. *Immunity* 2002; 16:441–451.

36. Hirsch E, Bosco O, Tropel P, et al. Resistance to thromboembolism in PI3-Kγ-deficient mice. *FASEB J* 2001; 15:2019–2021.

37. Crackower MA, Oudit GY, Kozieradzki I, et al. Regulation of myocardial contractility and cell size by distinct PI3-K-PTEN signaling pathways. *Cell* 2002; 110:737–749.

38. Vlahos CJ, McDowell SA, Clerk A. Kinases as therapeutic targets for heart failure. *Nat Rev Drug Discov* 2003; 2:99–113.

39. Susuki H, Terauchi Y, Fujiwara M, et al. Xid-like immunodefiency in mice and disruption of the p85α subunit of phosphoinositide 3-kinase. *Science* 1999; 283:390–392.

40. Fruman DA, Snapper SB, Yballe CM, et al. Impaired B cell development and proliferation in absence of phosphoinositide 3-kinase p85α. *Science* 1999; 283:393–397.

41. Ueki K, Yballe CM, Brachmann SM, et al. Increased insulin sensitivity in mice lacking p85beta subunit of phosphoinositide 3-kinase. *Proc Natl Acad Sci USA* 2002; 99:419–424.

42. Luo J, Manning BD, Cantley LC. Targeting the PI3-K-Akt pathway in human cancer: rationale and promise. *Cancer Cell* 2003; 4:257–262.

43. Roymans D, Slegers H. Phosphatidylinositol 3-kinases in tumor progression. *Eur J Biochem* 2001; 268:487–498.

44. Vivanco I, Sawyers CL. The phosphatidylinositol 3-kinase-Akt pathway in human cancer. *Nat Rev Cancer* 2002; 2:489–501.

45. Whitman M, Kaplan DR, Schaffhausen B, Cantley L, Roberts TM. Association of phosphatidylinositol kinase activity with polyoma middle-T competent for transformation. *Nature* 1985; 315:239–242.

46. Courtneidge SA, Heber A. An 81 kd protein complexed with middle T antigen and pp60c-src: a possible phosphatidylinositol kinase. *Cell* 1987; 50:1031–1037.

47. Chang HW, Aoki M, Fruman D, et al. Transformation of chicken cells by the gene encoding the catalytic subunit of PI3-K. *Science* 1997; 276:1848–1850.

48. Jimenez C, Jones DR, Rodriguez-Viciana P, et al. Identification and charaterisation of a new oncogene derived form the regulatory subunit of phosphoinositide 3-kinase. *EMBO J* 1998; 17:743–753.

49. Borlado RL, Redondo C, Alvarez B, et al. Increased phosphoinositide 3-kinase activity induces a lymphoproliferative disorder and contributes to tumor generation in vivo. *FASEB J* 2000; 14:895–903.

50. Jucker M, Sudel K, Horn S, et al. Expression of a mutated form of the p85alpha regulatory subunit of phosphatidylinositol 3-kinase in a Hodgkin's lymphoma-derived cell line (CO). *Leukemia* 2002; 16:894–901.

51. Shayesteh L, Lu Y, Kuo WL, et al. *PIK3CA* is implicated as an oncogene in ovarian cancer. *Nature Genetics* 1999; 21:99–102.

52. Ma YY, Wei S-J, Lin Y-C, et al. *PIK3CA* as an oncogene in cervical cancer. *Oncogene* 2000; 19:2739–2744.

53. Woenckhaus J, Steger K, Werner E, et al. Genomic gain of PI3-KCA and increased expression of p110alpha are associated with progression of dysplasia into invasive squamous cell carcinoma. *J Pathol* 2002; 198:335–342.

53a. Samuels Y, Wang Z, Bardelli A, et al. High frequency of mutations of the *PIK3CA* gene in human cancers. *Science* 2004; 304:554.

53b. Bachman KE, Argani P, Samuels Y, et al. The *PIK3CA* gene is mutated with high frequeny in human breast cancers. *Cancer Biol. Ther.* 2004; 8:772–775.

53c. Campbell IG, Russell SE, Choong DY, et al. Mutation of the *PIK3CA* gene in ovarian and breast cancer. *Cancer Res.* 2004; 64:7678–7681.

53d. Kang S, Bader AG, Vogt PK. Phosphatidylinositol 3-K mutations identified in human cancer are oncogenic. *Proc. Natl. Acad. Sci. USA* 2005; 102:802–807.

54. Testa JR, Bellacosa A. AKT plays a central role in tumorigenesis. *Proc Natl Acad Sci USA* 2002; 98:10983–10985.

55. Bonneau D, Longy M. Mutations of the human PTEN gene. *J Med Genet* 2000; 16:109–122.

56. Zhou X-P, Gimm O, Hampel H, Niemann T, Walker MJ, Eng C. Epigenetic PTEN silencing in malignant melanomas without PTEN mutation. *Am J Pathol* 2000; 157:1123–1128.

57. Soria JC, Lee HY, Lee JL, et al. Lack of PTEN expression in non-small cell lung cancer could be related to promoter methylation. *Clin Cancer Res* 2002; 8:1178–1184.

58. Roche S, Koegl M, Courtneidge SA. The phosphatidylinositol 3-kinase alpha is required for DNA synthesis induced by some but not all growth factors. *Proc Natl Acad Sci USA* 1994; 91:9185–9189.

59. Liang J, Zubovitz J, Petrocelli T, et al. PKB/Akt phosphorylates p27, impairs nuclear import of p27 and opposes p27-mediated G1 arrest. *Nat Med* 2002; 8:1153–1160.

60. Shin I, Yakes FM, Rojo F, et al. PKB/Akt mediates cell-cycle progression by phosphorylation of p27(Kip1) at threonine 157 and modulation of its cellular localization. *Nat Med* 2002; 8:1145–1152.

61. Viglietto G, Motti ML, Bruni P, et al. Cytoplasmic relocalization and inhibition of the cyclin-dependent kinase inhibitor p27(Kip1) by PKB/Akt-mediated phosphorylation in breast cancer. *Nat Med* 2002; 8:1136–1144.

62. Diehl JA, Cheng M, Roussel MF, Sherr CJ. Glycogen synthase kinase-3beta regulates cyclin D1 proteolysis and subcellular localisation. *Genes Dev* 1998; 12:3499–3511.

63. Sears R, Nuckolls F, Haura E, Taya Y, Tamai K, Nevins JR. Multiple Ras-dependent phosphorylation pathways regulate Myc protein stability. *Genes Dev* 2000; 14:2501–2514.

64. Burgering BM, Medema RH. Decisions on life and death: FOXO forkhead transcription factors are in command when PKB/Akt is off duty. *J Leukoc Biol* 2003; 73:689–701.

65. Dudek H, Datta SR, Franke TF, et al. Regulation of neuronal survival by the serine-threonine kinase Akt. *Science* 1997; 275:661–665.

66. Philpott KL, McCarthy MJ, Klippel A, Rubin LL. Activated phosphatidylinositol 3-kinase and Akt kinase promote survival of superior cervical neurons. *J Cell Biol* 1997; 139: 809–815.

67. Datta SR, Dudek H, Tao X, et al. Akt phosphorylation of BAD couples survival signals to the cell-intrinsic death machinery. *Cell* 1997; 91:231–241.

68. Cardone MH, Roy N, Stennicke HR, et al. Regulation of cell death protease caspase-9 by phosphorylation. *Science* 1998; 282:1318–1321.

69. Brunet A, Bonni A, Zigmond MJ, et al. Akt promotes cell survival by phosphorylating and inhibiting a forkhead transcription factor. *Cell* 1999; 96:857–868.

70. Mayo LD, Donner DB. The PTEN, Mdm2, p53 tumor suppressor-oncoprotein network. *Trends Biochem Sci* 2002; 27:462–467.

71. Fingar DC, Salama S, Tsou C, Harlow E, Blenis J. Mammalian cell size is controlled by mTOR and its downstream targets S6K1 and 4EBP1/eIF4E. *Genes Dev* 2002; 16: 1472–1487.

72. Manning BD, Cantley LC. United at last: the tuberous sclerosis complex gene products connect the phosphoinositide 3-kinase/Akt pathway to mammalian target of rapamycin (mTOR) signalling. *Biochem Soc Trans* 2003; 31:573–578.

73. Manning BD, Cantley LC. Rheb fills a GAP between TSC and TOR. *Trends Biochem Sci* 2003; 28:573–576.

74. Shi Y, Gera J, Hu L, Hsu JH, Bookstein R, Li W, Lichtenstein A. Enhanced sensitivity of multiple myeloma cells containing PTEN mutations to CCI-779. *Cancer Res* 2002; 62:5027–5034.

75. Neshat MS, Mellinghoff IK, Tran C, et al. Enhanced sensitivity of PTEN-deficient tumors to inhibition of FRAP/mTOR. *Proc Natl Acad Sci USA* 2001; 98:10314–10319.

76. Wang D, Huang HJ, Kazlauskas A, Cavenee WK. Induction of vascular endothelial growth factor expression in endothelial cells by platelet-derived growth factor through the activation of phosphatidylinositol 3-kinase. *Cancer Res* 1999; 59:1464–1472.

77. Jiang BH, Zheng JZ, Aoki M, Vogt PK. Phosphatidylinositol 3-kinase signaling mediates angiogenesis and expression of vascular endothelial growth factor in endothelial cells. *Proc Natl Acad Sci USA* 2000; 97:1749–1753.

78. Rak J, Mitsuhashi Y, Sheenan C, et al. Oncogenes and tumor angiogenesis: differential modes of vascular endothelial growth factor up-regulation in Ras-transformed epithelial cells and fibroblasts. *Cancer Res* 2000; 60:490–498.

79. Shiojima I, Walsh K. Role of Akt signalling in vascular homeostasis and angiogenesis. *Circ Res* 2002; 90:1243–1250.

80. Zhong H, Chiles K, Feldser D, et al. Modulation of hypoxia-inducible factor 1α expression by the epidermal growth factor/phosphatidylinositol 3-kinase/PTEN/Akt/FRAP pathway in human prostate cancer cells: implications for tumor angiogenesis and therapeutics. *Cancer Res* 2000; 60:1541–1545.

81. Semenza GL. Signal transduction to hypoxia-inducible factor 1. *Biochem Pharmacol* 2002; 64:993–998.

82. Giaccia A, Siim B, Johnson RS. HIF-1 as a target for drug development. *Nat Rev Drug Discov* 2003; 2:803–811.

83. Staller P, Sulitkova J, Lisztwan J, Moch H, Oakeley EJ, Krek W. Chemokine receptor CXCR4 downregulated by von Hippel-Lindau tumour suppressor pVHL. *Nature* 2003; 425:307–311.

84. Stephens L, Ellson C, Hawkins P. Roles of PI3-Ks in leucocyte chemotaxis and phagocytosis. *Curr Opin Cell Biol* 2002; 14:203–213.

85. Bourne HR, Weiner O. A chemical compass. *Nature* 2002; 419:21.

86. Price JT, Tiganis T, Agarwal A, Djakiew D, Thompson EW. Epidermal growth factor promotes MDA-MB-231 breast cancer cell migration through a phosphatidylinositol 3´-kinase and phospholipase C-dependent mechanism. *Cancer Res* 1999; 59:5475–5478.

87. Muller A, Homey B, Soto H, et al. Involvement of chemokine receptors in breast cancer metastasis. *Nature* 2001; 410:50–56.

88. Curnock AP, Sotsios Y, Wright KL, Ward SG. Optimal chemotactic responses of leukemic T cells to stromal cell–derived factor-1 requires the activation of both class IA and IB phosphoinositide 3-kinases. *J Immunol* 2003; 170(8):4021–4030.

89. West KA, Castillo S, Dennis PA. Activation of the PI3-K/Akt pathway and chemotherapeutic resistance. *Drug Resist Updat* 2002; 5:234–248.

90. Schultz RM, Merriman RL, Andis SL, et al. In-vitro and in-vivo anti-tumor activity of the phosphatidylinositol-3-kinase inhibitor, wortmannin. *Anticancer Res* 1995; 15:1135–1139.

91. Lemke LE, Paine-Murrieta GD, Taylor CW, Powis G. Wortmannin inhibits the growth of mammary tumors despite the existence of a novel wortmannin-insensitive phosphatidylinositol-3-kinase. *Cancer Chemother Pharmacol* 1999; 44:491–497.

92. Boehle AS, Kurdow R, Boenicke L, et al. Wortmannin inhibits growth of human non-small-cell lung cancer in vitro and in vivo. *Langenbeck's Arch Surg* 2002; 387:234–239.

93. Hosoi Y, Miyachi H, Matsumoto Y, et al. A phosphatidylinositol 3-kinase inhibitor wortmannin induces radioresistant DNA synthesis and sensitizes cells to bleomycin and ionizing radiation. *Int J Cancer* 1998; 78:642–647.

94. Edwards E, Geng L, Tan J, Onishko H, Donnelly E, Hallahan, DE Phosphatidylinositol 3-kinase/Akt signalling in the response of vascular endothelium to ionizing radiation. *Cancer Res* 2002; 62:4671–4677.

95. Semba S, Itoh N, Ito M, Harada M, Yamakawa M. The in vitro and in vivo effects of 2-(4-Morpholinyl)-8-phenyl-chromone (LY294002), a specific inhibitor of phosphatidylinositol 3'-kinase, in human colon cancer cells. *Clin Cancer Res* 2002; 8:1957–1963.

96. Bondar VM, Sweeney-Gotsch B, Andreeff M, Mills GB, McConkey DJ. Inhibition of the phosphatidylinositol 3'-kinase-Akt pathway induces apoptosis in pancreatic carcinoma cells in vitro and in vivo. *Mol Cancer Ther* 2002; 1:989–997.

97. Brognard J, Clark AS, Ni Y, Dennis PA. Akt/protein kinase b is constitutively active in non-small cell lung cancer cells and promotes cellular survival and resistance to chemotherapy and radiation. *Cancer Res* 2001; 61:3986–3997.

98. Clark AE, West K, Streicher S, Dennis PA. Constitutive and inducible Akt activity promotes resistance to chemotherapy, trastuzumab and tamoxifen in breast cancer cells. *Mol Cancer Ther* 2002; 1:707–717.
99. Gupta AK, Cerniglia GJ, Mick R, et al. Radiation sensitization of human cancer cells in vivo by inhibiting the activity of PI3-K using LY294002. *Int J Radiat Oncol Biol Phys* 2003; 56:846–853.
100. Davies MA, Kim SJ, Parikh NU, Dong Z, Bucana CD, Gallick GE. Adenoviral-mediated expression of MMAC/PTEN inihibits proliferation and metastasis of human prostate cancer cells. *Clin Cancer Res* 2002; 8:1695–1698.
101. Saito Y, Swanson X, Mhashikar AM, et al. Adenoviral-mediated transfer of the PTEN gene inhibits human colorectal cancer growth in vitro and in vivo. *Gene Ther* 2003; 10:1961–1969.
102. Tanaka M, Grossman HB. In vivo gene therapy of human bladder cancer with PTEN suppresses tumor growth, downregulates phosphorylated Akt, and increases sensitivity to doxorubicin. *Gene Ther* 2003; 10:1636–1642.
103. Czauderna F, Fechtner M, Aygun H, et al. Functional studies of the PI(3)-kinase signalling pathway employing synthetic and expressed siRNA. *Nucleic Acids Res* 2003; 31:670–682.
104. Czauderna F, Santel A, Hinz M, et al. Inducible shRNA expression for application in a prostate cancer model. *Nucleic Acids Res* 2003; 31:e127.
105. Ward S, Sotsios Y, Dowden J, Bruce I, Finan P. Therapeutic potential of PI3-K inhibitors. *Chem Biol* 2003; 10:207–213.
106. Fry MJ. Phosphoinositide 3-kinase signalling in breast cancer: how big a role might it play? *Breast Cancer Res* 2001; 3:304–312.

4 Src as a Target for Pharmaceutical Intervention
Potential and Limitations

Mira Šuša, PhD, Martin Missbach, PhD, Rainer Gamse, MD, Michaela Kneissel, PhD, Thomas Buhl, PhD, Jürg A. Gasser, PhD, Markus Glatt, PhD, Terence O'Reilly, PhD, Anna Teti, PhD, and Jonathan Green, PhD

CONTENTS

SRC AS A TARGET FOR TREATMENT OF CANCER AND BONE LOSS
SRC INHIBITORS: OVERVIEW
SRC INHIBITORS: ACTIVITY IN VITRO
SRC INHIBITORS: MECHANISM OF ACTION IN OSTEOCLASTS
SRC INHIBITORS: ACTIVITY IN VIVO IN BONE LOSS MODELS
SRC INHIBITORS: ACTIVITY IN CANCER MODELS
SRC INHIBITORS: LIMITATIONS FOR PHARMACEUTICAL
    APPLICATION
OUTLOOK: SRC AS A TARGET FOR OTHER INDICATIONS, MORE
    SELECTIVE AND DUAL SRC INHIBITORS
REFERENCES

1. SRC AS A TARGET FOR TREATMENT OF CANCER AND BONE LOSS

Src is a nonreceptor type protein tyrosine kinase and a prototype of a family consisting of eight members in vertebrates: Src, Yes, Fyn, Fgr, Lyn, Hck, Lck, Blk. Src is an evolutionary well-conserved gene in vertebrates with homologs

From: *Cancer Drug Discovery and Development:*
Protein Tyrosine Kinases: From Inhibitors to Useful Drugs
Edited by: D. Fabbro and F. McCormick © Humana Press Inc., Totowa, NJ

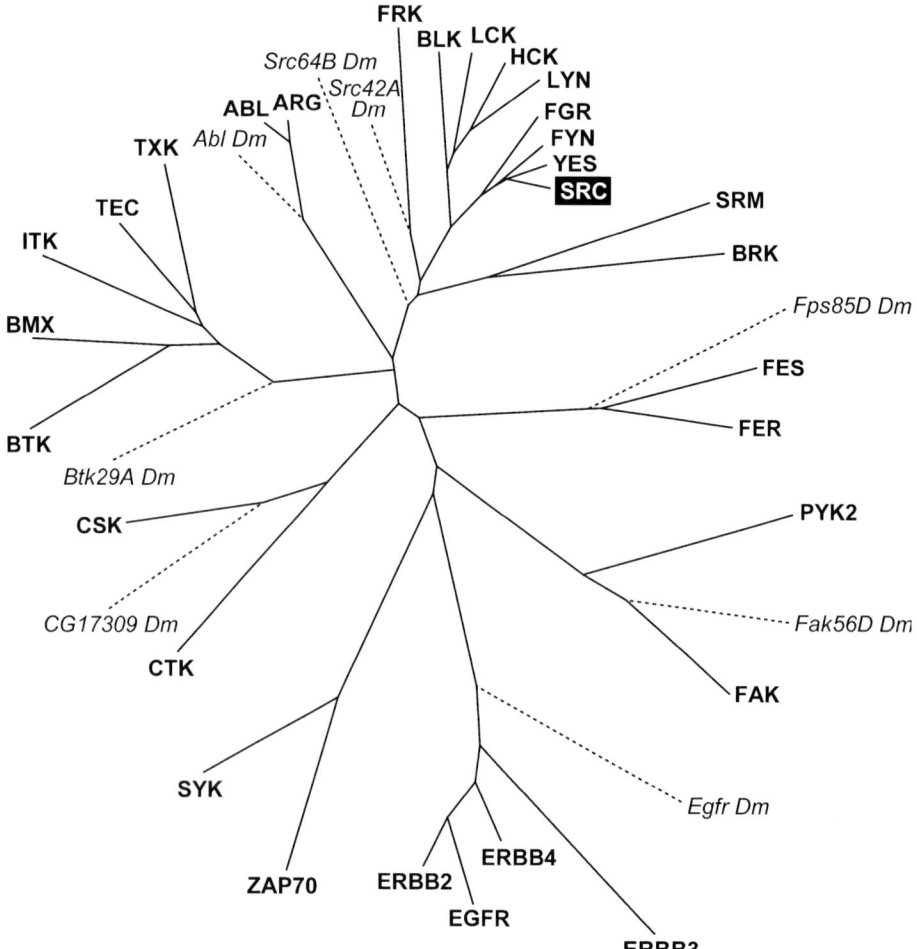

Fig. 1. Relationship of the Src family with other tyrosine kinases. The kinase domains of representative human nonreceptor tyrosine kinases and the epidermal growth factor receptor family were aligned with *Drosophila melanogaster* homologs. The 30 human genes (**bold**) are part of 90 tyrosine kinases from a total of 518 putative protein kinases in the human genome. The 8 fly genes (*italic Dm*) belong to 32 tyrosine kinases of 240 fly kinases (*34*). The neighbor-joining tree reveals higher sequence conservation in the Src family than in other families in humans, but more divergence in fly Src homologs (*35*). (This image was generated by J. Spring.)

expressed in invertebrates. When compared with other nonreceptor type 1 tyrosine kinases in the human genome, the Src family is the largest, with its members very closely related to each other, particularly those within two Src family subgroups (Fig. 1). Such a high degree of similarity is also present in receptor tyrosine kinase families, e.g., the epidermal growth factor receptor (EGFR) family, which is the most similar to nonreceptor tyrosine kinases (Fig. 1). Interestingly, the presence of many close family members within the Src family corresponds to their biological activity, which is linked to physical and functional association with receptor tyrosine kinases at the plasma membrane *(1)*.

The main structural and functional domains in Src are: (a) a catalytic tyrosine kinase domain or SH1, (b) a noncatalytic domain SH2, involved in phosphotyrosine-mediated protein–protein interactions, and (c) a noncatalytic domain SH3, involved in proline-rich sequence-mediated protein–protein interactions. Consequently, the principal functions of Src are its enzymatic activity as a tyrosine kinase and its adapter molecule capability *(1)*. The N- and C-terminal sequences do not form defined domains, but participate in regulation of subcellular localization and Src activity.

Src originally raised interest as a proto-oncogene in the 1980s and has been extensively studied in cellular models of transformation and animal models of carcinogenesis. The main Src substrates in cellular transformation systems are Fak, p130 Cas, Shc, phospholipase C, and phosphatidylinositol 3-kinase, all components of growth factor-induced intracellular signaling networks *(2)*. Mutated and/or constitutively active Src (Tyr527Phe mutant, viral form v-Src or polyoma middle T-activated c-Src) has the ability to transform cells to a malignant phenotype in vitro and to cause tumors in vivo *(3,4)*. Such mutations have not been well documented in human cancers, but instead, nonmutated Src is overexpressed in certain tumors and cooperates with receptor tyrosine kinases such as c-Met and the EGFR family *(5)*, in some cases via signal transducer and activator of transcription 5b (STAT5b) *(6)*. Human colon, breast, and lung cancer are most strongly associated with changes in Src activity and expression. More recent evidence further strengthens this notion, inasmuch as the Src substrate cortactin potentiates bone metastases arising from human breast cancer cells *(7)* and activation of Src in human colorectal carcinoma indicates a poor clinical prognosis *(8)*.

The interest for Src within the bone metabolism field was stirred in 1990s by the finding of an osteopetrotic phenotype (excess bone mass owing to osteoclast failure) in Src knockout mice *(9)*. Src is highly expressed in bone-resorbing cells osteoclasts, where it regulates adhesion, exocytosis, and survival. The main Src substrate in osteoclasts is Pyk2, a nonreceptor tyrosine kinase involved in adhesion *(10,11)*. None of the human osteopetrotic mutations has yet been linked to Src.

These data were a basis to start research programs to identify Src inhibitors for the pharmacological treatment of cancer and diseases characterized by elevated bone loss, such as osteoporosis. Typically such programs involved high-throughput screening for inhibitors of Src kinase activity in enzymatic in vitro assays, then profiling such compounds in cellular Src assays, followed by evaluation of an inhibitor's activity in functional assays, such as osteoclast-mediated bone resorption or tumor cell proliferation. Selectivity of compounds was then tested in a battery of assays with other kinases, in both the enzymatic and cellular context. Finally, the most active compounds with acceptable pharmacokinetic properties were tested in appropriate in vivo models (bone loss or tumor xenografts) and in vivo side effects were evaluated.

2. SRC INHIBITORS: OVERVIEW

Thorough reviews of Src inhibitor classes from the chemical and biological perspective are already available (e.g., refs. *12–15*). Here, we will briefly summarize the updated knowledge on the most promising Src inhibitory compounds and then focus on new unpublished data on Src mechanism of action and in vivo activity. Src inhibitors fall into two broad classes: inhibitors of the tyrosine kinase activity (adenosine triphosphate [ATP] binding domain-mediated) and inhibitors of protein–protein (SH2-, SH3-, or substrate binding domain-mediated) interactions. Worth mentioning are the nonpeptidic Src SH2 domain ligands, designed to accumulate in bone via binding of a phosphate group to bone mineral (recently reviewed in ref. *15*). However, a big improvement in cellular potency and pharmacokinetic properties of such compounds, linked to their chemical properties, remains a challenge. The kinase inhibitors are still the most promising group in terms of potency, selectivity, and therapeutic application. Most potent and selective compounds with cellular activity belong to the related classes of heterocyclic ATP analogs: pyrazolo-, pyrido-, and pyrrolopyrimidines (Fig. 2). Another similar compound class is purine-derived olomoucines (Fig. 2). Further variations in condensed heterocyclic rings are allowed, as exemplified by indolinones *(16)* and quinolinecarbonitriles (Fig. 2). The enzymatic and cellular activities of two pyrrolopyrimidines are shown in Tables 1 and 2. Recently it has been reported that addition of a bisphosphonate moiety to known kinase inhibitor classes, such as pyrrolopyrimidines and olomoucines, produced inhibitors targeted to bone *(17)*. A critical issue with these inhibitors will be cell penetration and bioavailability after oral application, which may be low because of the charged phosphonate moiety.

Modeling based on the crystal structure of Src family kinases, as well as experimental data with Src crystals, has confirmed the binding mode of pyrrolopyrimidines within the ATP binding domain, exhibiting typical features of a tyrosine kinase inhibitor (hydrogen bond pairs between the compound and the backbone of the kinase, and an interaction with the hydrophobic region not occupied by ATP) *(12)*. The pharmacokinetic properties of the pyrrolopyrimidines were satisfactory for oral application, and compounds showed the expected in vivo oral activity in animal models of bone loss *(18; see* Headings 5 and 6). The pharmacokinetic properties of several pyrrolopyrimidine and olomoucine Src inhibitors are shown in Table 3.

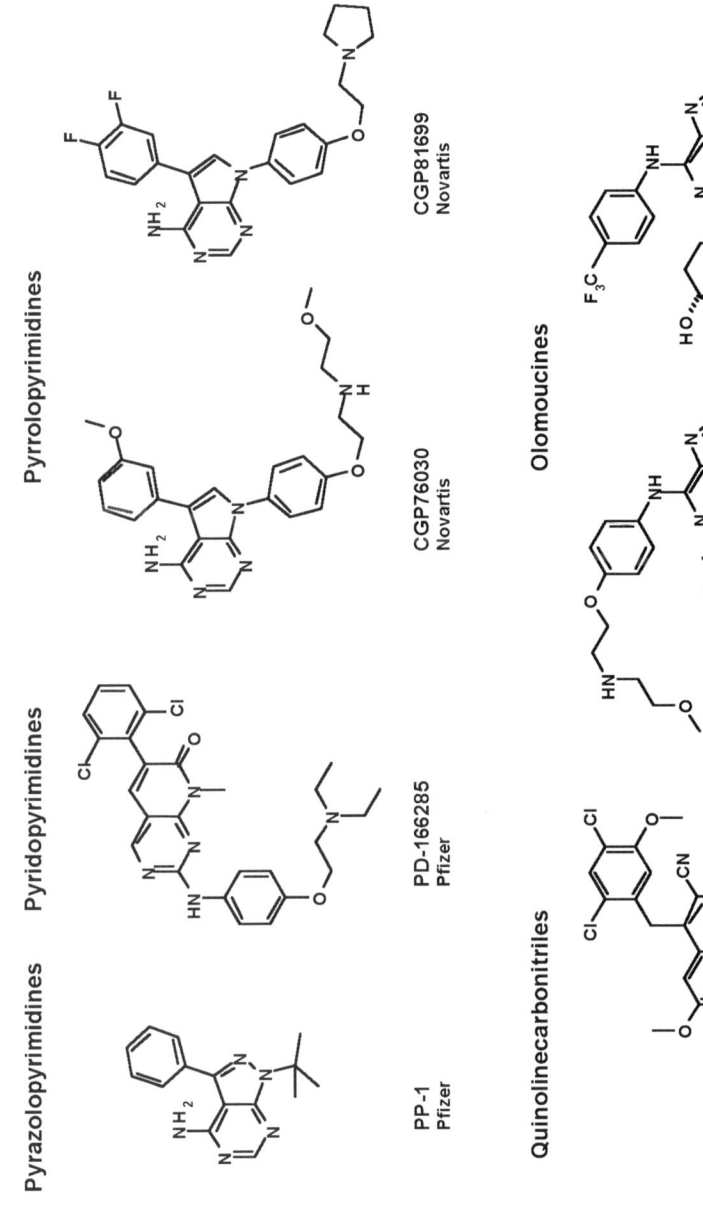

Fig. 2. Structures of the most interesting Src inhibitors. Representative compounds from different classes of Src inhibitors mentioned in the text are shown.

Table 1
Enzymatic Selectivity of Pyrrolopyrimidine Src Inhibitors: IC_{50} Values in nM

| | Compounds | |
| | CGP76030 | CGP77675 |
Protein kinase	Pyrrolopyrimidines	
Src tyrosine kinase family		
Src	5	6
Yes	2	7
Fgr	13	Nd
Lck	250	290
Lyn	53	Nd
Nonreceptor tyrosine kinases		
Abl	180	310
Fak	Nd	>1000
Pyk2	500	>1000
Syk2	>1000	Nd
Csk	560	Nd
Receptor tyrosine kinases		
EGFR[a]	260	150
Met	20,000	Nd
KDR	2700	1000
Flt-1	3500	>1000
Serine/threonine kinases		
PKC-α	>10,000	Nd
Cdc2	>10,000	>10,000

[a]Epidermal growth factor receptor inhibition in vitro was not confirmed in cells (Table 2). Recombinant purified or immunoprecipitated protein kinases were used in in vitro kinase reactions with increasing compound concentrations to determine IC_{50} values. The data are derived from ref. *18* or are our unpublished data. Nd, not determined.

3. SRC INHIBITORS: ACTIVITY IN VITRO

We will now describe the in vitro activity of pyrrolopyrimidines and olomoucines, for which we have the most extensive information available. In vitro and in vivo effects of the pyrrolopyrimidine CGP77675 on Src activity and bone resorption have been previously reported *(18)*. Briefly, CGP77675 inhibits human Src in enzymatic assays and cellular Src in an Src-overexpressing cell line with IC_{50} values of 20 and 200 nM, respectively. This compound was active in assays of in vitro osteoclast bone resorption at 0.1–1 μ*M* as well as in mice and rat models of bone resorption at 5–50 mg/kg after twice daily administration by mouth. Another pyrrolopyrimidine, CGP76030, displayed a similar profile. A summary of enzymatic and cellular activities for the two Src inhibitors

Table 2
Cellular Selectivity of Pyrrolopyrimidine Src Inhibitors: IC_{50} Values in nM

| | Compounds | |
| | CGP76030 | CGP77675 |
Cellular protein kinase	Pyrrolopyrimidines	
Src	80	200
Nonreceptor tyrosine kinases		
Csk	2600	5700
Jak-2	Nd	>1000
Receptor tyrosine kinases		
EGFR	>10,000	10,000
PDGFR	190	150
Kit	<100	Nd
FGFR	>2000	>2000
IGFR	>2000	>2000
Serine/threonine kinases		
PKC	~2000	>2000
Erk	>2000	>2000

Cells were treated with increasing compound concentrations to determine IC_{50} values. Cellular kinase activities were measured by immunoblotting of kinase-specific immunoprecipitates or whole-cell lysates with phosphotyrosine of phospho-kinase-specific antibodies. The data are derived from refs. *18, 37,* or are our unpublished data. Nd, not determined; EGFR, epidermal growth factor receptor; PDGFR, platelet-derived growth factor receptor; FGFR, fibroblast growth factor receptor; IGFR, insulin-like growth factor receptor; PKC, protein kinase C.

from the pyrrolopyrimidine class is shown in Tables 1 and 2. CGP76030 was also shown to inhibit thrombin-stimulated tyrosine phosphorylation in human platelets, cells expressing very high amounts of Src, which is activated upon thrombin treatment (our unpublished data). This result indicated that compounds are active in human cells and this assay could serve as a surrogate marker to follow the in vivo activity of compounds in animal models and in clinical trials. The olomoucine NVP-AAK980 inhibited Src in enzymatic and cellular assays (IC_{50} values of 3 and 220 nM, respectively), as well as in bone resorption assays (at submicromolar concentrations).

4. SRC INHIBITORS: MECHANISM OF ACTION IN OSTEOCLASTS

The mechanism of action of pyrrolopyrimidines in bone resorption models has been studied in primary mouse, rabbit, and human osteoclasts by biochemical and microscopic methods. The pyrrolopyrimidines CGP76030 and CGP77675 were

Table 3
Pharmacokinetic Parameters for Several Src Inhibitors

	Pyrrolopyrimidines		Olomoucines	
Compound no.	CGP76030	CGP81699	CGP79883	NVP-AAK980
CL (mL/min·kg)	5.6	11.6	47.6	29.7
V_{SS} (L/kg)	1.1	7.3	8.7	2.9
$t_{1/2}$ (h)	2.5	7.4	2.4	1.4
F (%)	25	36	97	13
C_{max}/dose (μM)	0.34	0.11	0.10	0.03
T_{max} (h)	2.0	1	6.3	2.8

Pharmacokinetic parameters were measured in one-in-one or cassette dosing experiments after i.v. (1–10 mg/kg) and p.o. (3–10 mg/kg) dosing to conscious rats (mean values, $n = 3–4$).

CL, total clearance; V_{ss}, volume of distribution at steady state; $t_{1/2}$, terminal half-life for elimination; F, absolute oral bioavailability; C_{max}/dose, dose-normalized maximal plasma concentration after oral administration; T_{max}, time of maximal plasma concentration after oral administration.

Total clearance for the pyrrolopyrimidines CGP76030 and CGP81699 in rats was rather low, whereas that of the olomoucines CGP79883 and NVP-AAK980 was moderate to high, suggesting higher metabolic stability of the two pyrrolopyrimidines. This was also reflected by the higher terminal plasma half-life of CGP81699 (>7 h) vs CGP79883 (2.4 h) and NVP-AAK980 (1.4 h). Because of the small volume of distribution, and despite its low clearance, the terminal half-life of CGP076030 was also only 2.5 h, which, similar to CGP79883 and NVP-AAK980, would require multiple daily dosing in rats to maintain high plasma levels more than 24 h. CGP81699 and CGP79883 had a relatively high volume of distribution, suggesting high drug levels also in peripheral tissues. Absolute oral bioavailability in the rat was moderate to very good for CGP76030 (25%), CGP81699 (36%) and CGP79883 (97%), whereas it was quite limited for NVP-AAK980 (13%).

shown to inhibit phosphorylation of the adhesion tyrosine kinase Pyk2, a member of the focal adhesion kinase Fak family (Fig. 3). This inhibition correlated with the break up of the actin rings and with the dissociation of Pyk2 from patches of polymerized actin (Fig. 3 and data not shown). Interestingly, two other cytoskeletal proteins, p130 Cas and vinculin, remained associated with patches of polymerized actin. These data suggested that Pyk2 physically and possibly functionally associates with supramolecularly organized actin structures, but not with less organized actin polymers. These molecular events correlated with reduced

Fig. 3. (Opposite Page) Src inhibitors of the pyrrolopyrimidine class diminish Pyk2 tyrosine phosphorylation and induce disruption of the actin ring in osteoclasts. Mature rabbit osteoclasts are large multinucleated cells (upper panel, left) that resorb bone slices (upper panel,

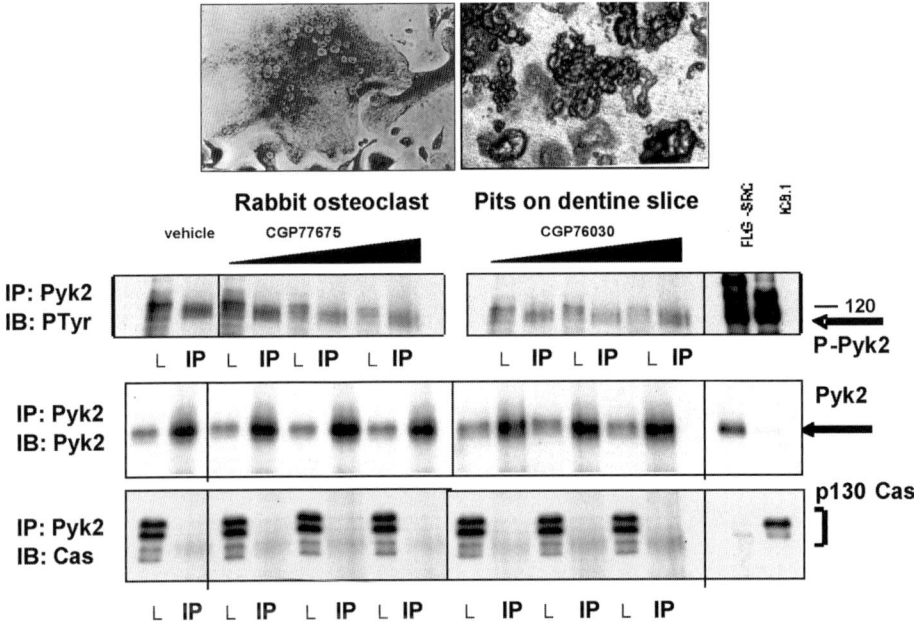

Pyk2 phosphorylation and lack of association with p130 Cas

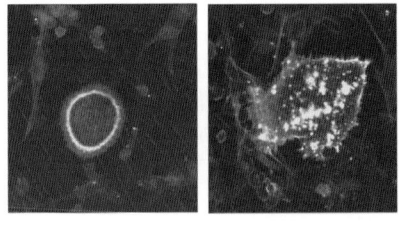

CGP77675:   None   1 µM
Actin ring: Intact   Disrupted

Fig. 3. *(Continued)* right). Treatment of rabbit osteoclasts with Src inhibitors (0.1–1 µ*M*) induces a decrease in tyrosine phosphorylation of the tyrosine kinase Pyk2 (middle panels), as shown by immunoprecipitating Pyk2 with specific antibodies and immunoblotting with anti-phospho-tyrosine antibodies (middle panels, top; L, lysate; IP, immunoprecipitate). Equivalent amounts of Pyk2 were immunoprecipitated, as shown by Pyk2 immunoblotting of the same Pyk2 immunoprecipitates (middle panels, middle). Pyk2 is much less sensitive to direct inhibition by the compounds than Src (Table 1), thus the effect on Pyk2 is indirect. Pyk2 is known to be involved in osteoclast adhesion. No association of Pyk2 with p130 Cas (middle panels, bottom, IP lanes) was detected, in contrast to a previous report *(36)*. Positive controls for immunoblotting are shown on the right (FLG-SRC and IC8.1 cell lysates). Treatment with Src inhibitors induced disruption of the actin ring structure (lower panel), but not actin depolmerization, because patches of actin remained visible after phalloidin staining (lower panel, right). The pictures in the lower panels were generated by P. Lehenkari and M. Muzylak.

bone resorption activity of osteoclasts (ref. *19* and our unpublished data). These data suggested that Src inhibition affects osteoclast morphology, movement, and resorption via inhibition of Pyk2 kinase phosphorylation.

Another aspect of Src inhibitor action in osteoclasts is inhibition of osteoclast formation (osteoclastogenesis) in stromal cell-dependent and independent systems and induction of osteoclast apoptosis. These phenotypical changes correlated with the stimulation of the mitogen-activated protein (MAP) kinases Erk and p38, but not Jnk, in osteoclasts (ref. *19* and our unpublished data). This effect of Src inhibitors was surprising in two aspects: first, Src knockout mice had a normal number of osteoclasts and, second, Erk is a kinase commonly induced during cell growth. A difference in the Src inhibitor effect from Src knockout data is likely to stem from inhibition of other Src family members by the inhibitory compounds. An Src inhibitor-induced increase in Erk phosphorylation is clearly an osteoclast-specific effect, as we previously showed that the same compounds did not affect Erk phosphorylation in osteoblastic cells *(18)*. Erk activation by Src inhibitor in osteoclasts occurs at late time points, similarly to Erk activation by DNA damage-inducing agents, which has been shown to mediate apoptosis *(20)*. In conclusion, we suggest that Src inhibitors exert their main effect in osteoclasts on the Src activity, followed by a change in phosphorylation and activity of other kinases, such as Pyk2, leading to inhibition of adhesion, movement, and bone resorption, and Erk and p38 MAP kinase, leading to cell death by apoptosis.

5. SRC INHIBITORS: ACTIVITY IN VIVO IN BONE LOSS MODELS

The in vivo activity of the pyrrolopyrimidine CGP77675 was previously reported in bone loss models *(19)*. Here we present new data with other compounds in several bone loss models: young growing and skeletally mature female ovariectomized (OVX) rats and young female and male retinoid-treated intact or thyroparathyroidectomized (TPTX) rats. Table 4 gives details on treatment regimen and methods of measurement and summarizes activities of several Src inhibitors in these models. The highest activity was displayed by two compounds: the pyrrolopyrimidine CGP76030 and the olomoucine NVP-AAK980. Two other compounds—CGP81699 and CGP79833, also a pyrrolopyrimidine and an olomoucine—showed less activity and more side effects (*see* Heading 6). Various activities and side-effect profiles within the same chemical classes suggested that the effects were not linked to the core compound structure, but rather to different biological selectivity profiles of each compound.

The most extensively studied compound was the pyrrolopyrimidine CGP76030, which showed activity in all models, even at doses as low as 10 mg/kg by mouth once or twice daily (Table 4). The compound fully prevented hypercalcemia and bone loss in retinoid-treated male TPTX and intact

Table 4
In Vivo Activity of Several Src Inhibitors in Models of Bone Loss

Pyrrolopyrimidines

CGP76030

Models:	8- and 40 to 44-wk-old female OVX Sprague-Dawley or Wistar rats[a]
	6- or 12-wk-old retinoid-treated, TPTX, and intact Wistar rats[b]
Treatments:	10 to 100 mg/kg, p.o., administered once or twice daily, 6–12 wk[a]
	10 to 100 mg/kg, p.o., once daily for 4 d[b]
Measurements:	Chemical analyses[a], DEXA[a,b], pQCT[a], serum calcium[b]
Results/bone:	At 10 mg/kg, twice daily, significant prevention of trabecular bone loss in 8- and 40-wk-old OVX rats (chemical analyses)
	Nonsignificant protection at some sites by in 40 to 44-wk-old OVX rats (pQCT)
	No effects by DEXA in OVX model
	Dose-dependent protection in retinoid model:
	Calcium protection 29% at 10 mg/kg, significant 53% at 50 mg/kg, and full at 100 mg/kg, once daily
	Full, significant bone loss protection at 100 mg/kg (DEXA)

CGP81699

Model:	44-wk-old female OVX Sprague-Dawley rats
Treatment:	3, 10, and 30 mg/kg p.o., daily, for 12 wk
Measurements:	pQCT in proximal tibial metaphysis, DEXA in femur and lumbar vertebral bodies
Results/bone:	No protection of bone by pQCT or DEXA

Olomoucines

NVP-AAK980

Model:	6 to 7-wk-old male Wistar rats, thyroxin- and retinoid-treated
Treatment:	10, 30, and 75 mg/kg, daily, for 4 d
Measurement:	Serum calcium, 3 h after the last daily injection
Results/bone:	Inhibition of serum calcium: none at 10 mg/kg, significant 39% at 30 mg/kg, and full at 100 mg/kg, ED_{50} of 34 mg/kg

CGP79883

Model:	6 to 7-wk-old male Wistar rats, thyroxin- and retinoid-treated
Treatment:	30 and 75 mg/kg, daily, for 4 d
Measurement:	Serum calcium, 3 h after the last daily injection
Results/bone:	Inhibition of serum calcium: 30% at 30 mg/kg, significant 37% at 100 mg/kg

[a] and [b] indicate the corresponding models, treatments, and measurements.

OVX, ovariectomized; TPTX, thyroparathyroidectomized; DEXA, dual energy X-ray absorptiometry; pQCT, peripheral quantitative computed tomography.

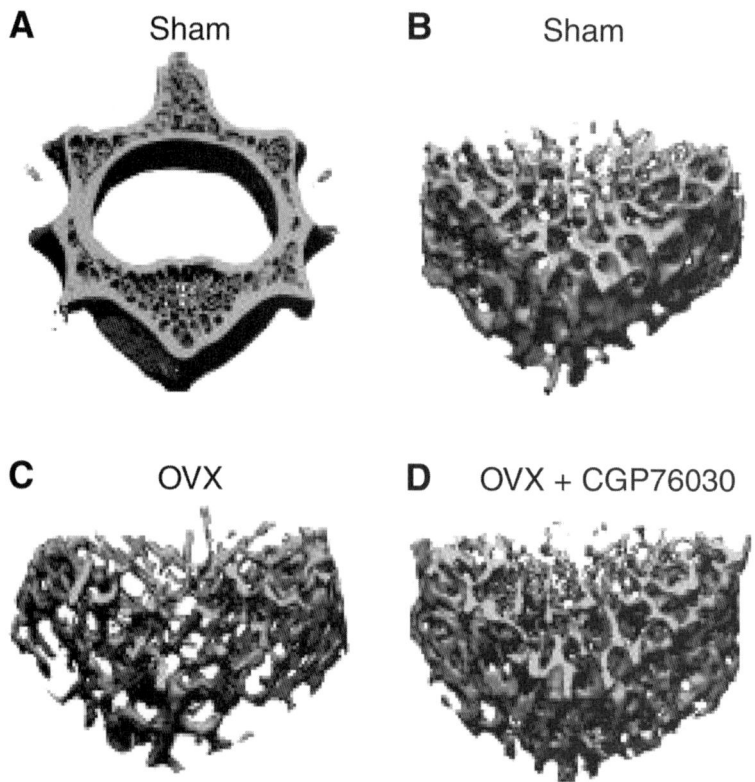

Fig. 4. Protective effect of CGP76030 on vertebral trabecular bone. Three-dimensional microcomputed tomographic reconstruction of rat lumbar vertebrae (LV2). (**A**) The central portion (1.3-mm length) of a complete vertebra of a sham-operated rat is shown. (**B**) Enlarged reconstruction of the isolated vertebral body from vertebrae shown in A. (**C,D**) Vertebral bodies of 8-wk-old ovariectomized rats without (C) or with (D) treatment with 50 mg/kg orally CGP76030 twice daily. The compound partially prevented microarchitectural changes caused by ovariectomy. OVX, ovariectomized.

female rats at 100 mg/kg by mouth on day 4, but could not be tested at such high doses in a long-term (12 wk) model of OVX-induced bone loss in skeletally mature rats owing to side effects (*see* Heading 6). Deaths occurred already at 30 mg/kg/day in 40 to 44-wk-old OVX female rats, but not up to 50 mg/kg/twice daily in young 8-wk-old rats. The reasons for this difference are probably connected to overproportional exposure at higher doses, particularly in old rats, and were facilitated by the long treatment period (12 wk). In OVX rat models, the protective effects of CGP76030 on bone measured by peripheral quantitative computed tomography and dual energy X-ray absorptionmetry were partial and often not significant, which can be attributed to relatively low bone loss by OVX. However, chemical analyses of bone content by measurement of bone calcium and hydroxyproline detected significant prevention of bone loss at 10–20 mg/kg by mouth in both young and old OVX rats. In addition, OVX-induced loss in trabecular bone microarchitecture was partially prevented by CGP76030, as illustrated in Fig. 4.

The olomoucine NVP-AAK980 was tested only in the short-term retinoid-treated rat model. The compound was active at 30 and 100 mg/kg by mouth, with the latter dose showing a full protection of hypercalcemia. Further tests were not done because of an insufficient therapeutic window (*see* Heading 6).

In conclusion, the pyrrolopyrimidine CGP76030 and the olomoucine NVP-AAK980 were active in preventing bone loss in the OVX and retinoid-treated rat models (from 10 and 30 to 100 mg/kg/by mouth). In the retinoid-treated rat model, they showed a full protection against hypercalcemia at higher doses (both compounds) and bone loss (only CGP76030 tested), an effect that is matched only by bisphosphonates and calcitonin, recognized bone resoprtion inhibitors already used in osteoporosis therapy.

6. SRC INHIBITORS: ACTIVITY IN CANCER MODELS

There is little information available on the effects of Src inhibitors in cancer models. The pyrrolopyrimidine CGP76030 and olomoucine NVP-AAK980 potently inhibited tyrosine phosphorylation in colon carcinoma cells (Fig. 5A) and the pyrrolopyrimidine CGP77675 inhibited PC3 cell migration and invasion at submicrolmolar concentrations *(21)* (Fig. 6). Both CGP76030 and NVP-AAK980 were active at inhibiting the growth of human colon cancer cell lines in vitro, with similar potency to the clinically used compound 5-fluorouracil (Fig. 5B); however the IC_{50} values of the Src inhibitors did not perfectly align with Src expression levels in these cell lines. Both CGP76030 and NVP-AAK980 were orally active, and possessed pharmacokinetic profiles demonstrating superior drug levels in tumor tissue as compared to plasma (Fig. 5C and data not shown). Subcutaneous HT29 cell tumor xenografts in female BALB/c nude mice were subsequently used to test for anti-tumor activity. NVP-AAK980 was inactive in vivo at 50 mg/kg, by mouth, once per day, which may be explained in part by low tumor drug levels (data not shown). However, CGP76030 was also apparently inactive in inhibiting the growth of subcutaneous HT29 tumors (Fig. 5D) despite drug penetration into tumor well in excess of the in vitro IC_{50}. No potentiation of the anti-tumor activity of 5-fluorouracil by CGP76030 was observed when the compounds were administered concomitantly (data not shown). A limitation with the in vivo testing of Src inhibitors is the identification of an appropriate Src-dependent cancer model with significance for human disease. Experimental and clinical evidence strongly implicates Src in tumor progression and metastases of colon cancer *(8)*. In the experiments described above, a model system that does not involve the formation of metastases was

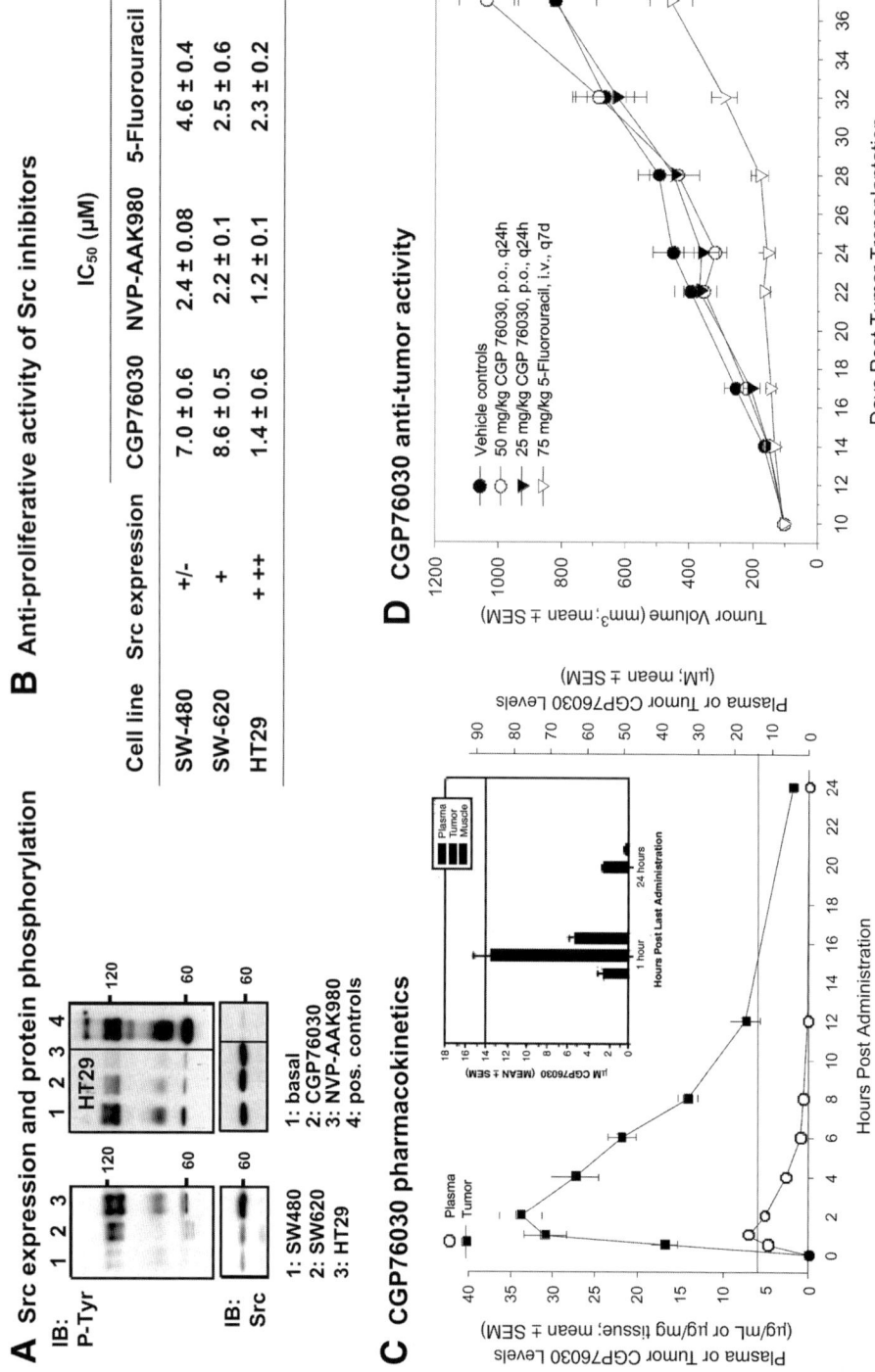

A Src expression and protein phosphorylation

IB:
P-Tyr

HT29
1 2 3 4

1 2 3

IB:
Src

1: SW480
2: SW620
3: HT29

1: basal
2: CGP76030
3: NVP-AAK980
4: pos. controls

B Anti-proliferative activity of Src inhibitors

Cell line	Src expression	IC$_{50}$ (µM)			
		CGP76030	NVP-AAK980	5-Fluorouracil	
SW-480	+/-	7.0 ± 0.6	2.4 ± 0.08	4.6 ± 0.4	
SW-620	+	8.6 ± 0.5	2.2 ± 0.1	2.5 ± 0.6	
HT29	+++	1.4 ± 0.6	1.2 ± 0.1	2.3 ± 0.2	

C CGP76030 pharmacokinetics

Plasma or Tumor CGP76030 Levels
(µg/mL or µg/mg tissue; mean ± SEM)

○ Plasma
■ Tumor

Hours Post Administration

Plasma or Tumor CGP76030 Levels
(µM; mean ± SEM)

Plasma
Tumor
Muscle

Hours Post Last Administration

1 hour   24 hours

µM CGP76030 (MEAN ± SEM)

D CGP76030 anti-tumor activity

Tumor Volume (mm³; mean ± SEM)

● Vehicle controls
○ 50 mg/kg CGP 76030, p.o., q24h
▲ 25 mg/kg CGP 76030, p.o., q24h
▽ 75 mg/kg 5-Fluorouracil, i.v., q7d

Days Post Tumor Transplantation

used, so that inhibition of the formation of metastases was not evaluable. Furthermore, despite the accumulated in vitro evidence, we have not yet unequivocally demonstrated Src-dependence of HT29 tumor growth in vivo. Lastly, inhibition of Src may be overridden by mechanisms operating in vivo in subcutaneous tumors that are not functioning in tumor cells in vitro. In light of these reservations, based on association of Src with human cancers and the available in vitro and in vivo data, it would be worth examining various colon, lung, breast, and

---

Fig. 5. The effect of Src inhibitors on the colon cancer cell lines and on HT29 xenografts in nude mice. **(A)** Src expression and protein phosphorylation. Equivalent amounts of protein from three cell lines related to colon cancer (SW480, SW620, HT29) were analyzed by antiphosphotyrosine and anti-Src immunoblotting (IB). HT29 cells expressed highest amounts of Src and phosphotyrosine phosphorylated proteins. Human HT29 colon cancer cells were treated with Src inhibitors for 2 h (CGP76030 at 0.5 μM and NVP-AAK980 at 1 μM). The cellular protein was extracted and equal amounts were analyzed by antiphosphotyrosine or anti-Src immunoblotting. Positive control lanes contain extracts from cells known to have high tyrosine phosphorylation (top: chicken c-Src expressing fibroblast cell line) or to express human Src (bottom: human osteoclastic cell line). HT29 cells expressed more Src than the positive control. Molecular marker weights in kilodaltons are shown on the right. Both inhibitors suppressed tyrosine phosphorylation in HT29 cells without affecting Src levels. **(B)** Antiproliferative activity of Src inhibitors. Each cell line was added to 96-well plates and incubated for 24 h. Subsequently, each compound was added in a twofold dilution series, and the cells were reincubated for 3 d. Methylene blue staining was performed at day 4 and the amount of bound dye (proportional to the number of surviving cells, which bind the dye) determined. IC_{50} values were determined using the SoftmaxPro program. CGP76030 showed the best correlation between antiproliferative activity and Src expression in three cell lines. **(C)** CGP76030 pharmacokinetics. Three millions of HT-29 tumor cells were injected subcutaneously into the left flank of each female nude mouse ($n = 4$/group). A single per os treatment of 50 mg/kg was administered when the tumors were approx 250 mm^3. High performance liquid chromatography-mass spectrometry (HPLC-MS) was used to determine the concentrations of CGP76030 in plasma and tumors. The horizontal line represents the concentrations of compound that is 10 times the antiproliferation IC_{50} for HT-29 cells in vitro. For CGP76030 the plasma AUC was 1525 mg.ml/h and the tumor AUC was 17,636 mg.g/h. Tumor drug levels were above 10 times the IC_{50} for approx 15 h postadministration, and above the IC_{50} for 24 h. (C inset) Mice bearing HT-29 tumors established from transplanted fragments were treated with CGP76030 for 27 d (*see* panel **[D]**). On the last treatment day, mice were sacrificed 1 and 24 h postadministration ($n = 4$/group). At this time, the tumors were approx 900 mm^3. HPLC-MS was used to determine the concentrations of CGP76030 in plasma, thigh muscle, and tumors. The horizontal line represents the concentration of compound that is 10 times the antiproliferation IC_{50} for HT-29 cells in vitro. Tumor drug levels did not exceed 10 times IC_{50} levels, but remained above the IC_{50} for 24 h. **(D)** CGP76030 anti-tumor activity. HT-29 tumor fragments of approx 25 mg were implanted into the left flank of each female nude mouse ($n = 8$/group). Treatments were started on day 10 following tumor transplantation. Tumor volumes were determined according to the formula: Length $\times$ Diameter$^2 \times \pi/6$. *designates $p < 0.05$ vs controls. Both 25- and 50-mg/kg doses of Src inhibitor were inactive, whereas the control compound 5-flurouracil showed the expected activity. CGP76030 was well tolerated.

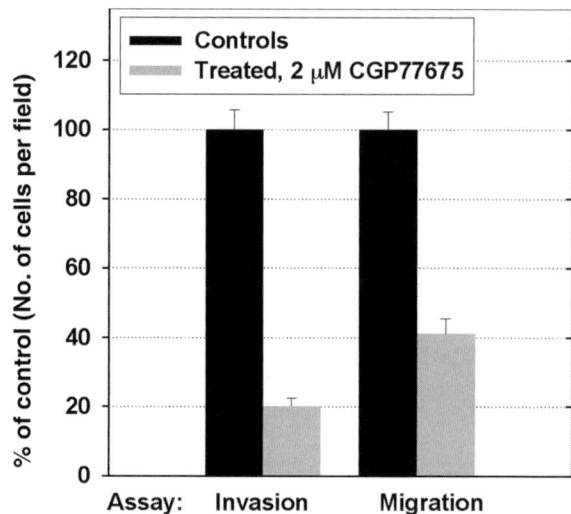

Fig. 6. Src inhibitor of the pyrrolopyrimidine class reduces prostrate cancer cell migration and invasion. PC3 prostate cancer cells were cultured in Boyden's chambers on porous membranes coated with gelatin or Matrigel and either left untreated (black bars) or treated with 2 μM CGP77675 (gray bars). The conditioned medium from NIH3T3 cells was used as chemoattractant. The number of cells migrating through the pores of the membrane were counted to estimate cell invasion (Matrigel coating) and migration (gelatin coating). The results are presented as means ± SEM of three experiments done in triplicates and show the inhibition of invasion and migration. The invasion was inhibited at a higher degree. The data were generated by I. Recchia and N. Rucci.

prostate cancer in vivo models that involve metastases *(5,21)* and establishing their dependence on Src. Tumor angiogenesis models are also worth examining, as there is in vitro and in vivo evidence for regulation of vascular endothelial growth factor-induced vascular permeability and angiogenesis by Src *(22)*.

Another strategy could be to develop dual inhibitors of Src and specific receptor tyrosine kinases, such as ErbB2 or c-Met. The effectiveness of dual kinase inhibition approach was shown by co-treatment by inhibitors of Src and ErbB2, which suppressed growth of Ras-induced sarcomas in mice *(23)*. In conclusion, two crucial factors for development of Src inhibitors for cancer are better definition of optimal in vivo tumor models and innovative ideas for the desired inhibitor profile.

7. SRC INHIBITORS: LIMITATIONS FOR PHARMACEUTICAL APPLICATION

Given the in vivo activity of Src inhibitors in models of bone resorption, what are the chances for the development of therapeutic agents from these compounds? A main indication for the pharmacological inhibition of bone loss is

Table 5
Side Effects of Src Inhibitors in the Rat

	Olomoucines		Pyrrolopyrimidines	
Compound	NVP-AAK980	CGP79883	CGP76030	CGP81699
Doses and side effects	25 mg/kg None 75 mg/kg Thrombopenia, LDH, GOT ↑, glucose ↑	25, 75 mg/kg Enlarged liver, anemia	25, 75 mg/kg Throat swelling, ovarian cysts, anemia, LDH ↑	25, 75 mg/kg Enlarged liver, enlarged spleen, swollen eyes, ovarian cysts, anemia, leukocytosis, thrombopenia, food intake ↓, deaths
Lack of specificity	Yes	Yes Cdc2 Kit	Yes PDGFR Kit	Yes PDGFR EGFR Kit

The experiments were performed with adult (9-mo-old) female Wistar rats during 2 wk. The compounds were administered daily by mouth. If not noted, body wt and food intake were not changed. Lack of specificity refers to enzymatic and cellular kinase activities, which were measured in vitro, as indicated in Tables 1 and 2 and designates IC_{50} ratio smaller than 10 between indicated kinase and Src. LDH, lactate dehydrogenase; GOT, aspartate aminotransferase; PDGFR, platelet-derived growth factor receptor; EGFR, epidermal growth factor receptor.

osteoporosis, a prevalent disease of old age, especially in postmenopausal women. In order to prevent the slow, progressive bone loss in such a population, a long-term, extremely well tolerated therapy is required. Furthermore, the patients need to take the therapy before they experience any symptoms, as bone fractures are an indication of already advanced osteoporosis. These factors put a very high requirement on tolerability of the potential therapy and Src inhibitors have not been adequately examined. The majority of compounds from both the pyrrolopyrimidine and olomoucine classes did not show overt toxicity in rats, detectable as a reduction of body weight and food consumption. However, in 9-mo-old Wistar rats treated for 2 wk at 25 and 75 mg/kg/d by mouth, the pyrrolopyrimidines CGP76030 and CGP81699 induced a dose-dependent decrease in red blood cells (anemia), affected the ovaries, and induced edema and erythema of nose, eyes, and ears (Table 5). The range and severity of symptoms and pathological findings were much more pronounced with CGP81699, which also caused reduced food intake and deaths (Table 5). As a more promising compound, CGP76030 was examined in another, more detailed study in

8-mo-old female Wistar rats at doses of 15, 30, and 60 mg/kg/d by mouth. A slight, but dose-dependent decrease in red blood cell count was confirmed, which appeared to be a consequence of bone marrow effects rather than direct hemolysis. The effect on red blood cells was smaller in younger, 8-wk-old rats. In this study, ovarian cysts were found at 30 and 60 mg/kg and hemorrhagic or necrotic corpora lutea in individual animals even at 15 mg/kg. A high dose of CGP76030 (60 mg/kg) induced a number of other changes, as evidenced by hematological, clinical chemical, and histopathological examinations, but no edema and erythema. A study in BALB/c mice for immune response to allergy, and for general activity and behavior did not show any adverse effects. Also, no adverse effects were seen with therapeutically relevant doses on cardiovascular and respiratory function in Wistar rats. The compound was not mutagenic in the Ames test and was weakly clastogenic in the micronucleus test. Therefore, owing primarily to changes in red blood cells and ovaries, which although mild, were nevertheless observed even at 15 mg/kg by mouth, it was not possible to define a "no toxic effect level" for the pyrrolopyrimidine CGP76030, as this dose was too close to 10 mg/kg by mouth, the lowest dose necessary to produce beneficial effects on bone. Thus, the side-effect profile limited further development of CGP76030 for the osteoporosis indication. It is possible that this side-effect profile would not be limiting for other indications or if the active dose were lower than 10 mg/kg. It should also be noted that this side-effect profile might be different in other species and in humans.

The olomoucine NVP-AAK980 at 25 mg/kg/d by mouth did not show any side effects, but caused thrombopenia at 75 mg/kg/d, whereas another olomoucine, CGP79883, caused dose-dependent anemia and effects on liver (Table 5). Because both compounds showed effects on a bone resorption surrogate (serum calcium) in TPTX retinoid-treated rats only at doses higher than 30 mg/kg/d, no therapeutic window could be defined for these compounds as well.

The severity of the side effects among two pyrrolopyrimidines and two olomoucines was in the order NVP-AAK980 (the mildest effects), CGP76030, CGP79883, and CGP81699 (most severe effects). A rough comparison of enzymatic selectivity of Src inhibitors with the number and severity of side effects shows an interesting correlation. The pyrrolopyrimidine CGP81699, the least tolerated compound, showed a lack of selectivity against three receptor tyrosine kinases (PDGFR, EGFR, and c-Kit). By contrast, the olomoucine NVP-AAK980, the best tolerated compound, was selective against these three receptor tyrosine kinases. All compounds were not selective against Yes (ref. 24, and our unpublished data), a Src family member that is most closely related to Src, and most compounds showed only a limited selectivity against other Src family members. Therefore, the most plausible explanation for the side effects of Src inhibitors is their limited enzymatic selectivity, which was not correlated to the compound's core structure and which led to inhibition of other kinases in nontarget tissues (bone marrow, ovaries). As cell cycle is of crucial importance in both of these tissues, it is possible that a common, protein kinase-dependent mechanism caused all observed side effects. Our recent unpublished study in BALB/C nu/nu mice model of breast cancer bone metastases showed positive effects of 100 mg/kg/d of CGP76030 on the incidence of metastases and lethality without any apparent side effects for 38 d (N. Rucci et al.)

8. OUTLOOK: SRC AS A TARGET FOR OTHER INDICATIONS, MORE SELECTIVE AND DUAL SRC INHIBITORS

The application of Src inhibitors for inhibition of benign bone loss in postmenopausal osteoporosis remains questionable. High safety requirements for this indication and unacceptable side effects of the compounds examined seem to preclude further development of current compounds.

Apart from a few isolated studies, the potential of Src inhibitors for inhibition of different tumors has not yet been extensively examined. A lack of enthusiasm in this area is certainly caused by a lack of convincing epidemiological studies implicating Src in the etiology of human tumors. Our preliminary studies with HT29 colon cancer xenografts in nude mice were not very encouraging (Fig. 5). However, recent in vitro data on potent effects of Src pyrrolopyrimidine inhibitors on prostate cell migration and invasion would warrant examination of existing compounds in suitable in vivo cancer metastases models. In accord with this notion, the Src inhibitors PP1 and herbimycin A have been shown to suppress collagen type I/III-dependent decrease in cadherin E, coupled to changes in cell–cell adhesion, proliferation, and migration of pancreatic carcinoma cells (25). In another tumor type, malignant peripheral nerve sheath tumors, there is a correlation between Src activity, inhibition by a Src inhibitor, and tumor cell migration and invasion. These aggressive malignancies arise with increased incidence in patients with neurofibromatosis type 1, in which Ras and EGFR activity is deregulated. A recent study by Sherman and colleagues (26) indicates that invasion and metastatic potential depend on CD44 expression, which is regulated by Src and can be inhibited by submicromolar concentrations of the pyrrolopyrimidine CGP77675. In particular, bone metastases models of breast and other cancers are also worth examining, since Src inhibitors would have a possibility to inhibit both tumor cells and tumor cell-stimulated osteoclasts that cause painful bone destruction (13). Thus, in vitro studies demonstrate that Src inhibitors may be most efficacious in models of tumor metastases, including those homing to bone. Our recent study in mouse models of bone metastases provides data supporting this idea.

Another useful feature of Src inhibitors in treatment of malignancies may be their ability to induce apoptosis of BCR-Abl kinase-transformed hematopoietic cells that are resistant to the BCR-Abl inhibitor Glivec (STI-571) (27,28). The beneficial effect of CGP76030 appears to be mediated by inhibition of Lyn and Hck, Src family kinases upregulated in STI-571-resistant cells. Furthermore, recent in vivo experiments in a murine BCR-Abl-induced B-cell leukemia model showed a positive effect of CGP76030, alone or after co-treatment with STI-571 *(29)*. Wyeth has also recently reported in vitro and in vivo anti-tumor activity in leukemia models with the compound SKI-606 (identical to compound 31a in Fig. 2), which inhibited both Abl and Src kinases *(30)*. Thus, development of a potent, dual BCR-Abl/Src inhibitor is an attractive possibility.

What are other potential indications for Src inhibitors? Here we will mention three of the most promising ones: stroke, graft rejection, and infection. A recent study showed that Src-deficient mice or mice treated with pyrazolopyrimidine PP1 (0.15–1.5 mg/kg, intraperitoneally) are partially protected from cerebral ischemic injury in stroke *(31)*. The mechanism of action of Src inhibitors involves modulation of vascular permeability induced by VEGF, which is dependent on Src *(22)*. In the murine cardiac allograft model, the pyrrolopyrimidine CGP77675 delayed graft rejection *(32)*. Src activity was induced in rejected grafts and was reduced upon treatment with the Src inhibitor. The mechanism of action may involve inhibition of the fibroblast growth factor (FGF)-induced proliferation of endothelial cells and of IL-2-dependent T cells. Finally, studies with internalization of the bacterium *Staphylococcus aureus* into host cells showed that pyrrolopyrimidine blocks this process and, thus, has potential as an anti-infection agent *(33)*.

Further development of Src inhibitors as therapeutic agents depends on their therapeutic index (ratio of efficacy vs tolerability). Of the known inhibitors, olomoucine derivatives, such as NVP-AAK980, seem to have a good potential since they show better in vivo tolerability than other compounds, while retaining high activity in a bone loss model. Development of more selective Src kinase inhibitors or precisely targeted dual inhibitors (Src and Abl, Src and EGFR or Erbb2, Src and Met) would provide a new impetus in the area. In addition, strategies to selectively target Src inhibitors to a certain tissue and to define optimal cancer models could prove useful.

ACKNOWLEDGMENTS

We thank Drs. P. Ulrich and U. W. Laengle from Preclinical Toxicology, Novartis Pharma AG, Basel, for the toxicology studies on CGP76030. Drs. L. Widler, E. Altmann, and R. Beerli, Arthritis and Bone Metabolism Disease Area, NIBR, are acknowledged for the synthesis of Src inhibitors. Numerous colleagues from the Oncology Disease Area, NIBR, are acknowledged for profiling compounds in several enzymatic and cellular assays, in particular Drs. T. Meyer, E. Buchdunger, and D. Fabbro. I. Recchia, and N. Rucci from Prof. A. Teti's group (L'Aquila, Italy) are acknowledged for their molecular studies of osteoclasts. We are grateful to Drs. P. Lehenkari, M. Muzylak, and Prof. M. Horton (London, UK) for their microscopy studies with osteoclasts. We thank Drs. L. Sherman, (Beaverton, Oregon), G. Miller (Nashville, Tennessee), and T. Fowler (Houston, Texas) for sharing their unpublished data on application of Src inhibitors. We thank Dr. J. Spring (Basel, Switzerland) for the generation of Fig. 1.

REFERENCES

1. Neet K, Hunter T. Vertebrate non-receptor protein tyrosine kinase families. *Genes Cells* 1996; 1:147–169.
2. Brown MT, Cooper JA. Regulation, substrates and functions of src. *Biochim Biophys Acta* 1996; 1287:121–149.
3. Thomas SM, Brugge JS. Cellular functions regulated by Src family kinases. *Annu Rev Cell Dev Biol* 1997; 13:513–609.
4. Guy CT, Muthuswamy SK, Cardiff RD, Soriano P, Muller WJ. Activation of the c-Src tyrosine kinase is required for the induction of mammary tumors in transgenic mice. *Genes Dev* 1994; 8:23–32.
5. Biscardi JS, Tice DA, Parsons SJ. c-Src, receptor tyrosine kinases, and human cancer. *Adv Cancer Res* 1999; 76:61–119.
6. Kloth MT, Laughlin KK, Biscardi JS, Boerner JL, Parsons SJ, Silva CM. STAT5b, a mediator of synergism between c-Src and the epidermal growth factor receptor. *J Biol Chem* 2003; 278:1671–1679.
7. Li Y, Tondravi M, Liu J, et al. Cortactin potentiates bone metastasis of breast cancer cells. *Cancer Res* 2001; 61:6906–6911.
8. Aligayer H, Boyd DD, Heiss MM, Abdalla EK, Curley SA, Gallick GE. Activation of Src kinase in primary colorectal carcinoma: an indicator of poor clinical prognosis. *Cancer* 2002; 94:344–351.
9. Soriano P, Montgomery C, Geske R, Bradley A. Targeted disruption of the c-src proto-oncogene leads to osteopetrosis in mice. *Cell* 1991; 64:693–702.
10. Duong LK, Lakkakorpi PT, Nakamura I, Machwate M, Nagy RM, Rodan GA. PYK2 in osteoclasts is an adhesion kinase, localized in the sealing zone, activated by ligation of alpha(v)beta3 integrin, and phosphorylated by src kinase. *J Clin Invest* 1998; 102:881–892.
11. Jeschke M, Brandi ML, Susa M. Expression of Src family kinases and their putative substrates in the human preosteoclastic cell line FLG 29.1. *J Bone Miner Res* 1998; 13:1880–1889.
12. Susa M, Missbach M, Green J. Src inhibitors: drugs for the treatment of osteoporosis, cancer or both? *Trends Pharmacol Sci* 2000; 21:489–495.
13. Susa M, Teti A. Tyrosine kinase Src inhibitors: potential therapeutic applications. *Drug News Perspect* 2000; 13:169–175.
14. Altmann E, Widler L, Missbach M. N(7)-substituted-5-aryl-pyrrolo[2,3-d]pyrimidines represent a versatile class of potent inhibitors of the tyrosine kinase c-Src. *Med Chem* 2002; 2:201–208.
15. Metcalf CA III, van Schrevendijk MR, Dalgarno DC, Sawyer TK. Targeting protein kinases for bone disease: discovery and development of Src inhibitors. *Curr Pharm Des* 2002; 8:2049–2075.
16. Blake RA, Broome MA, Liu X, et al. SU6656, a selective Src family kinase inhibitor, used to probe growth factor signaling. *Mol Cell Biol* 2000; 20:9018–9027.
17. Shakespeare WC. Src tyrosine kinase inhibitors for bone diseases. Abstracts of Papers, 223rd American Chemical Society National Meeting, April 7–11, 2002, Orlando.
18. Missbach M, Jeschke J, Feyen J, et al. A novel inhibitor of the tyrosine kinase Src suppresses phosphorylation of its major cellular substrates and reduces bone resorption in vitro and in rodent models in vivo. *Bone* 1999; 24:437–449.

19. Recchia I, Rucci N, Funari A, et al. Reduction of c-Src activity by substituted 5,7-diphenyl-pyrrolo[2,3-*d*]- pyrimidines induces osteoclast apoptosis in vivo and in vitro. Involvement of ERK 1/2 pathway. *Bone* 2004; 34:65–79.

20. Tang D, Wu D, Hirao A, et al. ERK activation mediates cell cycle arrest and apoptosis after DNA damage independently of p53. *J Biol Chem* 2002; 277:12710–12717.

21. Recchia I, Rucci N, Festuccia C, et al. Pyrrolopyrimidine c-Src inhibitors reduce prostate cancer cell activity in vitro. *Eur J Cancer* 2003; 39:1927–1935.

22. Eliceiri BP, Paul R, Schwartzberg PL, Hood JD, Leng J, Cheresh DA. Selective requirement for Src kinases during VEGF-induced angiogenesis and vascular permeability. *Mol Cell* 1999; 4:915–924.

23. He H, Hirokawa Y, Manser E, Lim L, Levitzki A, Maruta H. Signal therapy for RAS-induced cancers in combination of AG 879 and PP1, specific inhibitors for ErbB2 and Src family kinases, that block PAK activation. *Cancer J* 2001; 7:191–202.

24. Susa M, Luong-Nguyen N-H, Crespo J, Maier R, Missbach M, McMaster G. Active recombinant human tyrosine kinase c-Yes: expression in baculovirus system, purification, comparison to c-Src, and inhibition by a c-Src inhibitor. *Protein Expr Purif* 2000; 19:99–106.

25. Menke A, Philippi C, Vogelmann R, et al. Down-regulation of E-cadherin gene expression by collagen type I and type III in pancreatic cancer cell lines. *Cancer Res* 2001; 61:3508–3517.

26. Su W, Sin M, Darrow A, Sherman L. Malignant peripheral nerve sheath tumor cell invasion is facilitated by Src and aberrant CD44 expression. *Glia* 2003; 42:350–358.

27. Donato NJ, Wu JY, Stapley J, et al. BCR-ABL independence and LYN kinase overexpression in chronic myelogenous leukemia cells selected for resistance to STI571. *Blood* 2003; 101:690–698.

28. Warmuth M, Simon N, Mitina O, et al. Dual-specific Src and Abl kinase inhibitors, PP1 and CGP76030, inhibit growth and survival of cells expressing imatinib mesylate-resistant Bcr-Abl kinases. *Blood* 2003; 101:664–672.

29. Li S, Hu Y. Src kinase inhibitor CGP 76030 synergizes with STI571 in the treatment of B-cell acute lymphoblastic leukemia induced by the BCR/ABL oncogene in mice. The American Society of Hematology, 45th Annual Meeting, December 6–9, 2002, San Diego, California.

30. Golas JM, Arndt K, Etienne C, et al. SKI-606, a 4-anilino-3-quinolinecarbonitrile dual inhibitor of Src and Abl kinases, is a potent antiproliferative agent against chronic myelogenous leukemia cells in culture and causes regression of K562 xenografts in nude mice. *Cancer Res* 2003; 63:375–381.

31. Paul R, Zhang ZG, Eliceiri BP, et al. Src deficiency or blockade of Src activity in mice provides cerebral protection following stroke. *Nat Med* 2001; 7:222–227.

32. Fairchild R, Azimzadeh A, Pierson R, Miller G. Activation of Src family members in allografts. American Transplant Congress, April 26–May 1, 2002, Washington, DC.

33. Fowler T, Johansson S, Wary KK, Hook M. Src kinase has a central role in in vitro cellular internalization of *Staphylococcus aureus. Cell Microbiol* 2003; 5:417–426.

34. Manning G, Whyte DB, Martinez R, Hunter T, Sudarsanam S. The protein kinase complement of the human genome. *Science* 2002; 298:1912–1934.

35. Gibson TJ, Spring J. Genetic redundancy in vertebrates: polyploidy and persistence of genes encoding multidomain proteins. *Trends Genet* 1998; 14:46–49.

36. Lakkakorpi PT, Nakamura I, Nagy RM, Parsons TJ, Rodan GA, Duong LT. Stable association of PYK2 and p130(Cas) in osteoclasts and their co-localization in the sealing zone. *J Biol Chem* 1999; 274:4900–4907.

37. Olayiole MA, Beuvink I, Horsch K, Daly JM, Hynes NE. ErbB receptor–induced activation of Stat transcription factors is mediated by Src tyrosine kinases. *J Biol Chem* 1999; 274:17209–17218.

5 Activated FLT3 Receptor Tyrosine Kinase as a Therapeutic Target In Leukemia

Blanca Scheijen and James D. Griffin

CONTENTS

INTRODUCTION
STRUCTURE AND FUNCTION OF FLT3 AND ITS LIGAND
    IN NORMAL HEMATOPOIETIC CELLS
FLT3 EXPRESSION IN LEUKEMIAS
ACTIVATING *FLT3* MUTATIONS IN MYELOID LEUKEMIA
BIOLOGICAL CONSEQUENCES OF FLT3 SIGNALING
FLT3 KINASE INHIBITORS
FUTURE DIRECTIONS
REFERENCES

1. INTRODUCTION

Acute myeloid leukemia (AML) is a heterogeneous neoplastic disease characterized by deregulated proliferation of myeloid precursor cells combined with an arrest in the differentiation process. The abnormal survival advantage of transformed immature myeloid precursors further contributes to a marked impairment of the functional maturation of the various myeloid cell lineages. Intensive studies over the past 20 yr have resulted in the identification of leukemia-specific cytogenetic abnormalities, which provided considerable insight into the underlying pathogenesis of AML. This has also resulted in the prognostic stratification of AML patients into three different risk groups: those with favorable, intermediate or standard, and poor risk disease. Patients with favorable cytogenetics, i.e., those with translocations t(15;17), t(8;21), and inversion inv(16), have particularly benefited from an improved understanding of the

From: *Cancer Drug Discovery and Development:*
Protein Tyrosine Kinases: From Inhibitors to Useful Drugs
Edited by: D. Fabbro and F. McCormick © Humana Press Inc., Totowa, NJ

molecular pathology of their disease, which has resulted in the identification of potential therapeutic targets. For example, the addition of all-*trans* retinoic acid during induction chemotherapy for acute promyelocytic leukemia (APL) patients with the PML/RARα (promyelocytic leukemia/retinoic acid receptor-α) fusion gene in t(15;17) has increased the 5-yr survival a further 20–30% compared with chemotherapy alone *(1,2)*.

The more recent observation that about 30% of all AML patients show activating mutations in the macrophage colony stimulating receptor-like tyrosine kinase receptor-3 (FLT3) has resulted in a new and interesting opportunity to design drugs for targeted therapy in AML patients with a normal karyotype and intermediate/standard or poor-risk disease. Here we review our current knowledge about the biological functions of wild-type FLT3 receptor and its ligand in hematopoiesis and the role of activated FLT3 receptors in leukemia, with particular emphasis on the regulatory pathways stimulated by mutant FLT3 and recent efforts to develop selective small-molecule FLT3 inhibitors.

2. STRUCTURE AND FUNCTION OF FLT3 AND ITS LIGAND IN NORMAL HEMATOPOIETIC CELLS

2.1. FLT3 Receptor

The FLT3 *(3,4)*, also known as fetal liver kinase-2 *(5,6)* or human stem cell kinase-1 *(7)*, is a member of the class III receptor tyrosine kinase (RTK) family that includes the macrophage colony-stimulating factor receptor, Steel factor receptor (KIT), and the receptors for the platelet-derived growth factor receptors α and β (PDGFR-α and PDGFR-β). Characteristic features of type III RTK are five immunoglobulin-like extracellular domains, a transmembrane domain, a juxtamembrane (JM) domain, and two intracellular tyrosine kinase domains (TKDs) linked by a kinase-insert region *(8)*. Each of the members of the RTK subclass III fulfills important roles in the regulation of proliferation, differentiation, and survival of normal hematopoietic cells and are known targets for activating mutations or chromosomal translocations in hematological malignancies *(9)*.

The human *FLT3* gene consists of 24 exons and is located on chromosome 13q12 *(10)*, has 85% amino acid sequence homology with mouse *Flt3 (11)*, and is expressed as a major 160 kDa protein that is glycosylated at *N*-linked glycosylation sites in the extracellular domain and localized to the plasma membrane, and a minor 135–140 kDa unglycosylated protein that is not membrane bound *(12–14)*. Within the hematopoietic compartment, expression of FLT3 occurs on the surface of hematopoietic stem cells (HSCs), uncommitted lymphoid and myeloid progenitors, as well as dendritic cells (DCs) and CD14[+] monocytes, but not on granulocytes, megakaryocytes, erythroid cells, and mast cells *(3,15–17)*. *FLT3* expression has also been detected in placenta, gonads, brain, and retina *(12,18,19)*, but its exact role in these tissues is far from understood.

A number of studies have demonstrated that FLT3 is expressed on a fraction of human adult bone marrow (40–60%) and cord blood (30–40%) CD34$^+$ progenitor cells *(17,20)*, which corresponds with long-term culture-initiating cells that are capable of in vivo lymphomyeloid reconstitution of nonobese diabetic/severe combined immunodeficiency mice *(21,22)*. In contrast, the fraction of murine Lin$^-$Sca1$^+$c-KIT$^+$ HSC that expresses FLT3 fails to long-term reconstitute myelopoiesis in vivo, as opposed to the sustained reconstitution activity of Lin$^-$Sca1$^+$c-KIT$^+$-FLT3$^-$ HSC *(23,24)*. Thus, in contrast to human candidate HSCs, mouse long-term reconstituting HSCs are negative for FLT3 expression during steady-state hematopoiesis and upregulation of FLT3 expression within the HSC compartment is accompanied by loss of self-renewal capacity. Interestingly, c-KIT displays an opposite expression pattern, where mouse long-term resulting HSCs show high expression of c-KIT *(25)*, whereas c-KIT ligand or stem cell factor (SCF) has limited capacity to support the in vitro survival of human CD34$^+$CD38$^-$ candidate HSCs *(26)*, which have been shown to express low levels of c-KIT *(27,28)*.

FLT3 has an important role in the early stages of B-cell development, corresponding with FLT3 expression on common lymphoid progenitor, pre-pro-B-cells (B220$^+$, CD43$^+$, AA4.1$^+$, CD19$^-$), and pro-B-cells (B220$^+$, CD43$^+$, AA4.1$^+$, CD19$^+$), whereas pre-B-cells, immature, and mature B-cells are devoid of FLT3 expression *(29,30)*. FLT3-deficient mice have reduced levels of predominantly primitive B-lymphoid progenitor cells, whereas normal numbers of functional B-cells are present in the periphery *(31)*. The population of multipotential and myeloid colony-forming progenitors is not affected in *Flt3$^{-/-}$* mice, and only competitive repopulation experiments reveal that FLT3-deficient stem cells have a defect in reconstitution of the lymphoid lineage, and to a lesser extent the myeloid lineage *(31)*.

2.2. FLT3 Ligand

In contrast to the rather restricted expression pattern of FLT3, FLT3 ligand (FL) RNA is expressed in multiple isoforms by a wide variety of tissues and most types of hematopoietic cells *(32–34)*. The predominant isoform of human FL is a cell surface transmembrane protein type I that can be proteolytically processed and released as a soluble protein that is also biologically active. The natural occurring soluble FL protein exists of a 65-kDa nondisulfide-linked homodimeric glycoprotein composed of 30 kDa subunits, each containing up to 12 kDa of N- and O-linked sugars *(35)*. Unlike human FL, the most abundant isoform of mouse FL is a 220-amino acid membrane associated protein *(32,36)*. This membrane-associated isoform results from a failure to splice out an intron, creating a change in the reading frame, which terminates in a stretch of hydrophobic amino acids that anchors the ligand to the cell surface. An alternatively spliced sixth exon generates a different soluble FL isoform, which has been detected in both human and mouse tissues *(36,37)*.

The supply of FL for hematopoiesis in humans is tightly regulated by a process specific for this cytokine *(38)*. This regulation is based on the intracellular retention of preformed FL and its release from intracellular stores depending on the status of the stem cell compartment *(39)*. During steady-state hematopoiesis, FL is expressed constitutively, but little of this cytokine is released by cells. Preformed FL can be detected in the cytoplasm of bone marrow stroma fibroblasts, T-cells, B-cells, and $CD34^+$ progenitors *(39)*. In fibroblasts and T-cells, intracellular FL colocalizes with proteins present in the Golgi and trans-Golgi compartments *(40)*. Release of FL can be triggered in T-cells through signaling via the common receptor-γ chain (interleukin (IL)-2, IL-4, IL-7, and IL-15) *(40)*, and in vivo, by stem cell deficiency in the bone marrow, as is the case with Fanconi anemia, acquired aplastic anemia, and cytoreductive chemotherapy *(41,42)*.

FL is usually not highly efficient when used as a single cytokine to stimulate progenitor cell expansion in tissue culture. However, FL acts as a potent stimulator of human HSCs when used in combination with thrombopoietin or SCF and IL-3 *(43,44)*, and is therefore often used to promote HSC ex vivo expansion in the course of retroviral-mediated gene transfer *(45,46)*. FL strongly enhances the myeloid colony-stimulating activity on hematopoietic progenitor cells in synergy with granulocyte-CSF, granulocytic-monocytic-CSF, or M-CSF. FL synergizes with IL-3, IL-7, and IL-11 to stimulate B lymphopoiesis in vitro and with IL-15 to drive the development of natural-killer (NK) cells. FL is also able to promote the expansion of primitive thymic progenitors in combination with IL-3, IL-6, and IL-7 *(47)*. In contrast, FL has no growth-stimulating activity on progenitor cells committed to the erythrocyte, megakaryocyte, eosinophil, or mast-cell lineages *(48–50)*. No restrictions in species specificity has been observed for the effects of FL, as murine and human FL are active on cells from both species.

In vivo administration of FL alone increases multipotent colony-forming unit granulocytic-erythroid-monocytic-megakaryocytic and colony-forming unit granulocytic-monocytic (CFU-GM) colonies as well as B-cell colony-forming unit, but few or no CFUs or mature cells of the erythroid or megakaryocytic lineage *(51)*. Mice transplanted with bone marrow cells that express transmembrane FL from a retroviral vector develop leukocytosis with abnormal cell infiltration into many organs and are predisposed to leukemic transformation of B-cell and myeloid-cell lineages *(52,53)*.

Most notably, FL as a single cytokine is able to induce all major DC populations from whole bone marrow cells *(54)*. In vivo treatment of mice with FL results in dramatic increase of DCs in all primary and secondary lymphoid tissues *(55)*, and FL stimulates DC development from CLPs, common myeloid progenitors, and granulocyte/macrophage progenitors *(56–58)*. In humans, FL induces both $CD11c^+$ and $CD11c^-$ DC subsets, and FL enhances the production of myeloid-type DCs from $CD34^+$ bone marrow progenitor cells in combination with GM-CSF, tumor necrosis factor-α, and IL-4 *(59–60)*. Because DCs are the most efficient antigen-presenting cells for T-cells, FL administration has been shown to inhibit tumor growth and promote tumor regression and immunization in experimental cancer models *(61–64)*.

Studies from gene-targeted mice confirm the important role for FL signaling in B-cell and DC development and differentiation. As opposed to *Flt3*$^{-/-}$ animals, *Flt3L*$^{-/-}$ mice display an overt reduction in leukocyte counts of bone marrow, spleen, lymph nodes, and peripheral blood *(65)*. Absolute numbers of CFU-GM are slightly reduced, whereas B-cell precursors show a significant reduction, similar to FLT3 receptor-deficient mice. In addition, *Flt3L*$^{-/-}$ mice have 4- to 14-fold reduced numbers of myeloid-related (CD8α$^-$CD11c^{hi}) and lymphoid-related (CD8α$^+$CD11c^{hi}) DCs and also show about fivefold lower numbers of NK cells. The finding that mice doubly deficient for FL and IL-7 receptor α (IL-7Rα) completely lack mature B-cells and all stages of committed B-cell progenitors during embryogenesis and in adult mice, indicates that complementary signaling through FLT3 and IL-7Rα is indispensable for both fetal and adult B-cell development *(30)*.

3. FLT3 EXPRESSION IN LEUKEMIAS

Given its expression on the cell surface of myeloid and lymphoid progenitors, it is not surprising that FLT3 receptor is expressed in the vast majority of human B-lineage acute lymphocytic leukemias (ALL) and most myeloid leukemias throughout the different morphological subtypes (M0–M7), in both cell lines as well as primary leukemia blasts *(66–68)*. A smaller fraction of T-lineage cases (less than 30%) are positive for FLT3 expression *(14,69)*. More detailed analyses have shown that the mean mRNA expression level of FLT3 is higher in AML than in normal mononuclear cells, but there is a wide variation between individual AML samples *(70)*. Virtually all primary AML samples and leukemic-derived cell lines display co-expression of FL and FLT3 *(66,68)*. Often this results in constitutive phosphorylation of wild-type FLT3 receptor in leukemic cells, even though in most cases FLT3 receptor autophosphorylation can be further augmented by the addition of exogenous FL *(71)*. This is in line with the finding that FL has anti-apoptotic effects and can induce proliferation in some FLT3-expressing primary leukemic cells and cell lines *(72,73)*. Thus, there is accumulating evidence that FL causes autocrine signaling in AML, which may contribute to the development of leukemia.

4. ACTIVATING *FLT3* MUTATIONS IN MYELOID LEUKEMIA

FLT3-activating mutations represent the most common known genetic lesion in *de novo* AML (~30%) and are also present in some patients with myelodysplastic syndromes (~5%) *(74–81)*. *FLT3* genetic aberrations have not been detected in patients with chronic myelogenous leukemia (CML), nonbiphenotypic ALL, chronic lymphoid leukemia, non-Hodgkin's lymphoma, or multiple myeloma. In 1996, the first type of *FLT3* mutation was identified as a genomic duplication (internal tandem duplication [ITD]) within the JM domain of FLT3 encoded by exons 14 and 15 *(74)*. The duplicated regions among the different AML patients are variable in size (between 3 and 400 bp) and exact location, but the resulting transcripts are always in frame and give rise to a duplicated stretch rich in tyrosine residues. It is thought that alteration of the length of the JM domain, rather than increase of tyrosine residues causes gain-of-function of FLT3 *(75–82)*. Furthermore, an activating point mutation has been found in the JM domain in leukemia cell lines MonoMac1 and MonoMac6, which results in a substitution of valine by alanine at position valine 592 (V592A) *(83)*, but this mutation has not yet been reported in primary AML blasts.

A different set of genetic alterations have been detected in the activation loop within the second tyrosine kinase domain of the FLT3 receptor that normally blocks the access of adenosine triphosphate and substrate to the kinase domain when the receptor is in an inactive state. Frequently, these involve mutations at aspartic acid 835 (D835) or isoleucine 836 (I836) in exon 20 *(79,80)*. Activating point mutations of the corresponding residues in the activation loop of other RTKs have been identified, including c-KIT, c-FMS, RET, and MET, which are also associated with human malignancies *(84–87)*. In addition, insertions and deletions in the codons surrounding amino acids 835/836 of FLT3 have also been found *(80,81,88)*. Most likely, additional genetic alterations in *FLT3* will be detected, perhaps in the first tyrosine kinase domain or C-terminal region as reported for c-FMS *(89,90)*, or in the extracellular domain as is the case in c-FMS and c-KIT *(90–92)*.

4.1. Internal Tandem Duplications

FLT3-ITD mutations can be detected in 20–25% of adult patients with AML, and correlate with higher leukocytosis at diagnosis, decreased disease-free survival, and increased relapse risk *(88,93–95)*. Moreover, concomitant loss of the FLT3-wild-type allele has been reported to be associated with significantly inferior clinical outcome and decreased overall survival in FLT3-ITD patients *(88,96)*. In pediatric AML patients, FLT3-ITDs are present in 11–16% of the cases and represent a strong adverse prognostic factor *(97–100)*. FLT3-ITD mutations occur more frequently in patients with APL (30–39%) harboring t(15;17), which generates the PML/RARα oncogene *(101–104)*. On the other hand, *RAS* mutations *(105–107)* and translocations involving mixed-lineage leukemia (MLL) or core binding factor proteins rarely coexist with FLT3-ITD mutations *(88,94,95)*. Thus, specific oncogenic gene aberrations may either cooperate or be redundant with FLT3-ITD mutations.

FLT3-ITD receptors display ligand-independent receptor dimerization and phosphorylation, but the exact mechanism whereby ITD mutations cause constitutive kinase activity has not yet been fully elucidated. Both elongation as well as shortening of the JM domain result in constitutively activated FLT3 receptor *(82)*. Based on the structural data of the EphB2 RTK *(108)*, it has been postulated that the JM domain takes up an α-helical conformation, which blocks activation of the kinase and may inhibit self-dimerization. ITD mutations would then disrupt steric hindrance of the kinase domain by the JM domain, leading to constitutive autophosphorylation. However, a negative regulatory function of the JM region through binding of an inhibitory protein has not been formally excluded.

FLT3-ITD receptors have been shown to confer growth-factor independence in cytokine-dependent-cell lines like Ba/F3 and 32D cells, and injection of 32D cells expressing FLT3-ITD receptors into syngeneic mice results in the rapid development of a leukemia-like disease *(109–111)*. Furthermore, transplantation of bone marrow cells retrovirally transduced with FLT3-ITD into recipient mice leads to an oligoclonal myeloproliferative disorder *(112)*, confirming the tumorigenic potential of FLT3-ITD mutations.

4.2. TKD Mutations

Activating TKD mutations occur in 7–14% of adults with AML, but D835 mutations seem to show no significant correlation with poor prognosis *(88,113, 114)*. About 3–8% of infants with AML have FLT3-TKD mutations *(115,116)*. Clustering algorithms have revealed that childhood acute leukemias carrying rearrangements involving the *MLL* gene on chromosome 11q23 show higher mRNA expression of wild-type FLT3, distinguishing them from conventional pre-B ALL and AML *(117,118)*. Interestingly, FLT3-TKD deletions involving codons D835/I836 occur more frequently in MLL patients *(119,120)*, and patient samples with these activating FLT3-TKD alterations express even higher levels of FLT3 transcripts *(120)*.

Like FLT3-ITD receptors, FLT3-TKD mutations induce ligand-independent receptor activation and confer growth factor independence in 32D cells *(80)*. However, it remains to be established whether FLT3-TKD receptors phosphorylate the same substrates as and signal in a similar manner to FLT3-ITD receptors. Given the fact that FLT3-D835 mutations are not associated with a significant decrease on the overall survival of AML patients, FLT3-TKD mutations might be less tumorigenic than FLT3-ITD mutations, even though FLT3-D835Y receptors display a higher level of intrinsic tyrosine kinase activity than FLT3-ITD receptors (Scheijen and Griffin, unpublished results).

5. BIOLOGICAL CONSEQUENCES OF FLT3 SIGNALING

Initial studies on FLT3 signaling have been performed using chimeric receptors containing the extracellular part of human c-FMS. These activated mouse FLT3 receptors induce direct interactions with adaptor protein GRB2 and the p85 subunit of PI3K *(121,122)*, and concomitant phosphorylation of SHIP,

SHC, Vav, RasGAP, and PLCγ *(121,123)*. FL-mediated triggering of human FLT3 receptor results in direct association with GRB2 and SOCS1, and promotes phosphorylation and complex formation of CBL, CBLB, SHC, SHIP, SHP2, GAB1, GAB2, and activation of mitogen-activated protein kinase (MAPK) pathway *(124–126)*. In contrast to murine FLT3, human FLT3 has no consensus SH2-domain binding site for p85-PI3K in the carboxyl terminus, nor does p85 seem to be tyrosine phosphorylated on FL binding *(127)*. Instead, p85-PI3K has been found to be associated with complexes containing tyrosine phosphorylated SHP2, SHIP, GAB1, GAB2, CBL, and CBLB *(126–128)*.

As expected, signal transduction downstream of FLT3-ITD receptors mimics to a large extent FL-induced activation of wild-type FLT3 receptors. FLT3-ITD signaling results in complex formation with SHC, CBL, Vav, and SHP2, and constitutive activation of signaling pathways such as RAS, MAPK, signal tranducers and activators of transcription (STAT)3, STAT5, and AKT/PKB *(82,109,110)*. However, there are accumulating data suggesting also differences between wild-type and mutant FLT3 signaling. For instance, FLT3-ITD mutants strongly induce STAT5 activation, which is hardly observed with FL stimulation of wild-type FLT3 receptors *(109,110,129)*. Anti-apoptotic and proliferative signaling of FLT3-ITD receptors specifically correlates with induction of STAT target genes, such as *Pim2*, *SOCS2/3*, and *Bcl-X$_L$ (130,131)*.

One characteristic feature of AML is a block in one of the different stages of myeloid differentiation, and several reports have shown that activated FLT3 signaling can directly contribute to inhibition of myeloid differentiation. Expression of FLT3-ITD prohibits G-CSF-induced differentiation of 32D cells *(132)*, and microarray analysis has demonstrated that FLT3-ITD receptor signaling represses expression of transcription factors PU.1 and C/EBPα, which are normally required to induce myeloid differentiation *(130,133)*.

6. FLT3 KINASE INHIBITORS

Recently, clinical success has been achieved with the tyrosine kinase inhibitor imatinib mesylate (Gleevec; STI571) for the treatment of patients with BCR/ABL-positive chronic mylogenous leukemia in both chronic phase and blast crises *(134,135)*, chronic myelomonocytic leukemias associated with constitutively activated TEL-PDGFRβ fusion *(136)*, as well as gastrointestinal stromal tumors carrying activating c-KIT mutations *(137)*. Based on these positive results, there has been a large effort to generate FLT3-specific kinase inhibitor molecules for potential therapeutic application in AML.

To date at least six different compounds have been reported that display reasonable good inhibitory activity toward FLT3 tyrosine kinase receptor. Many of those compounds share structural characteristics and show inhibitory activity against RTKs similar to FLT3, including c-FMS, c-KIT, PDGFRα, and PDGFRβ.

Therefore, it has to be taken under serious consideration that administration of such a compound to leukemia patients might prevent or delay recovery from chemotherapy- or leukemia-induced cytopenias. On the other hand, combined inhibition of both wild-type c-KIT and FLT3 tyrosine kinase activity of AML blasts, even in the absence of activating FLT3 mutations, could have a beneficial effect and may result in a clinical response. Thus, it will be important to find an optimum treatment regimen without development of prolonged marrow suppression or any other serious adverse effects.

There is a class of more nonspecific protein tyrosine kinase (PTK) inhibitors, such as herbimycin A, AG1295, and AG1296, which have shown efficacy to inhibit FLT3. Herbimycin A, which is a benzoquinone ansamycine known to target Hsp90, suppresses constitutive tyrosine phosphorylation of FLT3-ITD but not tyrosine phosphorylation of the ligand-stimulated wild-type FLT3, with 50% inhibition (IC_{50}) of FLT3-ITD tyrosine autophosphorylation in 32D cells between 100 and 300 nM *(138,139)*. Tyrphostins AG1295 and AG1296, originally defined as PDGFR and c-KIT inhibitors *(140,141)*, are selective cytotoxic for FLT3-ITD-transfected Ba/F3 cells as well as AML blasts harboring FLT3-ITD mutations, and IC_{50} of AG1296 is approx 1 µM for inhibition of FLT3-ITD autophosphorylation *(111,142,143)*. However, these compounds are expected to be quite toxic in humans, because they are rather nonspecific and no clinical studies have been reported with them.

Sugen has developed and tested several more selective compounds (SU5416, SU5614, SU11248, and SU11657) that block FLT3 activation. The indolinone compound SU5416 was originally developed as vascular endothelial growth factor receptor-2/kinase domain receptor (VEGFR-2/KDR) inhibitor *(144)*, and is currently being employed as an anti-tumor agent by its ability to inhibit angiogenesis in vivo *(145–147)*. SU5416 also targets c-KIT and is a potent inhibitor of both wild-type and mutant FLT3; IC_{50} of SU5416 is 100 nM for inhibiting FLT3-ITD autophosphorylation in AML cell line MV4;11 *(148)*. As a single agent, SU5416 has modest clinical activity in refractory AML and myelodyplastic syndromes, where in one study 5% of 55 patients achieved partial responses (reduction of blasts by at least 50% in bone marrow and peripheral blood) and 2% hematological improvement *(149)*. In another phase 2 clinical study involving 43 patients with refractory AML, SU5416 treatment resulted in 16% PR and one patient achieved morphological remission, defined as absence of blasts in the peripheral blood and less than 5% in the bone marrow but without normalization of neutrophil and platelet counts for a duration of 2 mo *(150)*. Reported side effects of SU5416 are various thromboembolic events, like ischemic stroke and deep venous thrombosis, and bone pain *(149–151)*.

Similar to SU5416, SU5614 inhibits VEGFR-2/KDR, c-KIT, as well as wild-type and mutant FLT3 *(148)*. IC_{50} of SU5614 is 10 nM for inhibition of FLT3-ITD autophosphorylation in MV4;11 AML cells *(148)*, whereas IC_{50} of SU5614 is 175 nM for inducing growth arrest in FLT3-ITD-transformed Ba/F3 cells *(83)*. Both SU11657 and SU11248 potently inhibit PDGFR, VEGFR, c-KIT, and FLT3 *(152,153)*. Introduction of an activated allele of *FLT3* into MRP8-*PML-RARα* mouse transgenic bone marrow accelerates the onset of PML-RARα-induced leukemia *(154)*. SU11657 cooperates with all-*trans* retinoic acid to rapidly cause regression of activated FLT3/PML-RARα leukemias in these mice *(154)*. IC_{50} of SU11248 is 50 nM to inhibit FLT3-ITD autophosphorylation in MV4;11 cells and in preclinical experiments, SU11248 exhibits dose-dependent efficacy in both an FLT3-ITD xenograft tumor model and a bone marrow engraftment model *(153)*. A phase 1 study on 15 patients with refractory AML showed 40% partial responses and 1 patient with morphological response after SU11248 treatment *(155)*.

The novel tyrosine kinase inhibitor guanosine triphosphate-14564 shares structural characteristics with the part of staurosporine that inhibits protein kinase C by competing for adenosine triphosphate. Guanosine triphosphate-14564 inhibits both wild-type FLT3 and FLT3-ITD autophosphorylation with the same efficiency as c-FMS and c-KIT with IC_{50} values of approx 300 nM *(156)*. Two *bis* (1*H*-2-indolyl)-1-methone derivatives D-64406 and D-65476 act as potent PDGFR and FLT3 inhibitors and they inhibit proliferation of Ba/F3 cells transfected with constitutive active TEL-FLT3 with IC_{50} values of 200–300 nM *(157,158)*. The piperazinyl quinazoline compound MLN518 (CT53518 from Millennium) suppresses PDGFR, c-KIT, and FLT3 kinase activity, inhibiting FLT3-ITD autophosphorylation in Ba/F3 cells with IC_{50} value of 10–100 nM *(159,160)*. In an FLT3-ITD-induced myeloproliferative disease using a nude mouse model and a murine bone marrow transplantation assay, MLN518 prolongs the disease latency period and increases survival of these mice *(160)*.

CEP-701 and CEP-5214 (Cephalon) are orally bioavailable PTK inhibitors derived from indolocarbazole, and have been described to act as RET and Trk tyrosine kinase inhibitors *(161–163)*. CEP-701 seems highly selective for inhibition of wild-type and constitutively activated FLT3 (IC_{50} for inhibition of FLT3-ITD autophosphorylation in Ba/F3 cells is ~5 nM), and is much less potent against c-KIT, c-FMS, and PDGFR *(164)*. CEP-701 exhibits a cytotoxic response in AML blasts harboring FLT3/ITD mutations, and prolongs survival of mice injected with Ba/F3 cells expressing FLT3-ITD *(164)*. In a phase 1/2 study, single agent CEP-701 shows clinical response in 5 out of 14 patients (35%) with refractory and relapsed AML *(165)*.

PKC412 (CGP 41251 from Novartis), a derivative of staurosporine (*N*-benzoyl-staurosporine), is a potent inhibitor of protein kinase C, VEGFR-2/KDR, c-KIT, PDGFR, and FLT3 *(166–168)*. PKC412 suppresses FLT3-ITD autophosphorylation in Ba/F3 cells with IC_{50} values of 10 nM, and PKC412 prolongs the survival of mice transplanted with FLT3-ITD-expressing bone marrow cells *(168)*. PKC412 inhibits constitutively active FIPL1-PDGFRα associated with hypereosinophilic syndrome *(169)*, and PKC412 has a cytotoxic effect on MLL lymphoblasts expressing either wild-type FLT3 or activating FLT3-TKD alterations *(119)*. The safety and pharmacokinetic characteristics of PKC412 have been determined in phase 1 clinical trial *(170)*, and recent data from two phase 2 clinical trials in patients with AML indicate that administration of 100–225 mg orally daily as a single agent leads to partial responses or reduction of blood and marrow blasts in a high fraction (>70%) of patients with mutant FLT3, and a much lower rate of clinical responses in patients with wild-type FLT3 *(171)*. Responses tended to be transient, lasting 1–6 mo, before emergence of progressive disease.

More recent studies have started to compare the sensitivity of the different FLT3 inhibitors to suppress distinct activating mutations of the FLT3 receptor. These included substitution of asparagine 835 for tyrosine (D835Y), deletion of isoleucine 836 (I836del), and exchange of isoleucine 836 for methionine plus an arginine insertion (I836M + R), all three FLT3-TKD mutations as well as FLT3-ITD mutation. For AG1296, the IC_{50} for inhibition of cell proliferation in Ba/F3 cells is D835Y >> I836M + R > I836del = ITD; for SU5614 IC_{50} values are I836M + R > D835Y = ITD > I836del, whereas for PKC412, IC_{50} values are ITD > I836del > D835Y = I836M + R *(172)*. These data indicate that it is very important to determine the exact nature of FLT3 mutation when applying RTK inhibitors in clinical trials.

7. FUTURE DIRECTIONS

In comparison to the optimistic results obtained in the clinic with imatinib mesylate (Gleevec; STI571) for the treatment of CML, the first reports on the effects of FLT3 tyrosine kinase inhibitors for the treatment of AML seem somewhat disappointing. However, it has to be taken into account that AML is a much more aggressive, heterogeneous, and multigene disease than CML. Successful treatment of AML using targeted therapy will require combination therapy and will benefit from a further understanding of critical targets downstream of activated FLT3 signaling, as well as identification of genetic alterations that cooperate with FLT3 in the process of leukemogenesis. This latter class of mutations will likely include transcription factors that are altered by genetic aberrations.

One important complication that has been observed for the treatment of CML with imatinib mesylate is the occurrence of multifactorial drug resistance, including more than 10 different mutations that can arise in the ABL TKD of patients with BCR/ABL-positive CML and ALL, which induce imatinib resistance *(173–175)*. Recent studies in vitro have demonstrated that specific mutations in the FLT3 TKD can also confer resistance to FLT3 PTK inhibitors *(176)*, and these FLT3 mutant phenotypes are likely to be presented in clinic as well.

REFERENCES

1. Burnett AK, Grimwade D, Solomon E, Wheatley K, Goldstone AH. Presenting white blood cell count and kinetics of molecular remission predict prognosis in acute promyelocytic leukemia treated with all-trans retinoic acid: result of the Randomized MRC Trial. *Blood* 1999; 93:4131–4143.
2. Tallman MS, Andersen JW, Schiffer CA, et al. All-trans-retinoic acid in acute promyelocytic leukemia. *N Engl J Med* 1997; 337:1021–1028.
3. Rosnet O, Marchetto S, deLapeyriere O, Birnbaum D. Murine Flt3, a gene encoding a novel tyrosine kinase receptor of the PDGFR/CSF1R family. *Oncogene* 1991; 6:1641–1650.
4. Rosnet O, Mattei MG, Marchetto S, Birnbaum D. Isolation and chromosomal localization of a novel FMS-like tyrosine kinase gene. *Genomics* 1991; 9:380–385.
5. Matthews W, Jordan CT, Gavin M, Jenkins NA, Copeland NG, Lemischka IR. A receptor tyrosine kinase cDNA isolated from a population of enriched primitive hematopoietic cells and exhibiting close genetic linkage to c-kit. *Proc Natl Acad Sci USA* 1991; 88:9026–9030.
6. Matthews W, Jordan CT, Wiegand GW, Pardoll D, Lemischka IR. A receptor tyrosine kinase specific to hematopoietic stem and progenitor cell-enriched populations. *Cell* 1991; 65:1143–1152.
7. Small D, Levenstein M, Kim E, et al. STK-1, the human homolog of Flk-2/Flt-3, is selectively expressed in CD34+ human bone marrow cells and is involved in the proliferation of early progenitor/stem cells. *Proc Natl Acad Sci USA* 1994; 91:459–463.
8. Blume-Jensen P, Hunter T. Oncogenic kinase signalling. *Nature* 2001; 411:355–365.
9. Scheijen B, Griffin JD. Tyrosine kinase oncogenes in normal hematopoiesis and hematological disease. *Oncogene* 2002; 21:3314–3333.
10. Abu-Duhier FM, Goodeve AC, Wilson GA, Care RS, Peake IR, Reilly JT. Genomic structure of human FLT3: implications for mutational analysis. *Br J Haematol* 2001; 113:1076–1077.
11. Rosnet O, Stephenson D, Mattei MG, et al. Close physical linkage of the FLT1 and FLT3 genes on chromosome 13 in man and chromosome 5 in mouse. *Oncogene* 1993; 8:173–179.
12. Maroc N, Rottapel R, Rosnet O, et al. Biochemical characterization and analysis of the transforming potential of the FLT3/FLK2 receptor tyrosine kinase. *Oncogene* 1993; 8:909–918.
13. Lyman SD, James L, Zappone J, Sleath PR, Beckmann MP, Bird T. Characterization of the protein encoded by the flt3 (flk2) receptor-like tyrosine kinase gene. *Oncogene* 1993; 8:815–822.
14. Carow CE, Levenstein M, Kaufmann SH, et al. Expression of the hematopoietic growth factor receptor FLT3 (STK-1/Flk2) in human leukemias. *Blood* 1996; 87:1089–1096.
15. Lyman SD, Jacobsen SE. c-kit ligand and Flt3 ligand: stem/progenitor cell factors with overlapping yet distinct activities. *Blood* 1998; 91:1101–1134.
16. Turner AM, Lin NL, Issarachai S, Lyman SD, Broudy VC. FLT3 receptor expression on the surface of normal and malignant human hematopoietic cells. *Blood* 1996; 88:3383–3390.
17. Rappold I, Ziegler BL, Kohler I, et al. Functional and phenotypic characterization of cord blood and bone marrow subsets expressing FLT3 (CD135) receptor tyrosine kinase. *Blood* 1997; 90:111–125.
18. deLapeyriere O, Naquet P, Planche J, et al. Expression of Flt3 tyrosine kinase receptor gene in mouse hematopoietic and nervous tissues. *Differentiation* 1995; 58:351–359.
19. Mitton KP, Swain PK, Khanna H, Dowd M, Apel IJ, Swaroop A. Interaction of retinal bZIP transcription factor NRL with Flt3-interacting zinc-finger protein Fiz1: possible role of Fiz1 as a transcriptional repressor. *Hum Mol Genet* 2003; 12:365–373.
20. Gotze KS, Ramirez M, Tabor K, Small D, Matthews W, Civin CI. Flt3[high] and Flt3[low] CD34+ progenitor cells isolated from human bone marrow are functionally distinct. *Blood* 1998; 91:1947–1958.

21. Sitnicka E, Buza-Vidas N, Larsson S, Nygren JM, Liuba K, Jacobsen SE. Human CD34+ hematopoietic stem cells capable of multilineage engrafting NOD/SCID mice express flt3: distinct flt3 and c-kit expression and response patterns on mouse and candidate human hematopoietic stem cells. *Blood* 2003; 102:881–886.

22. Ebihara Y, Wada M, Ueda T, et al. Reconstitution of human haematopoiesis in non-obese diabetic/severe combined immunodeficient mice by clonal cells expanded from single CD34+CD38- cells expressing Flk2/Flt3. *Br J Haematol* 2002; 119:525–534.

23. Adolfsson J, Borge OJ, Bryder D, et al. Upregulation of Flt3 expression within the bone marrow Lin(–)Sca1(+)c-kit(+) stem cell compartment is accompanied by loss of self-renewal capacity. *Immunity* 2001; 15:659–669.

24. Christensen JL, Weissman IL. Flk-2 is a marker in hematopoietic stem cell differentiation: a simple method to isolate long-term stem cells. *Proc Natl Acad Sci USA* 2001; 98:14541–14546.

25. Li CL, Johnson GR. Murine hematopoietic stem and progenitor cells: I. Enrichment and biologic characterization. *Blood* 1995; 85:1472–1479.

26. Borge OJ, Ramsfjell V, Cui L, Jacobsen SE. Ability of early acting cytokines to directly promote survival and suppress apoptosis of human primitive CD34+CD38- bone marrow cells with multilineage potential at the single-cell level: key role of thrombopoietin. *Blood* 1997; 90:2282–2292.

27. Kawashima I, Zanjani ED, Almaida-Porada G, Flake AW, Zeng H, Ogawa M. CD34+ human marrow cells that express low levels of Kit protein are enriched for long-term marrow-engrafting cells. *Blood* 1996; 87:4136–4142.

28. Sakabe H, Yahata N, Kimura T, et al. Human cord blood-derived primitive progenitors are enriched in CD34+c-kit– cells: correlation between long-term culture-initiating cells and telomerase expression. *Leukemia* 1998; 12:728–734.

29. Sitnicka E, Bryder D, Theilgaard-Monch K, Buza-Vidas N, Adolfsson J, Jacobsen SE. Key role of flt3 ligand in regulation of the common lymphoid progenitor but not in maintenance of the hematopoietic stem cell pool. *Immunity* 2002; 17:463–472.

30. Sitnicka E, Brakebusch C, Martensson IL, et al. Complementary signaling through flt3 and interleukin-7 receptor alpha is indispensable for fetal and adult B-cell genesis. *J Exp Med* 2003; 198:1495–1506.

31. Mackarehtschian K, Hardin JD, Moore KA, Boast S, Goff SP, Lemischka IR. Targeted disruption of the flk2/flt3 gene leads to deficiencies in primitive hematopoietic progenitors. *Immunity* 1995; 3:147–161.

32. Hannum C, Culpepper J, Campbell D, et al. Ligand for FLT3/FLK2 receptor tyrosine kinase regulates growth of haematopoietic stem cells and is encoded by variant RNAs. *Nature* 1994; 368:643–648.

33. Lyman SD, James L, Vanden Bos T, et al. Molecular cloning of a ligand for the flt3/flk-2 tyrosine kinase receptor: a proliferative factor for primitive hematopoietic cells. *Cell* 1993; 75:1157–1167.

34. Lyman SD, James L, Johnson L, et al. Cloning of the human homologue of the murine flt3 ligand: a growth factor for early hematopoietic progenitor cells. *Blood* 1994; 83:2795–2801.

35. McClanahan T, Culpepper J, Campbell D, et al. Biochemical and genetic characterization of multiple splice variants of the Flt3 ligand. *Blood* 1996; 88:3371–3382.

36. Lyman SD, James L, Escobar S, et al. Identification of soluble and membrane-bound isoforms of the murine flt3 ligand generated by alternative splicing of mRNAs. *Oncogene* 1995; 10:149–157.

37. Lyman SD, Stocking K, Davison B, Fletcher F, Johnson L, Escobar S. Structural analysis of human and murine flt3 ligand genomic loci. *Oncogene* 1995; 11:1165–1172.

38. Wodnar-Filipowicz A. Flt3 ligand: role in control of hematopoietic and immune functions of the bone marrow. *News Physiol Sci* 2003; 18:247–251.

39. Chklovskaia E, Jansen W, Nissen C, et al. Mechanism of flt3 ligand expression in bone marrow failure: translocation from intracellular stores to the surface of T lymphocytes after chemotherapy-induced suppression of hematopoiesis. *Blood* 1999; 93:2595–2604.

40. Chklovskaia E, Nissen C, Landmann L, Rahner C, Pfister O, Wodnar-Filipowicz A. Cell-surface trafficking and release of flt3 ligand from T lymphocytes is induced by common cytokine receptor gamma-chain signaling and inhibited by cyclosporin A. *Blood* 2001; 97:1027–1034.

41. Lyman SD, Seaberg M, Hanna R, et al. Plasma/serum levels of flt3 ligand are low in normal individuals and highly elevated in patients with Fanconi anemia and acquired aplastic anemia. *Blood* 1995; 86:4091–4096.

42. Wodnar-Filipowicz A, Lyman SD, Gratwohl A, Tichelli A, Speck B, Nissen C. Flt3 ligand level reflects hematopoietic progenitor cell function in aplastic anemia and chemotherapy-induced bone marrow aplasia. *Blood* 1996; 88:4493–4499.

43. Piacibello W, Sanavio F, Garetto L, et al. Extensive amplification and self-renewal of human primitive hematopoietic stem cells from cord blood. *Blood* 1997; 89:2644–2653.

44. Petzer AL, Zandstra PW, Piret JM, Eaves CJ. Differential cytokine effects on primitive (CD34+CD38–) human hematopoietic cells: novel responses to Flt3-ligand and thrombopoietin. *J Exp Med* 1996; 183:2551–2558.

45. Elwood NJ, Zogos H, Willson T, Begley CG. Retroviral transduction of human progenitor cells: use of granulocyte colony-stimulating factor plus stem cell factor to mobilize progenitor cells in vivo and stimulation by Flt3/Flk-2 ligand in vitro. *Blood* 1996; 88:4452–4462.

46. Dao MA, Hannum CH, Kohn DB, Nolta JA. FLT3 ligand preserves the ability of human CD34+ progenitors to sustain long-term hematopoiesis in immune-deficient mice after ex vivo retroviral-mediated transduction. *Blood* 1997; 89:446–456.

47. Moore TA, Zlotnik A. Differential effects of Flk-2/Flt-3 ligand and stem cell factor on murine thymic progenitor cells. *J Immunol* 1997; 158:4187–4192.

48. Hirayama F, Lyman SD, Clark SC, Ogawa M. The flt3 ligand supports proliferation of lymphohematopoietic progenitors and early B-lymphoid progenitors. *Blood* 1995; 85:1762–1768.

49. Hudak S, Hunte B, Culpepper J, et al. FLT3/FLK2 ligand promotes the growth of murine stem cells and the expansion of colony-forming cells and spleen colony-forming units. *Blood* 1995; 85:2747–2755.

50. Jacobsen SE, Okkenhaug C, Myklebust J, Veiby OP, Lyman SD. The FLT3 ligand potently and directly stimulates the growth and expansion of primitive murine bone marrow progenitor cells in vitro: synergistic interactions with interleukin (IL) 11, IL-12, and other hematopoietic growth factors. *J Exp Med* 1995; 181:1357–1363.

51. Brasel K, McKenna HJ, Morrissey PJ, et al. Hematologic effects of flt3 ligand in vivo in mice. *Blood* 1996; 88:2004–2012.

52. Hawley TS, Fong AZ, Griesser H, Lyman SD, Hawley RG. Leukemic predisposition of mice transplanted with gene-modified hematopoietic precursors expressing flt3 ligand. *Blood* 1998; 92:2003–2011.

53. Juan TS, McNiece IK, Van G, et al. Chronic expression of murine flt3 ligand in mice results in increased circulating white blood cell levels and abnormal cellular infiltrates associated with splenic fibrosis. *Blood* 1997; 90:76–84.

54. Brasel K, De Smedt T, Smith JL, Maliszewski CR. Generation of murine dendritic cells from flt3-ligand-supplemented bone marrow cultures. *Blood* 2000; 96:3029–3039.

55. Maraskovsky E, Brasel K, Teepe M, et al. Dramatic increase in the numbers of functionally mature dendritic cells in Flt3 ligand-treated mice: multiple dendritic cell subpopulations identified. *J Exp Med* 1996; 184:1953–1962.

56. Manz MG, Traver D, Miyamoto T, Weissman IL, Akashi K. Dendritic cell potentials of early lymphoid and myeloid progenitors. *Blood* 2001; 97:3333–3341.

57. Traver D, Akashi K, Manz M, et al. Development of CD8alpha-positive dendritic cells from a common myeloid progenitor. *Science* 2000; 290:2152–2154.
58. Karsunky H, Merad M, Cozzio A, Weissman IL, Manz MG. Flt3 ligand regulates dendritic cell development from Flt3+ lymphoid and myeloid-committed progenitors to Flt3+ dendritic cells in vivo. *J Exp Med* 2003; 198:305–313.
59. Maraskovsky E, Daro E, Roux E, et al. In vivo generation of human dendritic cell subsets by Flt3 ligand. *Blood* 2000; 96:878–884.
60. Pulendran B, Banchereau J, Burkeholder S, et al. Flt3-ligand and granulocyte colony–stimulating factor mobilize distinct human dendritic cell subsets in vivo. *J Immunol* 2000; 165:566–572.
61. Chen K, Braun S, Lyman S, et al. Antitumor activity and immunotherapeutic properties of Flt3-ligand in a murine breast cancer model. *Cancer Res* 1997; 57:3511–3516.
62. Lynch DH, Andreasen A, Maraskovsky E, Whitmore J, Miller RE, Schuh JC. Flt3 ligand induces tumor regression and antitumor immune responses in vivo. *Nat Med* 1997; 3:625–631.
63. Chakravarty PK, Alfieri A, Thomas EK, et al. Flt3-ligand administration after radiation therapy prolongs survival in a murine model of metastatic lung cancer. *Cancer Res* 1999; 59:6028–6032.
64. Pawlowska AB, Hashino S, McKenna H, Weigel BJ, Taylor PA, Blazar BR. In vitro tumor-pulsed or in vivo Flt3 ligand-generated dendritic cells provide protection against acute myelogenous leukemia in nontransplanted or syngeneic bone marrow-transplanted mice. *Blood* 2001; 97:1474–1482.
65. McKenna HJ, Stocking KL, Miller RE, et al. Mice lacking flt3 ligand have deficient hematopoiesis affecting hematopoietic progenitor cells, dendritic cells, and natural killer cells. *Blood* 2000; 95:3489–3497.
66. Meierhoff G, Dehmel U, Gruss HJ, et al. Expression of FLT3 receptor and FLT3-ligand in human leukemia-lymphoma cell lines. *Leukemia* 1995; 9:1368–1372.
67. DaSilva N, Hu ZB, Ma W, Rosnet O, Birnbaum D, Drexler HG. Expression of the FLT3 gene in human leukemia-lymphoma cell lines. *Leukemia* 1994; 8:885–888.
68. Brasel K, Escobar S, Anderberg R, de Vries P, Gruss HJ, Lyman SD. Expression of the flt3 receptor and its ligand on hematopoietic cells. *Leukemia* 1995; 9:1212–1218.
69. Drexler HG. Expression of FLT3 receptor and response to FLT3 ligand by leukemic cells. *Leukemia* 1996; 10:588–599.
70. Ozeki K, Kiyoi H, Hirose Y, et al. Biological and clinical significance of the FLT3 transcript level in acute myeloid leukemia. *Blood* 2003; 6:6.
71. Zheng R, Levis M, Piloto O, et al. FLT3 ligand causes autocrine signaling in acute myeloid leukemia cells. *Blood* 2003.
72. Drexler HG, Meyer C, Quentmeier H. Effects of FLT3 ligand on proliferation and survival of myeloid leukemia cells. *Leuk Lymphoma* 1999; 33:83–91.
73. Dehmel U, Zaborski M, Meierhoff G, et al. Effects of FLT3 ligand on human leukemia cells. I. Proliferative response of myeloid leukemia cells. *Leukemia* 1996; 10:261–270.
74. Nakao M, Yokota S, Iwai T, et al. Internal tandem duplication of the flt3 gene found in acute myeloid leukemia. *Leukemia* 1996; 10:1911–1918.
75. Kiyoi H, Towatari M, Yokota S, et al. Internal tandem duplication of the FLT3 gene is a novel modality of elongation mutation which causes constitutive activation of the product. *Leukemia* 1998; 12:1333–1337.
76. Abu-Duhier FM, Goodeve AC, Wilson GA, et al. FLT3 internal tandem duplication mutations in adult acute myeloid leukaemia define a high-risk group. *Br J Haematol* 2000; 111:190–195.
77. Horiike S, Yokota S, Nakao M, et al. Tandem duplications of the FLT3 receptor gene are associated with leukemic transformation of myelodysplasia. *Leukemia* 1997; 11:1442–1446.

78. Yokota S, Kiyoi H, Nakao M, et al. Internal tandem duplication of the FLT3 gene is preferentially seen in acute myeloid leukemia and myelodysplastic syndrome among various hematological malignancies. A study on a large series of patients and cell lines. *Leukemia* 1997; 11:1605–1609.

79. Abu-Duhier FM, Goodeve AC, Wilson GA, Care RS, Peake IR, Reilly JT. Identification of novel FLT-3 Asp835 mutations in adult acute myeloid leukaemia. *Br J Haematol* 2001; 113:983–988.

80. Yamamoto Y, Kiyoi H, Nakano Y, et al. Activating mutation of D835 within the activation loop of FLT3 in human hematologic malignancies. *Blood* 2001; 97:2434–2439.

81. Spiekermann K, Bagrintseva K, Schoch C, Haferlach T, Hiddemann W, Schnittger S. A new and recurrent activating length mutation in exon 20 of the FLT3 gene in acute myeloid leukemia. *Blood* 2002; 100:3423–3425.

82. Kiyoi H, Ohno R, Ueda R, Saito H, Naoe T. Mechanism of constitutive activation of FLT3 with internal tandem duplication in the juxtamembrane domain. *Oncogene* 2002; 21:2555–2563.

83. Spiekermann K, Dirschinger RJ, Schwab R, et al. The protein tyrosine kinase inhibitor SU5614 inhibits FLT3 and induces growth arrest and apoptosis in AML-derived cell lines expressing a constitutively activated FLT3. *Blood* 2003; 101:1494–1504.

84. Furitsu T, Tsujimura T, Tono T, et al. Identification of mutations in the coding sequence of the proto-oncogene c-kit in a human masT-cell leukemia cell line causing ligand-independent activation of c-kit product. *J Clin Invest* 1993; 92:1736–1744.

85. Jeffers M, Schmidt L, Nakaigawa N, et al. Activating mutations for the met tyrosine kinase receptor in human cancer. *Proc Natl Acad Sci USA* 1997; 94:11445–11450.

86. Santoro M, Carlomagno F, Romano A, et al. Activation of RET as a dominant transforming gene by germline mutations of MEN2A and MEN2B. *Science* 1995; 267:381–383.

87. Morley GM, Uden M, Gullick WJ, Dibb NJ. Cell specific transformation by c-fms activating loop mutations is attributable to constitutive receptor degradation. *Oncogene* 1999; 18:3076–3084.

88. Thiede C, Steudel C, Mohr B, et al. Analysis of FLT3-activating mutations in 979 patients with acute myelogenous leukemia: association with FAB subtypes and identification of subgroups with poor prognosis. *Blood* 2002; 99:4326–4335.

89. Ridge SA, Worwood M, Oscier D, Jacobs A, Padua RA. FMS mutations in myelodysplastic, leukemic, and normal subjects. *Proc Natl Acad Sci USA* 1990; 87:1377–1380.

90. Tobal K, Pagliuca A, Bhatt B, Bailey N, Layton DM, Mufti GJ. Mutation of the human FMS gene (M-CSF receptor) in myelodysplastic syndromes and acute myeloid leukemia. *Leukemia* 1990; 4:486–489.

91. Nakata Y, Kimura A, Katoh O, et al. c-kit point mutation of extracellular domain in patients with myeloproliferative disorders. *Br J Haematol* 1995; 91:661–663.

92. Gari M, Goodeve A, Wilson G, et al. c-kit proto-oncogene exon 8 in-frame deletion plus insertion mutations in acute myeloid leukaemia. *Br J Haematol* 1999; 105:894–900.

93. Rombouts WJ, Blokland I, Lowenberg B, Ploemacher RE. Biological characteristics and prognosis of adult acute myeloid leukemia with internal tandem duplications in the Flt3 gene. *Leukemia* 2000; 14:675–683.

94. Kottaridis PD, Gale RE, Frew ME, et al. The presence of a FLT3 internal tandem duplication in patients with acute myeloid leukemia (AML) adds important prognostic information to cytogenetic risk group and response to the first cycle of chemotherapy: analysis of 854 patients from the United Kingdom Medical Research Council AML 10 and 12 trials. *Blood* 2001; 98:1752–1759.

95. Schnittger S, Schoch C, Dugas M, et al. Analysis of FLT3 length mutations in 1003 patients with acute myeloid leukemia: correlation to cytogenetics, FAB subtype, and prognosis in the AMLCG study and usefulness as a marker for the detection of minimal residual disease. *Blood* 2002; 100:59–66.

96. Whitman SP, Archer KJ, Feng L, et al. Absence of the wild-type allele predicts poor prognosis in adult de novo acute myeloid leukemia with normal cytogenetics and the internal tandem duplication of FLT3: a cancer and leukemia group B study. *Cancer Res* 2001; 61:7233–7239.

97. Liang DC, Shih LY, Hung IJ, et al. Clinical relevance of internal tandem duplication of the FLT3 gene in childhood acute myeloid leukemia. *Cancer* 2002; 94:3292–3298.

98. Xu F, Taki T, Yang HW, et al. Tandem duplication of the FLT3 gene is found in acute lymphoblastic leukaemia as well as acute myeloid leukaemia but not in myelodysplastic syndrome or juvenile chronic myelogenous leukaemia in children. *Br J Haematol* 1999; 105:155–162.

99. Meshinchi S, Woods WG, Stirewalt DL, et al. Prevalence and prognostic significance of Flt3 internal tandem duplication in pediatric acute myeloid leukemia. *Blood* 2001; 97:89–94.

100. Zwaan CM, Meshinchi S, Radich JP, et al. FLT3 internal tandem duplication in 234 children with acute myeloid leukemia (AML): prognostic significance and relation to cellular drug resistance. *Blood* 2003; 19:19.

101. Noguera NI, Breccia M, Divona M, et al. Alterations of the FLT3 gene in acute promyelocytic leukemia: association with diagnostic characteristics and analysis of clinical outcome in patients treated with the Italian AIDA protocol. *Leukemia* 2002; 16:2185–2199.

102. Kainz B, Heintel D, Marculescu R, et al. Variable prognostic value of FLT3 internal tandem duplications in patients with de novo AML and a normal karyotype, t(15;17), t(8;21) or inv(16). *Hematol J* 2002; 3:283–289.

103. Kiyoi H, Naoe T, Yokota S, et al. Internal tandem duplication of FLT3 associated with leukocytosis in acute promyelocytic leukemia. Leukemia Study Group of the Ministry of Health and Welfare (Kohseisho). *Leukemia* 1997; 11:1447–1452.

104. Shih LY, Kuo MC, Liang DC, et al. Internal tandem duplication and Asp835 mutations of the FMS-like tyrosine kinase 3 (FLT3) gene in acute promyelocytic leukemia. *Cancer* 2003; 98:1206–1216.

105. Nakano Y, Kiyoi H, Miyawaki S, et al. Molecular evolution of acute myeloid leukaemia in relapse: unstable N- ras and FLT3 genes compared with p53 gene. *Br J Haematol* 1999; 104:659–664.

106. Stirewalt DL, Kopecky KJ, Meshinchi S, et al. FLT3, RAS, and TP53 mutations in elderly patients with acute myeloid leukemia. *Blood* 2001; 97:3589–3595.

107. Kiyoi H, Naoe T, Nakano Y, et al. Prognostic implication of FLT3 and N-RAS gene mutations in acute myeloid leukemia. *Blood* 1999; 93:3074–3080.

108. Wybenga-Groot LE, Baskin B, Ong SH, Tong J, Pawson T, Sicheri F. Structural basis for autoinhibition of the Ephb2 receptor tyrosine kinase by the unphosphorylated juxtamembrane region. *Cell* 2001; 106:745–757.

109. Hayakawa F, Towatari M, Kiyoi H, et al. Tandem-duplicated Flt3 constitutively activates STAT5 and MAP kinase and introduces autonomous cell growth in IL-3-dependenT-cell lines. *Oncogene* 2000; 19:624–631.

110. Mizuki M, Fenski R, Halfter H, et al. Flt3 mutations from patients with acute myeloid leukemia induce transformation of 32D cells mediated by the Ras and STAT5 pathways. *Blood* 2000; 96:3907–3914.

111. Tse KF, Allebach J, Levis M, Smith BD, Bohmer FD, Small D. Inhibition of the transforming activity of FLT3 internal tandem duplication mutants from AML patients by a tyrosine kinase inhibitor. *Leukemia* 2002; 16:2027–2036.

112. Kelly LM, Liu Q, Kutok JL, Williams IR, Boulton CL, Gilliland DG. FLT3 internal tandem duplication mutations associated with human acute myeloid leukemias induce myeloproliferative disease in a murine bone marrow transplant model. *Blood* 2002; 99:310–318.

113. Sheikhha MH, Awan A, Tobal K, Liu Yin JA. Prognostic significance of FLT3 ITD and D835 mutations in AML patients. *Hematol J* 2003; 4:41–46.

114. Frohling S, Schlenk RF, Breitruck J, et al. Prognostic significance of activating FLT3 mutations in younger adults (16 to 60 years) with acute myeloid leukemia and normal cytogenetics: a study of the AML Study Group Ulm. *Blood* 2002; 100:4372–4380.

115. Liang DC, Shih LY, Hung IJ, et al. FLT3-TKD mutation in childhood acute myeloid leukemia. *Leukemia* 2003; 17:883–886.

116. Meshinchi S, Stirewalt DL, Alonzo TA, et al. Activating mutations of RTK/ras signal transduction pathway in pediatric acute myeloid leukemia. *Blood* 2003; 102:1474–1479.

117. Armstrong SA, Staunton JE, Silverman LB, et al. MLL translocations specify a distinct gene expression profile that distinguishes a unique leukemia. *Nat Genet* 2002; 30:41–47.

118. Tsutsumi S, Taketani T, Nishimura K, et al. Two distinct gene expression signatures in pediatric acute lymphoblastic leukemia with MLL rearrangements. *Cancer Res* 2003; 63:4882–4887.

119. Armstrong SA, Kung AL, Mabon ME, et al. Inhibition of FLT3 in MLL. Validation of a therapeutic target identified by gene expression based classification. *Cancer Cell* 2003; 3:173–183.

120. Libura M, Asnafi V, Tu A, et al. FLT3 and MLL intragenic abnormalities in AML reflect a common category of genotoxic stress. *Blood* 2003; 102:2198–2204.

121. Dosil M, Wang S, Lemischka IR. Mitogenic signalling and substrate specificity of the Flk2/Flt3 receptor tyrosine kinase in fibroblasts and interleukin 3-dependent hematopoietic cells. *Mol Cell Biol* 1993; 13:6572–6585.

122. Rottapel R, Turck CW, Casteran N, et al. Substrate specificities and identification of a putative binding site for PI3K in the carboxy tail of the murine Flt3 receptor tyrosine kinase. *Oncogene* 1994; 9:1755–1765.

123. Marchetto S, Fournier E, Beslu N, et al. SHC and SHIP phosphorylation and interaction in response to activation of the FLT3 receptor. *Leukemia* 1999; 13:1374–1382.

124. Lavagna-Sevenier C, Marchetto S, Birnbaum D, Rosnet O. FLT3 signaling in hematopoietic cells involves CBL, SHC and an unknown P115 as prominent tyrosine-phosphorylated substrates. *Leukemia* 1998; 12:301–310.

125. Lavagna-Sevenier C, Marchetto S, Birnbaum D, Rosnet O. The CBL-related protein CBLB participates in FLT3 and interleukin-7 receptor signal transduction in pro-B-cells. *J Biol Chem* 1998; 273:14962–14967.

126. Zhang S, Broxmeyer HE. Flt3 ligand induces tyrosine phosphorylation of gab1 and gab2 and their association with shp-2, grb2, and PI3 kinase. *Biochem Biophys Res Commun* 2000; 277:195–199.

127. Zhang S, Broxmeyer HE. p85 subunit of PI3 kinase does not bind to human Flt3 receptor, but associates with SHP2, SHIP, and a tyrosine-phosphorylated 100-kDa protein in Flt3 ligand-stimulated hematopoietic cells. *Biochem Biophys Res Commun* 1999; 254:440–445.

128. Zhang S, Mantel C, Broxmeyer HE. Flt3 signaling involves tyrosyl-phosphorylation of SHP-2 and SHIP and their association with Grb2 and Shc in Baf3/Flt3 cells. *J Leukoc Biol* 1999; 65:372–380.

129. Spiekermann K, Bagrintseva K, Schwab R, Schmieja K, Hiddemann W. Overexpression and constitutive activation of FLT3 induces STAT5 activation in primary acute myeloid leukemia blast-cells. *Clin Cancer Res* 2003; 9:2140–2150.

130. Mizuki M, Schwable J, Steur C, et al. Suppression of myeloid transcription factors and induction of STAT response genes by AML-specific Flt3 mutations. *Blood* 2003; 101:3164–3173.

131. Minami Y, Yamamoto K, Kiyoi H, Ueda R, Saito H, Naoe T. Different anti-apoptotic pathways between wild-type and mutated FLT3: insights into therapeutic targets in leukemia. *Blood* 2003; 3:3.

132. Zheng R, Friedman AD, Small D. Targeted inhibition of FLT3 overcomes the block to myeloid differentiation in 32Dcl3 cells caused by expression of FLT3/ITD mutations. *Blood* 2002; 100:4154–4161.

133. Zheng R, Friedman AD, Levis M, Li L, Weir EG, Small D. Internal tandem duplication mutation of FLT3 blocks myeloid differentiation through suppression of C/EBPα expression. *Blood* 2003; 30:30.

134. Druker BJ, Talpaz M, Resta DJ, et al. Efficacy and safety of a specific inhibitor of the BCR-ABL tyrosine kinase in chronic myeloid leukemia. *N Engl J Med* 2001; 344:1031–1037.

135. Druker BJ. Imatinib alone and in combination for chronic myeloid leukemia. *Semin Hematol* 2003; 40:50–58.

136. Apperley JF, Gardembas M, Melo JV, et al. Response to imatinib mesylate in patients with chronic myeloproliferative diseases with rearrangements of the platelet-derived growth factor receptor beta. *N Engl J Med* 2002; 347:481–487.

137. Demetri GD, von Mehren M, Blanke CD, et al. Efficacy and safety of imatinib mesylate in advanced gastrointestinal stromal tumors. *N Engl J Med* 2002; 347:472–480.

138. Zhao M, Kiyoi H, Yamamoto Y, et al. In vivo treatment of mutant FLT3-transformed murine leukemia with a tyrosine kinase inhibitor. *Leukemia* 2000; 14:374–378.

139. Minami Y, Kiyoi H, Yamamoto Y, et al. Selective apoptosis of tandemly duplicated FLT3-transformed leukemia cells by Hsp90 inhibitors. *Leukemia* 2002; 16:1535–1540.

140. Gazit A, App H, McMahon G, Chen J, Levitzki A, Bohmer FD. Tyrphostins. 5. Potent inhibitors of platelet-derived growth factor receptor tyrosine kinase: structure–activity relationships in quinoxalines, quinolines, and indole tyrphostins. *J Med Chem* 1996; 39:2170–2177.

141. Kovalenko M, Gazit A, Bohmer A, et al. Selective platelet-derived growth factor receptor kinase blockers reverse sis-transformation. *Cancer Res* 1994; 54:6106–6114.

142. Levis M, Tse KF, Smith BD, Garrett E, Small D. A FLT3 tyrosine kinase inhibitor is selectively cytotoxic to acute myeloid leukemia blasts harboring FLT3 internal tandem duplication mutations. *Blood* 2001; 98:885–887.

143. Tse KF, Novelli E, Civin CI, Bohmer FD, Small D. Inhibition of FLT3-mediated transformation by use of a tyrosine kinase inhibitor. *Leukemia* 2001; 15:1001–1010.

144. Mendel DB, Schreck RE, West DC, et al. The angiogenesis inhibitor SU5416 has long-lasting effects on vascular endothelial growth factor receptor phosphorylation and function. *Clin Cancer Res* 2000; 6:4848–4858.

145. Abdollahi A, Lipson KE, Sckell A, et al. Combined therapy with direct and indirect angiogenesis inhibition results in enhanced antiangiogenic and antitumor effects. *Cancer Res* 2003; 63:8890–8898.

146. Lara PN, Jr., Quinn DI, Margolin K, et al. SU5416 plus interferon alpha in advanced renal cell carcinoma: a phase II California Cancer Consortium Study with biological and imaging correlates of angiogenesis inhibition. *Clin Cancer Res* 2003; 9:4772–4781.

147. Huss WJ, Barrios RJ, Greenberg NM. SU5416 selectively impairs angiogenesis to induce prostate cancer-specific apoptosis. *Mol Cancer Ther* 2003; 2:611–616.

148. Yee KW, O'Farrell AM, Smolich BD, et al. SU5416 and SU5614 inhibit kinase activity of wild-type and mutant FLT3 receptor tyrosine kinase. *Blood* 2002; 100:2941–2949.

149. Giles FJ, Stopeck AT, Silverman LR, et al. SU5416, a small molecule tyrosine kinase receptor inhibitor, has biologic activity in patients with refractory acute myeloid leukemia or myelodysplastic syndromes. *Blood* 2003; 102:795–801.

150. Fiedler W, Mesters R, Tinnefeld H, et al. A phase 2 clinical study of SU5416 in patients with refractory acute myeloid leukemia. *Blood* 2003; 102:2763–2767.

151. Giles FJ, Cooper MA, Silverman L, et al. Phase II study of SU5416—a small-molecule, vascular endothelial growth factor tyrosine-kinase receptor inhibitor—in patients with refractory myeloproliferative diseases. *Cancer* 2003; 97:1920–1928.

152. Mendel DB, Laird AD, Xin X, et al. In vivo antitumor activity of SU11248, a novel tyrosine kinase inhibitor targeting vascular endothelial growth factor and platelet-derived growth factor receptors: determination of a pharmacokinetic/pharmacodynamic relationship. *Clin Cancer Res* 2003; 9:327–337.

153. O'Farrell AM, Abrams TJ, Yuen HA, et al. SU11248 is a novel FLT3 tyrosine kinase inhibitor with potent activity in vitro and in vivo. *Blood* 2003; 101:3597–3605.
154. Sohal J, Phan VT, Chan PV, et al. A model of APL with FLT3 mutation is responsive to retinoic acid and a receptor tyrosine kinase inhibitor, SU11657. *Blood* 2003; 101:3188–3197.
155. O'Farrell AM, Foran JM, Fiedler W, et al. An innovative phase I clinical study demonstrates inhibition of FLT3 phosphorylation by SU11248 in acute myeloid leukemia patients. *Clin Cancer Res* 2003; 9:5465–5476.
156. Murata K, Kumagai H, Kawashima T, et al. Selective cytotoxic mechanism of GTP-14564, a novel tyrosine kinase inhibitor in leukemia cells expressing a constitutively active Fms-like tyrosine kinase 3 (FLT3). *J Biol Chem* 2003; 278:32892–32898.
157. Teller S, Kramer D, Bohmer SA, et al. Bis(1H-2-indolyl)-1-methanones as inhibitors of the hematopoietic tyrosine kinase Flt3. *Leukemia* 2002; 16:1528–1534.
158. Mahboobi S, Teller S, Pongratz H, et al. Bis(1H-2-indolyl)methanones as a novel class of inhibitors of the platelet-derived growth factor receptor kinase. *J Med Chem* 2002; 45:1002–1018.
159. Pandey A, Volkots DL, Seroogy JM, et al. Identification of orally active, potent, and selective 4-piperazinylquinazolines as antagonists of the platelet-derived growth factor receptor tyrosine kinase family. *J Med Chem* 2002; 45:3772–3793.
160. Kelly LM, Yu JC, Boulton CL, et al. CT53518, a novel selective FLT3 antagonist for the treatment of acute myelogenous leukemia (AML). *Cancer Cell* 2002; 1:421–432.
161. Strock CJ, Park JI, Rosen M, et al. CEP-701 and CEP-751 inhibit constitutively activated RET tyrosine kinase activity and block medullary thyroid carcinoma cell growth. *Cancer Res* 2003; 63:5559–5563.
162. Pinski J, Weeraratna A, Uzgare AR, Arnold JT, Denmeade SR, Isaacs JT. Trk receptor inhibition induces apoptosis of proliferating but not quiescent human osteoblasts. *Cancer Res* 2002; 62:986–989.
163. Miknyoczki SJ, Chang H, Klein-Szanto A, Dionne CA, Ruggeri BA. The Trk tyrosine kinase inhibitor CEP-701 (KT-5555) exhibits significant antitumor efficacy in preclinical xenograft models of human pancreatic ductal adenocarcinoma. *Clin Cancer Res* 1999; 5:2205–2212.
164. Levis M, Allebach J, Tse KF, et al. A FLT3-targeted tyrosine kinase inhibitor is cytotoxic to leukemia cells in vitro and in vivo. *Blood* 2002; 99:3885–3891.
165. Smith BD, Levis M, Beran M, et al. Single agent CEP-701, a novel FLT3 inhibitor, shows biologic and clinical activity in patients with relapsed or refractory acute myeloid leukemia. *Blood* 2004; 15:15.
166. Fabbro D, Ruetz S, Bodis S, et al. PKC412—a protein kinase inhibitor with a broad therapeutic potential. *Anticancer Drug Des* 2000; 15:17–28.
167. Fabbro D, Buchdunger E, Wood J, et al. Inhibitors of protein kinases: CGP 41251, a protein kinase inhibitor with potential as an anticancer agent. *Pharmacol Ther* 1999; 82:293–301.
168. Weisberg E, Boulton C, Kelly LM, et al. Inhibition of mutant FLT3 receptors in leukemia cells by the small molecule tyrosine kinase inhibitor PKC412. *Cancer Cell* 2002; 1:433–443.
169. Cools J, Stover EH, Boulton CL, et al. PKC412 overcomes resistance to imatinib in a murine model of FIP1L1-PDGFRalpha-induced myeloproliferative disease. *Cancer Cell* 2003; 3:459–469.
170. Propper DJ, McDonald AC, Man A, et al. Phase I and pharmacokinetic study of PKC412, an inhibitor of protein kinase C. *J Clin Oncol* 2001; 19:1485–1492.
171. Stone MR, De Angelo DG, Klimek V, et al. Patients with acute myeloid leukemin and an activating mutation in FLT3 respond to a small-molecule FLT3 tyrosine kinase inhibitor, PKC412. *Blood* 2005; 105:54–60.
172. Grundler R, Thiede C, Miething C, Steudel C, Peschel C, Duyster J. Sensitivity toward tyrosine kinase inhibitors varies between different activating mutations of the FLT3 receptor. *Blood* 2003; 102:646–651.

173. Gorre ME, Mohammed M, Ellwood K, et al. Clinical resistance to STI-571 cancer therapy caused by BCR-ABL gene mutation or amplification. *Science* 2001; 293:876–880.

174. Shah NP, Nicoll JM, Nagar B, et al. Multiple BCR-ABL kinase domain mutations confer polyclonal resistance to the tyrosine kinase inhibitor imatinib (STI571) in chronic phase and blast crisis chronic myeloid leukemia. *Cancer Cell* 2002; 2:117–125.

175. Branford S, Rudzki Z, Walsh S, et al. High frequency of point mutations clustered within the adenosine triphosphate-binding region of BCR/ABL in patients with chronic myeloid leukemia or Ph-positive acute lymphoblastic leukemia who develop imatinib (STI571) resistance. *Blood* 2002; 99:3472–3475.

176. Bagrintseva K, Schwab R, Kohl TM, et al. Mutations in the tyrosine kinase domain of FLT3 define a new molecular mechanism of acquired drug resistance to PTK inhibitors in FLT3-ITD transformed hematopoietic cells. *Blood* 2003; 6:6.

6 JAK Kinases in Leukemias, Lymphomas, and Multiple Myeloma

Renate Burger, PhD
and Martin Gramatzki, MD

CONTENTS

INTRODUCTION
STRUCTURAL FEATURES AND ACTIVATION OF JAKS
NEGATIVE REGULATION OF JAK SIGNALING
IMPLICATIONS OF ABERRANT JAK ACTIVATION
    IN HEMATOLOGICAL DISEASES
PHARMACOLOGICAL INHIBITION OF JAKS
CONCLUSIONS AND PERSPECTIVES
REFERENCES

1. INTRODUCTION

The Janus kinase or just another kinase (JAK) family comprises cytoplasmic receptor-associated protein tyrosine kinases that are involved in signal transduction pathways mediated by many cytokines and cytokine-like hormones *(1–4)*. Physiologically, these cytokines play a critical role in regulating normal cellular functions such as proliferation, survival, and differentiation. The importance of JAKs derives from the fact that they are the first proteins involved in the intracellular part of cytokine-induced signal transduction. The major signaling events downstream of JAKs are the activation of the signal transducer and activator of transcription (STAT) proteins, the Ras/Raf/mitogen-activated protein kinase (MAPK), and the phosphatidylinositol-3 kinase (PI3-K)/Akt pathways *(5)*. Dysregulated JAK activity has pathological implications: constitutive or enhanced JAK activation has been implicated in neoplastic transformation and abnormal cell proliferation in various hematological malignancies including multiple myeloma. Therefore, JAKs represent an attractive target for the

From: *Cancer Drug Discovery and Development:*
Protein Tyrosine Kinases: From Inhibitors to Useful Drugs
Edited by: D. Fabbro and F. McCormick © Humana Press Inc., Totowa, NJ

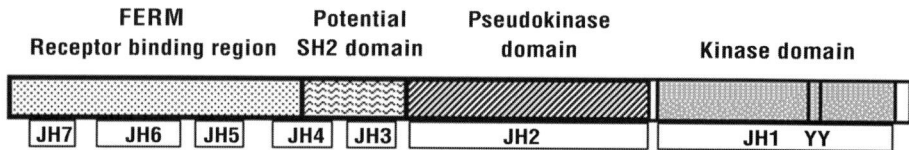

Fig. 1. General structure of Janus kinases (JAK). The seven JAK homology (JH) domains are regions of sequence similarity in the four known JAKs. Relevant tyrosine residues (Y) that become phosphorylated, are marked. The significance of the predicted SH2 domain remains unclear. FERM, four-point-one, ezrin, radixin, and moesin domain.

development of novel drugs that might selectively inhibit their activity. Although exciting breakthroughs have been achieved with other tyrosine kinase inhibitors, such as STI571/imatinib mesylate, pharmacological JAK inhibitors have not yet reached the clinical stage. This chapter focuses on what is known on the role of JAKs in hematological malignancies including multiple myeloma, and the development of pharmacological inhibitors for therapy of diseases associated with JAK activity.

2. STRUCTURAL FEATURES AND ACTIVATION OF JAKS

The JAK family of tyrosine kinases consists of four mammalian members: JAK1, JAK2, JAK3, and TYK2 (tyrosine kinase-2) (1). They are large proteins with more than a thousand amino acids in length and a molecular weight ranging from 120 to 140 kDa. The JAKs consist of seven regions with conserved homology (JAK homology domains JH1–JH7) (Fig. 1). JH1 at the carboxy-terminal end of the protein encodes the kinase domain that exhibits classical features of a protein tyrosine kinase (PTK) (6). This includes a lysine at the catalytically active site and two positionally conserved adjacent tyrosine residues in subdomain VII of the kinase domain. Autophosphorylation of these tyrosines regulates receptor-coupled autokinase activity of JAKs and may also increase substrate accessibility to the active site (7). Besides these two critical tyrosines, JAKs contain multiple additional phosphorylation sites that may mediate interaction with other intracellular proteins (4). JH2 represents a pseudokinase domain without any functional kinase activity, but it is suggested to regulate JH1 catalytic activity. The presence of this additional kinase domain is a unique feature of JAKs and distinguishes them from other classes of PTKs. The amino-terminal JH3–JH7 domains have been implicated in receptor association and also regulation of catalytic activity (8). JAKs are ubiquitously expressed with the exception of JAK3, which is mainly restricted to hematopoietic cells (5,9).

JAKs are involved in signaling pathways mediated by cytokines of the hematopoietin family, also called class II cytokines, and their respective receptors. JAK1 is an essential mediator for biological responsiveness to a

Table 1
Important Features of JAKs

JAK	Year identified/ cloned	Transcript size	Chromosomal location	Critical tyrosine residues	Cytokines by which kinase is activated	Phenotype of knock-out mice
TYK2	1990	4.4 kb	19p13.2	Y1054/Y1055	IFN-α/β, gp130-family, IL-10, IL-12, IL-13, TPO)	Increased pathogen susceptibility
JAK1	1991	5.4 kb	1p31.3	Y1038/Y1039	IFN-α/β, IFN-γ, gp130 family, G-CSF, IL-10, γC family	Early postnatal lethal
JAK2	1992	5.3/5.0 kb	9p24	Y1007/Y1008	IFN-γ, gp130 family, G-CSF, leptin, IL-12, IL-3 family, single chain family (EPO, GH, PRL, TPO), AT1, insulin	Embryonic lethal
JAK3	1994	4.2 kb	19p13.1	Y980/Y981	γC family	SCID-like

Reviewed in refs. 5 and 9.
IFN, interferon; TPO, thrombopoietin; G-CSF, granulocyte-colony stimulating factor; EPO, erythropoietin; GH, growth hormone; PRL, pro-lactin; AT, angiotensin; SCID, severe combined immunodeficiency.

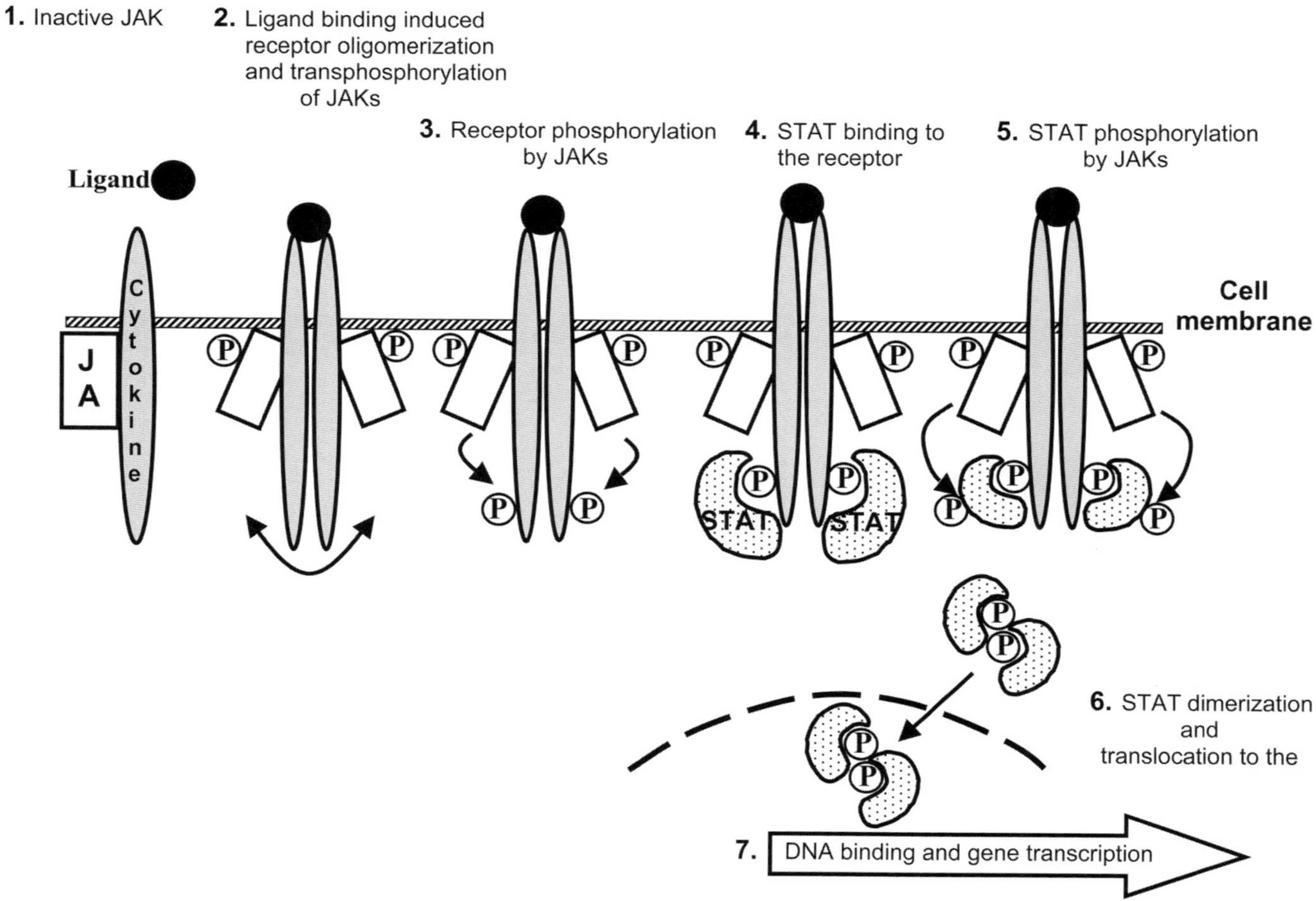

1. Inactive JAK

2. Ligand binding induced receptor oligomerization and transphosphorylation of JAKs

3. Receptor phosphorylation by JAKs

4. STAT binding to the receptor

5. STAT phosphorylation by JAKs

6. STAT dimerization and translocation to the

7. DNA binding and gene transcription

Cell membrane

Ligand

Cytokine

JAK

STAT

Fig. 2

118

broad range of cytokines, in particular the interferons (IFNs) and interleukin(IL)-2 receptor γ-chain utilizing cytokines (IL-2, IL-4, IL-7, IL-9, IL-15, IL-21), as well as the gp130 family (IL-6, IL-11, LIF, OSM, CNTF, CT-1, NNT-1/BSF-3) *(3,10)*. JAK2 is important for erythropoietin and IL-3-induced signal transduction, and JAK3 is essential for responses mediated by cytokines using the common γ-receptor subunit. TYK2 appears to be most important in mediating the biological response to IL-12 and lipopolysaccharide *(9)*. Thus, depending on the type of the cytokine receptor, specific JAK kinases may be preferentially activated, either alone or in combination with other JAKs. Although the JAK kinases do not seem to significantly contribute to signaling specificity, gene-targeting studies have identified characteristic signaling defects indicating clear nonredundant in vivo functions *(8)*. A summary of important features of the JAKs is given in Table 1.

The cytokine receptor complexes are composed of one or more individual proteins that, by themselves, lack intrinsic tyrosine kinase activity. However, the receptor chains contain specific, proline-rich recognition motifs (box 1/box 2) in their membrane-proximal region with which the JAKs are constitutively associated. The JAKs then are activated by binding of the cytokine to the extracellular domain of its cognate receptor. The precise mechanism by which ligand binding results in the activation of JAKs is not fully understood. It is believed that ligand binding triggers the homo- or hetero-oligomerization of cytokine receptor chains. The importance of receptor oligomerization for JAK activation has been documented in numerous studies (reviewed in ref. *5*). This oligomerization leads to a conformational change that brings receptor-associated JAKs into apposition. The JAKs then sequentially phosphorylate each other on specific tyrosine residues in the activation loop of the catalytic domain (Table 1). The transphosphorylation event induces the conversion of JAKs from an inactive enzyme to an active PTK. Notably, JAKs may form homo- or hetero-oligomers with a certain kinase hierarchy supporting the concept of transphosphorylation *(7)*. Upon activation, the JAKs phosphorylate specific tyrosine motifs within the cytoplasmic tail of the receptor chains, which then act as recruitment sites for various SH2-containing signaling molecules.

The probably most important and best characterized signaling pathway induced by JAKs is the activation of STAT proteins. Once recruited to the receptor, the STATs themselves become substrates for tyrosine phosphorylation by

Fig. 2. Model for the cytokine-induced activation of the Janus kinase-signal transducer and activator of transcription (JAK/STAT) signaling pathway. Ligand binding induces receptor oligomerization, activation of JAKs and subsequent phosphorylation of receptor chains and STAT transcription factors.

JAKs. Activated STATs are subsequently released from the receptor and form homo- or heterodimers through a reciprocal phosphotyrosine–SH2 domain interaction. Upon nuclear translocation, the STATs bind to specific sequence elements on the promoters of target genes and activate gene transcription (Fig. 2) (9,11).

Although the major function of JAKs is generally considered to be STAT activation, JAKs are also linked to other signaling processes depending on the cytokine receptor. The tyrosine sites on the receptors that are phosphorylated by JAKs can serve as docking sites for Src kinases, protein tyrosine phosphatases, PI3-K, and several adaptor proteins such as Shc, Grb2, and Cbl (12). Two important signaling pathways that are linked to JAK activation are the Ras/Raf/MAPK and the PI3-K/Akt pathways. Thus, JAK kinases provide important links between cytokine/hormone receptors and downstream effector proteins, ultimately resulting in transcriptional regulation of specific genes that mediate cellular responses.

JAK activation in hematopoietic cells can also be induced by oncogenic tyrosine kinases independent of cytokine stimulation. Murine pre-B lymphocytes transformed by the v-Abl oncogene of the Abelson murine leukemia virus harbor constitutively activated JAK1 and JAK3 and a direct association of v-Abl with JAK was found. This interaction was required for full activation of JAK1 (13,14). Although JAK1 activation seems to be essential for STAT phosphorylation and cytokine-independent proliferation, the role for JAK kinases in v-Abl-induced transformation and tumorigenicity is not clear (15,16). Even more controversial is the involvement of JAKs in Bcr-Abl-mediated STAT activation in chronic myeloid leukemia (CML). In some studies, JAK activation by Bcr-Abl could be demonstrated (17–20). Activation of JAK by the nucleophosmin/anaplastic lymption kinase (NPM-ALK) oncogenic fusion protein in a way similar to Bcr-Abl has been suggested recently. A constitutive phosphorylation of JAK2 was found in anaplastic large-cell lymphoma (ALCL) cell lines and in NPM-ALK-transformed hematopoietic cells. Moreover, JAK2 seemed to be physically associated with NPM-ALK, either directly or indirectly (21).

3. NEGATIVE REGULATION OF JAK SIGNALING

Under normal physiological conditions, JAK activation and downstream signaling are under control of negative regulation in order to prevent inappropriate gene expression. In this chapter, only those mechanisms that directly interfere with JAK activation will be briefly discussed. A more comprehensive coverage of negative regulation of JAK/STAT signaling is given in recent reviews (7,9,12,22–24).

3.1. Protein Tyrosine Phosphatases

Optimal activation of JAKs depends on the phosphorylation of critical tyrosine residues in the kinase-activating domain (Table 1). This phosphorylation is reversed by protein tyrosine phosphatases (PTPs), leading to reduced activation of subsequent signaling components. The SH2 domain-containing phosphatases SHP-1 and SHP-2 belong to a small, highly conserved subfamily of cytoplasmic PTPs *(25)*. They are recruited to the cytokine receptors by binding to specific phosphotyrosine motifs through their SH2 domains. These SHP–protein interactions lead to enzymic activation of the phosphatase by relieving the catalytic domain from N-terminal SH2 domain-mediated inhibition. On the other hand, SHPs have been shown to bind JAKs directly, and by themselves may be tyrosine phosphorylated *(26–29)*. SHP-1 (HCP, SHP, SH-PTP1, PTP1C) is expressed primarily in hematopoietic cells and has been shown to be physically associated with all JAK family members *(28,30–32)*. SHP-1 acts as negative regulator of intracellular signaling by many transmembrane receptors. It antagonizes JAK activation induced by IL-3 *(33)*, IFNs *(30)*, IL-2 *(34)*, IL-4 and IL-13 *(35)*, and EPO *(36,37)*. SHP-2 (SYP, SH-PTP2, PTP1D, PTP2C) is an ubiquitously expressed PTP generally recognized as a positive regulator of cytokine signaling by its function as an adaptor molecule *(10)*. However, there is also evidence for a role of SHP-2 as a negative regulator, e.g., in IFN *(38)*, IL-6 *(39)*, and leptin signaling *(40,41)*. The importance of negative regulation by SHPs in vivo is demonstrated by the motheaten phenotype of mice lacking functional SHP-1, which show hyperphosphorylation of JAKs and die of a disease with components of autoimmunity and inflammation *(42)*. SHP-2 signal-deficient mice display splenomegaly, lymphadenopathy, and autoimmune arthritis *(43,44)*.

Several other PTPs have also been reported to negatively regulate JAK/STAT signaling. The transmembrane CD45 protein, highly expressed in all hematopoietic cells, has been recently identified as a JAK phosphatase, directly binding to JAKs *(45)*. The cytoplasmic variant of PTPε, PTPεC, selectively inhibited IL-6- and leukemia inhibitory factor (LIF)-induced differentiation and apoptosis in murine M1 leukemic cells by reducing JAK1 and TYK2 phosphorylation levels *(46)*. JAK2 and TYK2 have been shown to be physiological substrates of PTP1B *(47)*.

3.2. The Supressor of Cytokine Signaling Protein Family

Negative regulation of cytokine signaling is also mediated by members of the suppressor of cytokine signaling (SOCS) family of proteins. The SOCS family consists of at least eight proteins, cytokine-indicible SH2-containing protein (CIS) and SOCS1–SOCS7, characterized by a conserved domain at the C-terminus referred to as the SOCS box (reviewed in refs. *22* and *23*). In addition, they contain a central SH2 domain enabling them to interact with other, tyrosine-phosphorylated proteins. Expression of SOCS proteins is rapidly induced by cytokines via the JAK/STAT pathway. Subsequently, they block STAT-mediated signal transduction, thus representing classical feedback inhibitors.

The exact mechanisms by which SOCS proteins antagonize JAK/STAT signaling seem to vary among the different family members, and still remain to be fully elucidated. Two mechanisms seem to be specifically relevant in this context. One is direct inhibition of JAKs by binding of the SOCS protein to activated JAKs and presentation of a pseudosubstrate motif. This has been well documented for SOCS1, also termed JAB or SSI-1. Two independent sites of SOCS1 are involved in this interaction: the SH2 domain specifically binds to the phosphorylated tyrosine residue (Y1007) in the activation loop of JAK2, and an additional N-terminal kinase inhibitory region, which binds to the catalytic pocket and is required for inhibition of JAK2 signaling and kinase activity (48,49). A similar mechanism accounts for the inhibitory function of SOCS3, which has been reported to bind and antagonize JAK2 (50,51). In contrast to SOCS1, but similar to CIS, SOCS3 associates with activated cytokine receptors, and it is not clear if receptor recruitment is a prereqisite for subsequent JAK binding (10). Another important mechanism by which SOCS proteins display their inhibitory function is targeting the JAK/receptor complex for regulated proteasomal degradation (see Section 3.3.).

3.3. Proteosomal Degradation

An important mechanism for modulating JAK activity and JAK-induced signaling pathways is proteasome-mediated degradation. In general, the ubiquitin-proteasome pathway mediates specific degradation of regulatory proteins and plays an important role in controlling a variety of cellular functions. This process involves conjugation of ubiquitin to the substrate, which then is recognized and degraded by the 26S proteasome (52).

Proteasome inhibitors have been shown to prolong IL-2- and IL-3-induced JAK activation (53,54). Some recent studies have brought more insight on how the proteasome deactivates JAK kinases. Members of the SOCS family of proteins seem to be involved in this process because they can associate with proteins that are linked to proteosomal degradation. The SOCS box promotes association with the elongin B/C complex, which binds cullin-2, an E3-like ubiquitin ligase (55,56). New findings report that SOCS1 can target JAK2 to degradation by recruiting ubiquitin ligase and promoting ubiquitination of JAK2. Phosphorylation of JAK2 on Y1007 in the activation loop was required for the interaction with SOCS1 (57). This mechanism also serves to downregulate cellular transformation by TEL/JAK (58,59). Consistent with these findings, dephosphorylation of JAK2 by SHP-2 resulted in JAK2 stabilization (60). In contrast, SHP-1-accelerated proteasome-mediated degradation of TYK2 and JAK1 in certain cancer cell lines (61). The precise role for these proteins in modulating proteasomal degradation of JAKs still needs to be defined.

4. IMPLICATIONS OF ABERRANT JAK ACTIVATION IN HEMATOLOGICAL DISEASES

The essential role that JAKs play in normal hematopoietic regulation has been shown by gene-targeting studies that have identified characteristic signaling defects. JAK1 and JAK2 knockout mice are perinatal or embryonic lethal, whereas JAK3 nullizygous mice show severe combined immunodeficiency (SCID) (Table 1) (described in more detail in refs. *9* and *62*). Consequently, inappropriate inhibition or absence of JAK activity causes immunosuppressive diseases. The prominent example is the presence of mutations in the JAK3 gene that leads to SCID in humans, characterized by an absence of peripheral T- and natural-killer cells and normal or slightly increased numbers of B-cells *(63–66)*. Other heritable human diseases caused by intrinsic inactivating defects in JAKs have not been reported. However, a variety of pathogens, such as human papilloma virus or cytomegalovirus, are able to inhibit JAK-mediated IFN signaling, thereby escaping immunosurveillance *(7,67,68)*.

On the other hand, constitutive or enhanced activation of JAK activity can result in malignant cell growth or even neoplastic transformation. The earliest evidence came from studies in nonhuman experimental systems. An amino acid substitution in the JAK homolog of *Drosophila,* HOP, owing to a point mutation in the hopscotch gene, was shown to produce a hyperactive kinase and caused leukemia-like abnormalities *(69,70)*. Elevated JAK activity has also been implicated in an increasing number of human hematological malignancies (Table 2) *(7,62,71,72)*. The mechanisms that may lead to increased JAK activation include intrinsic genetic defects like chromosomal translocations and gene amplification, phosphorylation by oncogenic tyrosine kinases, increased cytokine/growth factor production (either autocrine or paracrine), and disruption of normal negative regulation. So far, activating mutations resembling those in the *Drosophila* JAK homolog have not been found in human JAKs *(73)*. The consequences of elevated JAK activity are numerous, reflecting the pleiotropic biological activities of cytokines and growth factors, and they depend on the signal transduction pathways that are initiated in the given cellular context. Mainly, aberrant JAK activity delivers increased proliferation and survival signals and thereby contributes to the process of malignant transformation.

4.1. JAKs in Leukemias and Lymphomas

The first suggestion for a role of constitutively active JAK in human cancer came from a study of patients with B-precursor lymphoblastic acute leukemia (pre-B ALL) in relapse *(74)*. Tumor-derived cell lines harbored constitutively activated JAK2, and AG490, a PTK inhibitor shown to be effective against JAKs, prevented the tumor cells to engraft in SCID mice (*see* Section 5) *(75)*. Similar results were obtained with pre-B ALL cell lines harboring an 11q23 translocation

Table 2
JAK Associated Hematological Malignancies

Kinase	Type of altered activity	Disease implicated	References
Leukemias			
JAK2	Persistent activation (autocrine cytokine production, Btk, Bcr-Abl)	Precursor-B ALL	(74,76)
JAK2	TEL-JAK fusion protein (chromosomal translocation)	T-ALL, Precursor-B ALL, CML	(77,78)
JAK2	Bcr/JAK2 fusion protein (chromosomal translocation)	CML	(82)
JAK1	Constitutive activation		
JAK2	(cytokine production, Bcr-Abl)	Myeloid leukemia (cell lines)	(19,93)
JAK1	Constitutive activation	Adult T-cell	(34,91,92)
JAK3	(cytokine production, defect in negative regulation)	leukemia/ lymphoma (HTLV-1 transformed)	
Lymphomas			
JAK1	Constitutive activation (autocrine IL-10)	B-lymphoblastoid cell lines (EBV-positive)	(94)
JAK2	Increased expression (gene amplification)	Hodgkin's lymphoma	(83,84)
JAK2	Increased expression (gene amplification)	Diffuse large B-cell lymphoma	(85)
JAK2	Increased expression (gene amplification, IL-4?)	Mediastinal large B-cell lymphoma	(86,87,88)
JAK3	Constitutive activation	Cutaneous T-cell lymphoma	(89,90)
JAK1	Constitutive activation	LSTRA T-cell	(53)
JAK2	(Lck overexpression)	lymphoma (mouse cell line)	
JAK2	Constitutive activation	NPM/ALK positive	(21,95,96)
JAK3	(activated ALK)	anaplastic large cell lymphoma	

(Continued)

Table 2 *(Continued)*

Kinase	Type of altered activity	Disease implicated	References
TYK2	Defect in negative regulation	Hemophagocytic/ lymphohistiocytosis	*(32)*
Multiple myeloma			
JAK1 JAK2 TYK2	Persistent activation (cytokine/IL-6- mediated, defect in negative regulation)	Multiple myeloma cell lines and primary tumor cells	*(98,110,111, 112,118)*

JAK, janus kinase; ALL, acute lymphoblastic leukemia; CML, chronic myeloid leukemia; HTLV-1, human T-cell lymphotropic virus type-1; EBV, Epstein-Barr virus; ALK, anaplastic lymphoma kinase; NPM, nucleophosmin.

or the Philadelphia chromosome. In these cases, the Btk tyrosine kinase and Bcr-Abl were found to associate with JAK2 and suggested to be involved in JAK2 activation. AG490 inhibited JAK2 phosphorylation and cell growth *(76)*.

Direct evidence of the concept that dysregulation of JAKs can cause cellular transformation comes from the identification of TEL/JAK fusion proteins in lymphoid leukemias and a case of atypical CML *(77,78)*. In these cases, chromosomal translocations were involved that led to the production of a chimeric protein containing the oligomerization domain of the translocated ets leukemia (TEL) protein, a member of the ETS transcription factor family, fused to the catalytic JH1 domain of JAK2 (Fig. 3). As a result of this fusion, the kinase becomes constitutively activated *(79)*. Moreover, it has been demonstrated that the TEL/JAK chimera was able to transform Ba/F3 cells and render them factor-independent *(77,79,80)*. Mice transplanted with the retrovirus expressing the fusion protein develop a fatal mixed myeloproliferative and lymphoproliferative disorder *(80)*. Importantly, TEL/JAK transgenic mice develop T-cell leukemia with constitutive activation of STAT1 and STAT5 in leukemic tissues *(81)*. A similar chromosomal translocation, t(9;22)(p24;q11), involves JAK2 fused to the BCR region, and was found in cases of Bcr-Abl negative CML *(82)*.

Another genetic defect that leads to deregulated JAK expression involves gene amplification. In primary Hodgkin's lymphoma cells and cell lines, an increased copy number of chromosomal sequences spanning the *JAK2* gene were identified *(83,84)*. Genomic amplification of the *JAK2* gene was also evident in a case of diffuse large B-cell lymphoma, which resulted in a high expression of the respective JAK2 mRNA transcript *(85)*. In another aggressive non-Hodgin's lymphoma (NHL), mediastinal large B-cell lymphoma, gains in chromosome arm 9p that include *JAK2* gene amplification and high expression levels of JAK2 were found *(86–88)*.

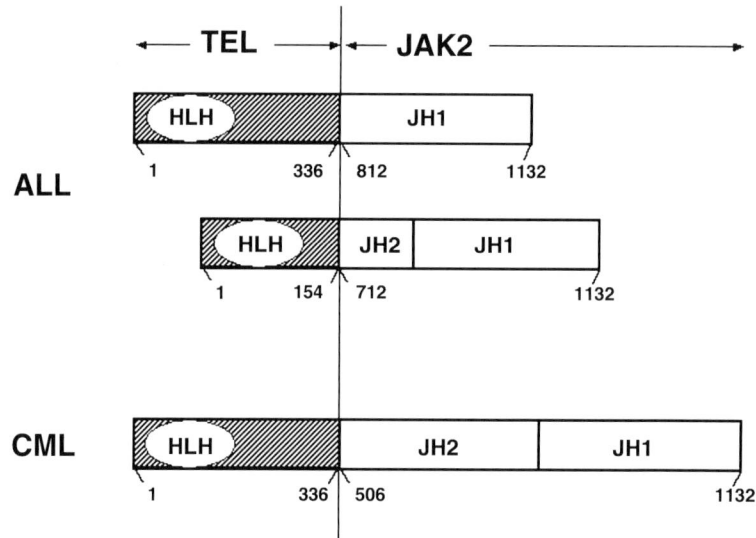

Fig. 3. Translocated ETS leukemia/Janus kinase-2 (TEL/JAK) fusion proteins in acute lymphoid leukemia (ALL) and chronic myeloid leukemia (CML). Expression of the TEL/JAK fusion protein is the result of a t(9;12)(p24;p13) chromosomal translocation. The amino acid positions of the fusions are indicated by numbers. HLH, Helix–Loop–Helix oligomerization domain. Adapted from ref. 62.

In many cases, however, the precise mechanisms underlying constitutive JAK activation have not been identified. In a study of cutaneous T-cell lymphomas, the data suggest that the progression from indolent to aggressive T-cell lymphomas may involve a switch from factor-dependent to constitutive JAK3 activation (89). In tumor cell lines from a patient with mycosis fungoides, a slowly migrating iso-form of STAT3 was found to be constitutively activated and associated with JAK3 (90). Similarly, the progression of adult T-cell leukemia/lymphoma (ATLL) seems to be correlated with a gradual onset of constitutive JAK1 and JAK3 acti-vation combined with loss of IL-2 dependence (91,92). Constitutive JAK phos-phorylation was also observed in myeloid and B-lymphoblastoid cell lines (19,93,94). For some of these cases, autocrine cytokine production has been sug-gested as the cause of JAK activation. Very recent studies suggest a role for JAK kinases in the pathophysiology of NPM/ALK positive ALCL. Constitutive JAK2 and JAK3 phosphorylation and physical association with NPM-ALK was observed in ALCL cells, suggesting a role for the ALK tyrosine kinase in JAK activation (21,95,96). However, the role of JAKs for STAT activation and NPM/ALK-mediated transformation is unclear and further studies are needed.

Defects in the negative regulation of JAKs may also contribute to malignant proliferation. Expression of the phosphatase SHP-1 is decreased in more than 90% of hematopoietic-related and some nonhematopoietic tumor cell lines and tissues including B- and T-cell lymphomas. The diminished or abolished SHP-1 expression could be because of a mutation of the *SHP-1* gene, methylation of the promoter region of the *SHP-1* gene or posttranscriptional regulation of SHP-1 protein synthesis (reviewed in ref. *97*). For example, expression of SHP-1 is downregulated in a number of IL-2-independent human T-cell lymphotrophic virus type-1 (HTLV-1) transformed T-cell lines that exhibit constitutive JAK/STAT, thereby contributing to HTLV-1-mediated T-cell transformation *(34)*. Another mechanism is the lack of interaction between the phosphatase and JAK, shown for SHP-1 and TYK2 in familial hemophagocytic lymphohistiocytosis *(32)*. Nonfunctional SOCS proteins owing to methylation were observed in multiple myeloma (MM) and acute myeloid leukemia *(98,99)*. Very recently, somatic SHP-2 mutations have been found in approx 30% of sporadic juvenile myelomonocytic leukemia *(100)*.

In many primary lymphoid and myeloid leukemia cells, constitutively activated STAT proteins have been found; however, the underlying mechanisms or the JAKs involved have not been identified *(101–104)*. It cannot be excluded that, at least in some cases, other PTKs than JAKs, i.e., kinases of the Src family, might be responsible for STAT phosphorylation *(12,105)*. Other studies, however, do suggest the involvement of JAKs by using the JAK inhibitor AG490. In large granular lymphocyte (LGL) leukemia and primary effusion lymphoma (PEL), AG490 induced apoptosis that was accompanied by a decrease or abrogation of STAT activation *(106,107)*.

4.2. JAKs in Multiple Myeloma

Recent studies indicate that the activation of the JAK/STAT pathway plays an important role in the pathophysiology of MM. In malignant plasma cells, STAT3 is strongly activated by IL-6, a major growth and survival factor, which is mainly produced in the myeloma bone marrow microenvironment *(108,109)*. This has been shown for a number of cell lines, but even more important, STAT3 was found to be constitutively activated in tumor samples from patients with MM *(110,111)*. Interestingly, constitutive JAK1/STAT3 phosphorylation was observed in IL-6-independent but not IL-6-dependent murine plasmacytomas and hybridomas, suggesting that acquirement of constitutive JAK/STAT signaling contributes to malignant progression *(112)*.

IL-6 belongs to the family of gp130 cytokines that exert their action via the signal transducer chain gp130 *(10)*. Binding of IL-6 to its specific receptor (gp80/CD126) leads to homodimerization of gp130, activation of JAKs, and gp130 phosphorylation. Phosphorylation of gp130 by JAKs occurs on six tyrosine residues in its cytoplasmic region (Fig. 4).

The JAKs known to be involved in IL-6/gp130 signaling are JAK1, JAK2, and/or TYK2 *(10,113,114)*. Activation of one or more of these JAKs in murine and human plasmacytoma cell lines as well as patient samples following

gp130

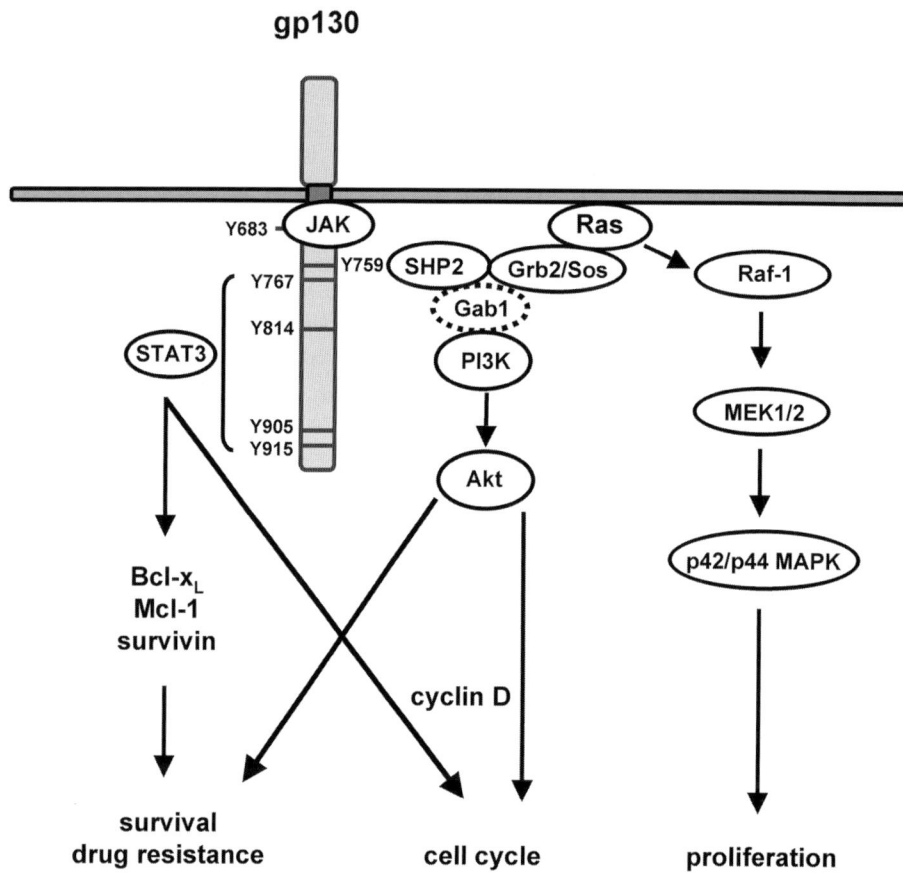

Fig. 4. Gp130/Janus kinase (JAK) mediated signaling pathways in malignant plasma cells. Three major signaling pathways are initiated through distinct tyrosine residues on gp130 that become phosphorylated by JAKs: tyrosines(Y)767, Y814, Y905, and Y915 represent recruitment sites for STAT3; Y759 links gp130 to the Ras-dependent mitogen-activated protein kinase and the PI3-K/Akt pathway via SHP2.

stimulation with IL-6 has been shown; however, the pattern of JAKs activated seems to vary *(112,113,115–119)*. Phosphorylation of JAKs by other cytokines in plasmacytoma was demonstrated for IL-11 *(115)*, Kaposi's sarcoma-associated virus-encoded IL-6 *(117)*, IL-21 *(120)*, and IFN-α *(119,121)*.

Four tyrosines in the carboxyl-terminus of gp130 function as the binding sites for STAT3 (Fig. 4). Activated STAT3 initiates multiple biological responses *(105,122)*. One of the mechanisms by which STAT3 signaling contributes to the pathogenesis of MM is the upregulation of proteins of the Bcl-2 family conferring resistance to apoptosis. This has been shown for Bcl-x_L in the

U266 cell line *(110)*, and for Mcl-1 in MDN cells *(123)*. Abrogation of constitutive STAT3 activation in U266 cells using an inhibitor of JAKs, AG490 (*see* Section 5), decreased Bcl-x$_L$ expression, induced apoptosis, and sensitized the cells to Fas- or drug-mediated cell death *(110,124,125)*. In MM1.S myeloma cells, IL-6 increased the expression of survivin, a member of the inhibitor of apoptosis protein family, which is suggested to be a target gene of STAT3 as well *(126,127)*. Survivin may play a role in IL-6-mediated protection from dexamethasone (Dex)-induced cell death. With the means of gene expression profiling, a higher expression of survivin was found in Dex-resistant MM1.R cells compared to Dex-sensitive MM1.S *(128)*. Our recent studies suggest that other, Bcl-2 family-independent mechanisms exist that mediate STAT3-triggered anti-apoptotic responses. In the IL-6-dependent INA-6 plasma cell line, among the anti-apoptotic members of the Bcl-2 family, only Mcl-1 was slightly induced by IL-6. However, overexpression studies demonstrated that IL-6 does not exert its survival effect primarily through this pathway. Importantly, activation of STAT3 is obligatory for survival of INA-6 cells as revealed by transfection with mutated gp130 receptor chimeras *(129)*. Together, these findings strongly support the important role for STAT3 in myeloma cell growth and survival.

Besides STAT activation, other pathways are induced by gp130/JAK signaling through tyrosine residue Y759 *(10,114)*. Numerous studies have shown that these pathways are activated by IL-6 in MM cells (reviewed in ref. *130*). Tyr759 mediates the interaction between gp130 and the tyrosine phosphatase *SHP-2*, which becomes phosphorylated in a JAK-dependent manner *(131)*. Activated *SHP-2* links gp130 to the Ras/Raf/MAPK signaling cascade via the Grb2 adaptor protein, and this pathway has been suggested to be important for myeloma cell proliferation *(116)*. In addition, *SHP-2* mediates the protective effect of IL-6 against Dex-induced apoptosis by dephosphorylating related adhesion focal tyrosine kinase (RAFTK) *(132)*. IL-6 also activates PI3-K and downstream protein kinase B/Akt in MM cells, where it is involved in protection against apoptosis and enhanced proliferation *(133–136)*. The molecular mechanism linking gp130 to PI3-K is not well understood and a role for the scaffolding adaptor protein Gab1 has been proposed *(137)*. In MM cells, IL-6 induces association of phosphorylated PI3-K with SHP2; however, no direct association with JAKs was found *(134)*. Fig. 4 summarizes gp130-induced signaling pathways in MM.

Taken together, JAKs play a critical role in the pathophysiology of MM primarily through their association with cytokine receptors. Because IL-6 is a major growth and survival factor for malignant plasma cells, JAKs associated with gp130, the signal transducer chain of the IL-6 receptor complex, are of particular importance. Activation of STAT3 seems to be an essential downstream signaling event. By triggering anti-apoptotic pathways, activated STAT3 not only mediates resistance to therapy but also directly contributes to the malignant progression of MM by allowing accumulation of long-lived plasma

Table 3
Features of JAK Inhibitors

Name	Chemical class	Molecular formula	Molecular weight	target	references
AG490/ tyrphostin B42	benzylidene malononitrile	$C_{17}H_{14}N_2O_3$	294.3	pan/JAK EGFR	(74,138,143)
Pyridone 6	tetracyclic pyridone	$C_{18}H_{16}FN_3O$	309.3	pan/JAK other PTKs	(174)
WHI-P131	dimethoxy-quinazoline	$C_{16}H_{15}N_3O_3$	297.3	JAK3	(162,163)
ZM39923	naphthyl ketone	$C_{23}H_{25}NO \cdot HCl$	367.9	JAK3	(165)
ZM449829	naphthyl ketone	$C_{13}H_{10}O$	182.2	JAK3	(165)
n.n.[a]	pyridyl oxindole	$C_{18}H_{13}N_3O \cdot$ $MeSO_3H$	383.4	JAK3	(166)
CR4	hydroxystyryl-acrylonitrile			JAK2 Bcr-Abl	(173)

[a] n.n., no specific name; EGFR, epidormal growth factor receptor.

cells *(105,122)*. However, at least two other signaling pathways downstream of JAK activation, the Ras/Raf/MAPK and the PI3-K/Akt pathway, are critical for myeloma cell growth and survival.

5. PHARMACOLOGICAL INHIBITION OF JAKS

Few selective JAK inhibitors are currently available, which are mostly used in cell culture studies to confirm the connection of the JAK-STAT pathway to certain biological functions such as cell growth, survival, differentiation, or drug resistance (features of selected JAK inhibitors are summarized in Table 3). Some studies have been performed in animal models. To our knowledge, there are no JAK inhibitors in clinical trials yet.

The probably best known and characterized JAK inhibitor is AG490 or tyrphostin (tyrosine phosphorylation inhibitor) B42 *(75,138)*. Tyrphostins are a class of low-molecular-weight PTK blockers designed on the basis of tyrosine, in which the benzylidene moiety of erbstatin, a natural compound with activity against the epidermal growth factor receptor (EGFR) and Src, was incorporated. The basic pharmacophore in tyrphostins is 3,4-dihydroxy-*cis*-cinnamonitrile; AG490 itself is an *N*-benzyl-3,4-dihydroxy-benzylidenecyanoacetamide *(138)* (Fig. 5). AG490 has often been designated as a JAK2-specific inhibitor. However, AG490 is not exclusively selective for JAK2, but also inhibits, at least partially, JAK1, JAK3, and TYK2 *(115,118,139–142)*. Originally, AG490 was

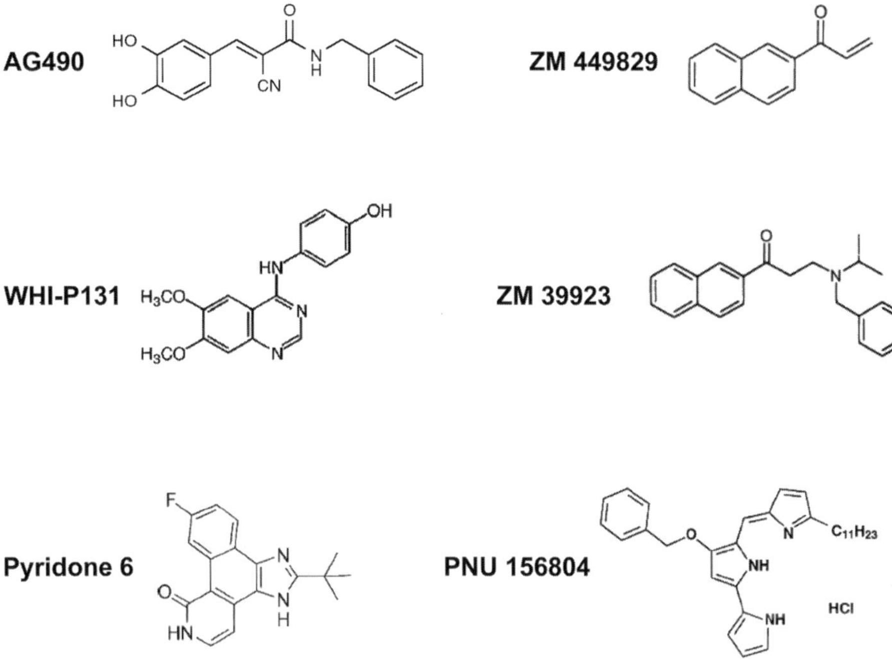

Fig. 5. Chemical structures of JAK inhibitors.

described as an inhibitor of members of the EGFR family, with an IC_{50} of 100 nM for receptor autophosphorylation *(143)*. The in vitro kinase activities of other PTKs, including Lck, Lyn, Btk, Syk, Src, and Zap70, are not affected by AG490 *(74,144,145)*.

AG490 has shown efficacy in vitro and in vivo as a JAK inhibitor in numerous studies. In pre-B ALL, the compound inhibited growth and induced apoptosis of cell lines and primary tumor cells *(74,76)*. This effect was exerted through inhibition of JAK2 kinase activity, which was found to be constitutively activated in these cells. AG490 at concentrations that inhibited ALL cell growth did not affect growth and maturation of hematopoietic progenitor cells as evaluated in colony-forming assays. Importantly, AG490 was able to prevent engraftment of a pre-B ALL cell line in SCID mice with no significant toxicity to normal blood cells *(74)*.

AG490 has growth-inhibiting effects also in MM. Inhibition of constitutive STAT3 signaling in the U266 cell line induced apoptosis by downregulating Bcl-X_L expression and rendered the cells sensitive to Fas-induced apoptosis *(110)*. A more recent study showed that AG490 treatment sensitized U266 cells to apoptosis by chemotherapeutic drugs as well, including cisplatin, fludarabine, adriamycin, and vinblastine *(125)*. In MDN cells, IL-6-mediated STAT3 phosphorylation and upregulation of Mcl-1 was inhibited by AG490 *(123)*. Besides inducing apoptosis through downregulation of Bcl-2 family proteins, JAK inhibition in myeloma could have effects on cell proliferation. Again in U266 cells, AG490 induced a cell cycle arrest actually diminishing the cytotoxic effects of DNA-damaging agents *(124)*. The cell cycle arrest was explained by a reduction of cyclin D1 expression, which is a STAT3 target gene *(122)*. AG490 may also decrease IL-6-induced cell proliferation by blocking ERK2 activation *(118)*. This signaling pathway has been shown to trigger proliferation in MM cells *(116)*. AG490 reduced STAT3 phosphorylation levels and cell growth induced by IL-6 and BSF-3, another cytokine of the gp130 family *(146)*. In vivo antimyeloma activity was demonstrated in a subcutaneous syngeneic mouse model, in which treatment with AG490 resulted in tumor regression. However, this was not complete because of regrowth of tumor cells after AG490 withdrawal. Notably, tumor specimens from AG490-treated mice were negative for STAT3 phosphorylation *(147)*.

Downregulation of STAT3 activation and Bcl-2 expression by AG490 might be responsible for growth inhibition in cell lines from a patient with mycosis fungoides *(90,148)*. Pre-B cells with a t(11q23) translocation or Philadelphia chromosome were found to be particularly sensitive to growth inhibition by AG490 *(76)*. AG490 induced apoptosis in primary LGL leukemia cells and PEL cell lines, which was accompanied by an abrogation of STAT activation and corresponding downregulation of Mcl-1 and survivin, respectively *(106,107)*. In Bcr-Abl expressing cell lines and primary leukemic cells from patients with CML, synergistic or additive effects could be achieved by a combination of AG490 with imatinib mesylate (STI571, Gleevec) *(149,150)*.

Novel analogs of AG490 have been screened recently from a chemical library in order to find more effective compounds that are useful also in animal studies. AG1801 and WP1066 turned out to effectively block IL-6-mediated STAT3 phosphorylation and induce apoptosis in numerous cell types including MM cells *(151,152)*.

Inhibition of JAKs has also been observed with the serine/threonine kinase inhibitor staurosporine at high concentrations, the immunosuppressive drug leflunomide, and octylamino-undecyl-dimethylxanthine derivatives *(153–156)*. Piceatannol, a resveratrol analog, is a natural stilbene byproduct from the seeds of Euphorbia lagascae and reported to inhibit the p72Syk kinase *(157)*. Recent studies showed inhibition of STAT3 and STAT5 activation by piceatannol *(158)*. An IL-10-dependent AIDS-related B-NHL cell line was sensitized by the compound to chemotherapeutic drug-induced apoptosis through inhibition of JAK1/STAT3 activation and Bcl-2 expression *(125)*. Curcumin (diferuloylmethane) is a plant product known for its antioxidant, antitumor, and anti-inflammatory activities *(159)*. In T-cells, curcumin has been shown to inhibit IL-12-induced phosphorylation of JAK2 and TYK2 *(160)*. Curcumin has recently been shown to effectively inhibit constitutive and IL-6-mediated STAT3 phosphorylation in MM cells with concomitant suppression of proliferation; however, the kinases inhibited by curcumin have not been identified *(161)*.

In contrast, highly selective JAK3 inhibitors can be found among different chemical classes. In a structure-based design approach, three dimethoxyquinazoline compounds, WHI-P131, WHI-P154, and WHI-P97, were developed *(162,163)*. The lead compound WHI-P131 (JANEX-1) showed potent JAK3 inhibitory activity without affecting JAK1, JAK2, or other PTKs like Syk, Btk, Lyn, or insulin receptor kinase. It was also inhibiting clonogenic growth of a number of JAK3 expressing leukemic B- and T-cell lines *(163)*. In a recent study, both WHI-P131 and WHI-P154 decreased constitutive phosphorylation of JAK3 and STAT3 associated with induction of apoptosis in ALK-positive ALCL cell lines *(96)*. In vivo toxicity and pharmacokinetic studies were performed in mice, Lewis rats, and cynomolgus monkeys, and no acute toxicity was observed *(164)*. Potent inhibition of JAK3 was also found for a series of naphthyl(β-aminoethyl)ketones *(165)*, a pyridyl oxindole compound *(166)*, the prodigiosin analog PNU156804 *(167–170)*, and for the compound CP-690550 *(171,172)*.

JAK3 is mainly found in lymphoid cells, and appears to be stimulated only by activation of cytokine receptors containing the γc subunit (IL-2, -4, -7, -9, -13, -15, -21 receptors). Therefore, inhibition of JAK3 results in immunosuppression by blocking the T-cell mitogenic signal. Besides their potential as antineoplastic agents, JAK3 inhibitors may find application as immunosuppressive drugs to control autoimmune diseases and transplant rejection.

Several new compounds with promising potential have been identified recently. JAK2 and Bcr-Abl were described to be the main targets of a new PTK inhibitor, CR4. This compound inhibited the growth and survival of both Philadelphia-positive and -negative ALL as well as AML cells. Alhough efficiently ablating leukemic cell growth, normal cell growth and differentiation remained unaffected *(173)*. As a pan-JAK inhibitor, the compound pyridone 6 inhibits each member of the JAK kinase family in the nanomolar range with specificity over numerous other kinases *(174)*. It also inhibits IL-2- and IL-4-driven proliferation of carmurine T-cell lymphoma cell line cells and phosphorylation of STAT5. Cucurbitacin I (JSI-124) is a natural plant product that has been used as a traditional herbal medicine for a long time. It was found recently to be highly selective in inhibiting JAK/STAT3 activation without affecting signaling pathways mediated by Akt, Erk1/2, or c-Jun NH_2-terminal kinase. However, the biochemical target of JSI-124 has not been identified *(139)*.

6. CONCLUSIONS AND PERSPECTIVES

Extensive research over the past decade has led to the identification of components of cellular signaling pathways as promising targets for therapeutic intervention. This knowledge resulted in the development of novel drugs that are now being studied in preclinical and clinical studies *(175,176)*. The importance of JAKs in regulating cytokine-dependent gene expression and cellular function is well established. Activation of JAKs is one of the first steps in cytokine receptor-mediated signaling and critical for virtually all subsequent downstream signaling cascades *(5)*. As outlined in this chapter, aberrant JAK activity, as a result of excessive cytokine signaling, chromosomal translocation, or disruption of negative regulation, has been shown to be involved in malignant disease. Therefore, pathologically active JAKs represent a potential and attractive target for novel drugs that may selectively inhibit them.

A direct apoptotic effect of JAK inhibitors may be seen on cells where growth factors play an important role in cell proliferation and survival. At least some of the effects of JAK inhibitors are the result of blocking downstream signaling by STAT proteins. However, because in most proliferative diseases more than one signaling pathway is involved, JAK inhibition by itself might not be sufficient to eradicate the tumor. A good example is the observed synergistic inhibitory effect of the JAK inhibitor AG490 in combination with imatinib mesylate (STI571, Gleevec) on Bcr-Abl-expressing cells *(149,150)*. In many cases, inhibition of JAK-dependent pathways may lead to sensitization of the tumor cells to other drugs, as it was observed in MM. Therefore, combining JAK inhibitors with cytotoxic drugs, other apoptosis-inducing agents, antibodies, or even other signaling inhibitors may have additive or synergistic effects.

Today, there is only a limited number of small-molecule JAK inhibitors, most of them being nonselective and none of them has yet entered the clinical arena. Tyrphostin AG490 belongs to the first generation of PTK inhibitors developed more than a decade ago. Blocking leukemic cell growth with AG490 in an SCID mouse model for pre-B ALL was considered a milestone in signal transduction therapy *(74,177)*. Structural studies of JAKs would offer a better understanding of their function and provide the basis for the development of selective inhibitors. A theoretical model for the JH1 kinase domain plus kinase-like domain (JH2) of JAK2 has been generated recently *(178)*. Using this structure as a starting point, the N-terminal region of JAK2 was also modeled, and the entire predicted three-dimensional structure of human JAK2, comprising all seven JAK homology domains, presented *(179)*. The availability of new high-throughput screening assay formats will allow to identify compounds that effectively block JAK enzymatic activity *(180)*. Such molecules can be tested for specificity and optimized into drug candidates. New JAK inhibitors are already under investigation and, without any doubt, will play an important role in cancer therapy in the future.

REFERENCES

1. Leonard WJ, O'Shea JJ. Jaks and STATs: biological implications. *Annu Rev Immunol* 1998; 16:293–322.
2. Ihle JN, Thierfelder W, Teglund S, et al. Signaling by the cytokine receptor superfamily. *Ann NY Acad Sci* 1998; 865:1–9.
3. Schindler C, Strehlow I. Cytokines and STAT signaling. *Adv Pharmacol* 2000; 47:113–174.
4. Ravandi F, Talpaz M, Kantarjian H, Estrov Z. Cellular signalling pathways: new targets in leukaemia therapy. *Br J Haematol* 2002; 116(1):57–77.
5. Rane SG, Reddy EP. Janus kinases: components of multiple signaling pathways. *Oncogene* 2000; 19(49):5662–5679.
6. Hubbard SR, Till JH. Protein tyrosine kinase structure and function. *Annu Rev Biochem* 2000; 69:373–398.
7. Duhe RJ, Wang LH, Farrar WL. Negative regulation of Janus kinases. *Cell Biochem Biophys* 2001; 34(1):17–59.
8. Schindler CW. Series introduction. JAK-STAT signaling in human disease. *J Clin Invest* 2002; 109(9):1133–1337.
9. Kisseleva T, Bhattacharya S, Braunstein J, Schindler CW. Signaling through the JAK/STAT pathway, recent advances and future challenges. *Gene* 2002; 285(1–2):1–24.
10. Heinrich PC, Behrmann I, Haan S, Hermanns HM, Muller-Newen G, Schaper F. Principles of interleukin (IL)-6-type cytokine signalling and its regulation. *Biochem J* 2003; 374(Pt 1):1–20.
11. Levy DE, Darnell JE, Jr. Stats: transcriptional control and biological impact. *Nat Rev Mol Cell Biol* 2002; 3(9):651–662.
12. Rane SG, Reddy EP. JAKs, STATs and Src kinases in hematopoiesis. *Oncogene* 2002; 21(21):3334–3358.
13. Danial NN, Losman JA, Lu T, et al. Direct interaction of Jak1 and v-Abl is required for v-Abl-induced activation of STATs and proliferation. *Mol Cell Biol* 1998; 18(11):6795–6804.
14. Danial NN, Pernis A, Rothman PB. Jak-STAT signaling induced by the v-abl oncogene. *Science* 1995; 269(5232):1875–1877.
15. Danial NN, Rothman P. JAK-STAT signaling activated by Abl oncogenes. *Oncogene* 2000; 19(21):2523–2531.
16. Sexl V, Kovacic B, Piekorz R, et al. Jak1 deficiency leads to enhanced Abelson-induced B-cell tumor formation. *Blood* 2003; 101(12):4937–4943.
17. Shuai K, Halpern J, ten Hoeve J, Rao X, Sawyers CL. Constitutive activation of STAT5 by the Bcr-Abl oncogene in chronic myelogenous leukemia. *Oncogene* 1996; 13(2):247– 254.
18. Henderson YC, Guo XY, Greenberger J, Deisseroth AB. Potential role of Bcr-Abl in the activation of JAK1 kinase. *Clin Cancer Res* 1997; 3(2):145–149.
19. Chai SK, Nichols GL, Rothman P. Constitutive activation of JAKs and STATs in Bcr-Abl-expressing cell lines and peripheral blood cells derived from leukemic patients. *J Immunol* 1997; 159(10):4720–4728.
20. Xie S, Wang Y, Liu J, et al. Involvement of Jak2 tyrosine phosphorylation in Bcr-Abl transformation. *Oncogene* 2001; 20(43):6188–6195.
21. Ruchatz H, Coluccia AM, Stano P, Marchesi E, Gambacorti-Passerini C. Constitutive activation of Jak2 contributes to proliferation and resistance to apoptosis in NPM/ALK-transformed cells. *Exp Hematol* 2003; 31(4):309–315.

22. Krebs DL, Hilton DJ. SOCS proteins: negative regulators of cytokine signaling. *Stem Cells* 2001; 19(5):378–387.

23. Kubo M, Hanada T, Yoshimura A. Suppressors of cytokine signaling and immunity. *Nat Immunol* 2003; 4(12):1169–1176.

24. Yasukawa H, Sasaki A, Yoshimura A. Negative regulation of cytokine signaling pathways. *Annu Rev Immunol* 2000; 18:143–164.

25. Neel BG, Gu H, Pao L. The "Shp"ing news: SH2 domain-containing tyrosine phosphatases in cell signaling. *Trends Biochem Sci* 2003; 28(6):284–293.

26. Stahl N, Farruggella TJ, Boulton TG, Zhong Z, Darnell JE, Jr., Yancopoulos GD. Choice of STATs and other substrates specified by modular tyrosine-based motifs in cytokine receptors. *Science* 1995; 267(5202):1349–1353.

27. Fuhrer DK, Feng GS, Yang YC. Syp associates with gp130 and Janus kinase 2 in response to interleukin-11 in 3T3-L1 mouse preadipocytes. *J Biol Chem* 1995; 270(42):24826–24830.

28. Jiao H, Berrada K, Yang W, Tabrizi M, Platanias LC, Yi T. Direct association with and dephosphorylation of Jak2 kinase by the SH2-domain-containing protein tyrosine phosphatase SHP-1. *Mol Cell Biol* 1996; 16(12):6985–6992.

29. Yin T, Shen R, Feng GS, Yang YC. Molecular characterization of specific interactions between SHP-2 phosphatase and JAK tyrosine kinases. *J Biol Chem* 1997; 272(2):1032–1037.

30. David M, Chen HE, Goelz S, Larner AC, Neel BG. Differential regulation of the alpha/beta interferon-stimulated Jak/Stat pathway by the SH2 domain–containing tyrosine phosphatase SHPTP1. *Mol Cell Biol* 1995; 15(12):7050–7058.

31. Yetter A, Uddin S, Krolewski JJ, Jiao H, Yi T, Platanias LC. Association of the interferon-dependent tyrosine kinase Tyk-2 with the hematopoietic cell phosphatase. *J Biol Chem* 1995; 270(31):18179–18182.

32. Tabrizi M, Yang W, Jiao H, et al. Reduced Tyk2/SHP-1 interaction and lack of SHP-1 mutation in a kindred of familial hemophagocytic lymphohistiocytosis. *Leukemia* 1998; 12(2):200–206.

33. Yi T, Mui AL, Krystal G, Ihle JN. Hematopoietic cell phosphatase associates with the interleukin-3 (IL-3) receptor beta chain and down-regulates IL-3-induced tyrosine phosphorylation and mitogenesis. *Mol Cell Biol* 1993; 13(12):7577–7586.

34. Migone TS, Cacalano NA, Taylor N, Yi T, Waldmann TA, Johnston JA. Recruitment of SH2-containing protein tyrosine phosphatase SHP-1 to the interleukin 2 receptor; loss of SHP-1 expression in human T-lymphotropic virus type I–transformed T cells. *Proc Natl Acad Sci USA* 1998; 95(7):3845–3850.

35. Haque SJ, Harbor P, Tabrizi M, Yi T, Williams BR. Protein-tyrosine phosphatase Shp-1 is a negative regulator of IL-4- and IL-13-dependent signal transduction. *J Biol Chem* 1998; 273(51):33893–33896.

36. Klingmuller U, Lorenz U, Cantley LC, Neel BG, Lodish HF. Specific recruitment of SH-PTP1 to the erythropoietin receptor causes inactivation of JAK2 and termination of proliferative signals. *Cell* 1995; 80(5):729–738.

37. Bittorf T, Seiler J, Zhang Z, Jaster R, Brock J. SHP1 protein tyrosine phosphatase negatively modulates erythroid differentiation and suppression of apoptosis in J2E erythroleukemic cells. *Biol Chem* 1999; 380(10):1201–1209.

38. You M, Yu DH, Feng GS. Shp-2 tyrosine phosphatase functions as a negative regulator of the interferon-stimulated Jak/STAT pathway. *Mol Cell Biol* 1999; 19(3):2416–2424.

39. Lehmann U, Schmitz J, Weissenbach M, et al. SHP2 and SOCS3 contribute to Tyr-759-dependent attenuation of interleukin-6 signaling through gp130. *J Biol Chem* 2003; 278(1):661–671.

40. Carpenter LR, Farruggella TJ, Symes A, Karow ML, Yancopoulos GD, Stahl N. Enhancing leptin response by preventing SH2-containing phosphatase 2 interaction with Ob receptor. *Proc Natl Acad Sci USA* 1998; 95(11):6061–6066.

41. Li C, Friedman JM. Leptin receptor activation of SH2 domain containing protein tyrosine phosphatase 2 modulates Ob receptor signal transduction. *Proc Natl Acad Sci USA* 1999; 96(17):9677–9682.

42. Shultz LD, Schweitzer PA, Rajan TV, et al. Mutations at the murine motheaten locus are within the hematopoietic cell protein-tyrosine phosphatase (Hcph) gene. *Cell* 1993; 73(7):1445–1454.

43. Ohtani T, Ishihara K, Atsumi T, et al. Dissection of signaling cascades through gp130 in vivo: reciprocal roles for STAT3- and SHP2-mediated signals in immune responses. *Immunity* 2000; 12(1):95–105.

44. Atsumi T, Ishihara K, Kamimura D, et al. A point mutation of Tyr-759 in interleukin 6 family cytokine receptor subunit gp130 causes autoimmune arthritis. *J Exp Med* 2002; 196(7):979–990.

45. Irie-Sasaki J, Sasaki T, Matsumoto W, et al. CD45 is a JAK phosphatase and negatively regulates cytokine receptor signalling. *Nature* 2001; 409(6818):349–354.

46. Tanuma N, Nakamura K, Shima H, Kikuchi K. Protein-tyrosine phosphatase PTPepsilon C inhibits Jak-STAT signaling and differentiation induced by interleukin-6 and leukemia inhibitory factor in M1 leukemia cells. *J Biol Chem* 2000; 275(36):28216–28221.

47. Myers MP, Andersen JN, Cheng A, et al. TYK2 and JAK2 are substrates of protein-tyrosine phosphatase 1B. *J Biol Chem* 2001; 276(51):47771–47774.

48. Yasukawa H, Misawa H, Sakamoto H, et al. The JAK-binding protein JAB inhibits Janus tyrosine kinase activity through binding in the activation loop. *Embo J* 1999; 18(5):1309–1320.

49. Nicholson SE, Willson TA, Farley A, et al. Mutational analyses of the SOCS proteins suggest a dual domain requirement but distinct mechanisms for inhibition of LIF and IL-6 signal transduction. *Embo J* 1999; 18(2):375–385.

50. Sasaki A, Yasukawa H, Suzuki A, et al. Cytokine-inducible SH2 protein-3 (CIS3/SOCS3) inhibits Janus tyrosine kinase by binding through the N-terminal kinase inhibitory region as well as SH2 domain. *Genes Cells* 1999; 4(6):339–351.

51. Sasaki A, Yasukawa H, Shouda T, Kitamura T, Dikic I, Yoshimura A. CIS3/SOCS-3 suppresses erythropoietin (EPO) signaling by binding the EPO receptor and JAK2. *J Biol Chem* 2000; 275(38):29338–29347.

52. Coux O, Tanaka K, Goldberg AL. Structure and functions of the 20S and 26S proteasomes. *Annu Rev Biochem* 1996; 65:801–847.

53. Yu CL, Burakoff SJ. Involvement of proteasomes in regulating Jak-STAT pathways upon interleukin-2 stimulation. *J Biol Chem* 1997; 272(22):14017–14020.

54. Callus BA, Mathey-Prevot B. Interleukin-3-induced activation of the JAK/STAT pathway is prolonged by proteasome inhibitors. *Blood* 1998; 91(9):3182–3192.

55. Kamura T, Sato S, Haque D, et al. The Elongin BC complex interacts with the conserved SOCS-box motif present in members of the SOCS, ras, WD-40 repeat, and ankyrin repeat families. *Genes Dev* 1998; 12(24):3872–3881.

56. Zhang JG, Farley A, Nicholson SE, et al. The conserved SOCS box motif in suppressors of cytokine signaling binds to elongins B and C and may couple bound proteins to proteasomal degradation. *Proc Natl Acad Sci USA* 1999; 96(5):2071–2076.

57. Ungureanu D, Saharinen P, Junttila I, Hilton DJ, Silvennoinen O. Regulation of Jak2 through the ubiquitin-proteasome pathway involves phosphorylation of Jak2 on Y1007 and interaction with SOCS-1. *Mol Cell Biol* 2002; 22(10):3316–3326.

58. Frantsve J, Schwaller J, Sternberg DW, Kutok J, Gilliland DG. Socs-1 inhibits TEL-JAK-mediated transformation of hematopoietic cells through inhibition of JAK2 kinase activity and induction of proteasome-mediated degradation. *Mol Cell Biol* 2001; 21(10):3547–3557.

59. Kamizono S, Hanada T, Yasukawa H, et al. The SOCS box of SOCS-1 accelerates ubiquitin-dependent proteolysis of TEL-JAK. *J Biol Chem* 2001; 276(16):12530–12538.

60. Ali S, Nouhi Z, Chughtai N. SHP-2 regulates SOCS-1-mediated Janus kinase-2 ubiquitination/degradation downstream of the prolactin receptor. *J Biol Chem* 2003; 278(52): 52021–52031.

61. Wu C, Guan Q, Wang Y, Zhao ZJ, Zhou GW. SHP-1 suppresses cancer cell growth by promoting degradation of JAK kinases. *J Cell Biochem* 2003; 90(5):1026–1037.

62. Ward AC, Touw I, Yoshimura A. The Jak-Stat pathway in normal and perturbed hematopoiesis. *Blood* 2000; 95(1):19–29.

63. Russell SM, Tayebi N, Nakajima H, et al. Mutation of Jak3 in a patient with SCID: essential role of Jak3 in lymphoid development. *Science* 1995; 270(5237):797–800.

64. Macchi P, Villa A, Giliani S, Sacco MG, Frattini A, Porta F, et al. Mutations of Jak-3 gene in patients with autosomal severe combined immune deficiency (SCID). *Nature* 1995; 377(6544):65–68.

65. Candotti F, Oakes SA, Johnston JA, et al. Structural and functional basis for JAK3-deficient severe combined immunodeficiency. *Blood* 1997; 90(10):3996–4003.

66. Cacalano NA, Migone TS, Bazan F, et al. Autosomal SCID caused by a point mutation in the N-terminus of Jak3: mapping of the Jak3-receptor interaction domain. *Embo J* 1999; 18(6):1549–1558.

67. Li S, Labrecque S, Gauzzi MC, et al. The human papilloma virus (HPV)-18 E6 oncoprotein physically associates with Tyk2 and impairs Jak-STAT activation by interferon-alpha. *Oncogene* 1999; 18(42):5727–5737.

68. Miller DM, Rahill BM, Boss JM, et al. Human cytomegalovirus inhibits major histocompatibility complex class II expression by disruption of the Jak/Stat pathway. *J Exp Med* 1998; 187(5):675–683.

69. Luo H, Hanratty WP, Dearolf CR. An amino acid substitution in the Drosophila hopTum-l Jak kinase causes leukemia-like hematopoietic defects. *Embo J* 1995; 14(7):1412–1420.

70. Harrison DA, Binari R, Nahreini TS, Gilman M, Perrimon N. Activation of a *Drosophila* Janus kinase (JAK) causes hematopoietic neoplasia and developmental defects. *Embo J* 1995; 14(12):2857–2865.

71. Benekli M, Baer MR, Baumann H, Wetzler M. Signal transducer and activator of transcription proteins in leukemias. *Blood* 2003; 101(8):2940–2954.

72. Verma A, Kambhampati S, Parmar S, Platanias LC. Jak family of kinases in cancer. *Cancer Metastasis Rev* 2003; 22(4):423–434.

73. Cools J, Peeters P, Voet T, et al. Genomic organization of human JAK2 and mutation analysis of its JH2-domain in leukemia. *Cytogenet Cell Genet* 1999; 85(3–4):260–266.

74. Meydan N, Grunberger T, Dadi H, et al. Inhibition of acute lymphoblastic leukaemia by a Jak-2 inhibitor. *Nature* 1996; 379(6566):645–648.

75. Levitzki A. Protein tyrosine kinase inhibitors as novel therapeutic agents. *Pharmacol Ther* 1999; 82(2-3):231–239.

76. Miyamoto N, Sugita K, Goi K, et al. The JAK2 inhibitor AG490 predominantly abrogates the growth of human B-precursor leukemic cells with 11q23 translocation or Philadelphia chromosome. *Leukemia* 2001; 15(11):1758–1768.

77. Lacronique V, Boureux A, Valle VD, et al. A TEL-JAK fusion protein with constitutive kinase activity in human leukemia. *Science* 1997; 278(5341):1309–1312.

78. Peeters P, Raynaud SD, Cools J, et al. Fusion of TEL, the ETS-variant gene 6 (ETV6), to the receptor-associated kinase JAK2 as a result of t(9;12) in a lymphoid and t(9;15;12) in a myeloid leukemia. *Blood* 1997; 90(7):2535–2540.

79. Ho JM, Beattie BK, Squire JA, Frank DA, Barber DL. Fusion of the ets transcription factor TEL to Jak2 results in constitutive Jak-Stat signaling. *Blood* 1999; 93(12):4354–4364.

80. Schwaller J, Frantsve J, Aster J, et al. Transformation of hematopoietic cell lines to growth-factor independence and induction of a fatal myelo- and lymphoproliferative disease in mice by retrovirally transduced TEL/JAK2 fusion genes. *Embo J* 1998; 17(18): 5321–5333.

81. Carron C, Cormier F, Janin A, et al. TEL-JAK transgenic mice develop T-cell leukemia. *Blood* 2000; 95(12):3891–3899.

82. Goldman JM, Melo JV. Chronic myeloid leukemia—advances in biology and new approaches to treatment. *N Engl J Med* 2003; 349(15):1451–1464.

83. Joos S, Kupper M, Ohl S, et al. Genomic imbalances including amplification of the tyrosine kinase gene JAK2 in CD30+ Hodgkin cells. *Cancer Res* 2000; 60(3):549–552.

84. Joos S, Granzow M, Holtgreve-Grez H, et al. Hodgkin's lymphoma cell lines are characterized by frequent aberrations on chromosomes 2p and 9p including REL and JAK2. *Int J Cancer* 2003; 103(4):489–495.

85. Wessendorf S, Schwaenen C, Kohlhammer H, et al. Hidden gene amplifications in aggressive B-cell non-Hodgkin lymphomas detected by microarray-based comparative genomic hybridization. *Oncogene* 2003; 22(9):1425–1429.

86. Joos S, Otano-Joos MI, Ziegler S, et al. Primary mediastinal (thymic) B-cell lymphoma is characterized by gains of chromosomal material including 9p and amplification of the REL gene. *Blood* 1996; 87(4):1571–1578.

87. Bentz M, Barth TF, Bruderlein S, et al. Gain of chromosome arm 9p is characteristic of primary mediastinal B-cell lymphoma (MBL): comprehensive molecular cytogenetic analysis and presentation of a novel MBL cell line. *Genes Chromosomes Cancer* 2001; 30(4):393–401.

88. Savage KJ, Monti S, Kutok JL, et al. The molecular signature of mediastinal large B-cell lymphoma differs from that of other diffuse large B-cell lymphomas and shares features with classical Hodgkin lymphoma. *Blood* 2003; 102(12):3871–3879.

89. Zhang Q, Nowak I, Vonderheid EC, et al. Activation of Jak/STAT proteins involved in signal transduction pathway mediated by receptor for interleukin 2 in malignant T lymphocytes derived from cutaneous anaplastic large T-cell lymphoma and Sezary syndrome. *Proc Natl Acad Sci USA* 1996; 93(17):9148–9153.

90. Nielsen M, Kaltoft K, Nordahl M, et al. Constitutive activation of a slowly migrating isoform of Stat3 in mycosis fungoides: tyrphostin AG490 inhibits Stat3 activation and growth of mycosis fungoides tumor cell lines. *Proc Natl Acad Sci USA* 1997; 94(13): 6764–6769.

91. Migone TS, Lin JX, Cereseto A, et al. Constitutively activated Jak-STAT pathway in T cells transformed with HTLV-I. *Science* 1995; 269(5220):79–81.

92. Takemoto S, Mulloy JC, Cereseto A, et al. Proliferation of adult T cell leukemia/lymphoma cells is associated with the constitutive activation of JAK/STAT proteins. *Proc Natl Acad Sci USA* 1997; 94(25):13897–13902.

93. Liu RY, Fan C, Garcia R, Jove R, Zuckerman KS. Constitutive activation of the JAK2/STAT5 signal transduction pathway correlates with growth factor independence of megakaryocytic leukemic cell lines. *Blood* 1999; 93(7):2369–2379.

94. Nepomuceno RR, Snow AL, Robert Beatty P, Krams SM, Martinez OM. Constitutive activation of Jak/STAT proteins in Epstein-Barr virus—infected B-cell lines from patients with posttransplant lymphoproliferative disorder. *Transplantation* 2002; 74(3):396–402.

95. Zamo A, Chiarle R, Piva R, et al. Anaplastic lymphoma kinase (ALK) activates Stat3 and protects hematopoietic cells from cell death. *Oncogene* 2002; 21(7):1038–1047.

96. Amin HM, Medeiros LJ, Ma Y, et al. Inhibition of JAK3 induces apoptosis and decreases anaplastic lymphoma kinase activity in anaplastic large cell lymphoma. *Oncogene* 2003; 22(35):5399–5407.

97. Wu C, Sun M, Liu L, Zhou GW. The function of the protein tyrosine phosphatase SHP-1 in cancer. *Gene* 2003; 306:1–12.

98. Galm O, Yoshikawa H, Esteller M, Osieka R, Herman JG. SOCS-1, a negative regulator of cytokine signaling, is frequently silenced by methylation in multiple myeloma. *Blood* 2003; 101(7):2784–2788.

99. Chen CY, Tsay W, Tang JL, et al. SOCS1 methylation in patients with newly diagnosed acute myeloid leukemia. *Genes Chromosomes Cancer* 2003; 37(3):300–305.

100. Tartaglia M, Niemeyer CM, Fragale A, et al. Somatic mutations in PTPN11 in juvenile myelomonocytic leukemia, myelodysplastic syndromes and acute myeloid leukemia. *Nat Genet* 2003; 34(2):148–150.

101. Weber-Nordt RM, Egen C, Wehinger J, et al. Constitutive activation of STAT proteins in primary lymphoid and myeloid leukemia cells and in Epstein-Barr virus (EBV)-related lymphoma cell lines. *Blood* 1996; 88(3):809–816.

102. Gouilleux-Gruart V, Gouilleux F, Desaint C, et al. STAT-related transcription factors are constitutively activated in peripheral blood cells from acute leukemia patients. *Blood* 1996; 87(5):1692–1697.

103. Hayakawa F, Towatari M, Iida H, et al. Differential constitutive activation between STAT-related proteins and MAP kinase in primary acute myelogenous leukaemia. *Br J Haematol* 1998; 101(3):521–528.

104. Xia Z, Baer MR, Block AW, Baumann H, Wetzler M. Expression of signal transducers and activators of transcription proteins in acute myeloid leukemia blasts. *Cancer Res* 1998; 58(14):3173–3180.

105. Bowman T, Garcia R, Turkson J, Jove R. STATs in oncogenesis. *Oncogene* 2000; 19(21):2474–2488.

106. Epling-Burnette PK, Liu JH, Catlett-Falcone R, et al. Inhibition of STAT3 signaling leads to apoptosis of leukemic large granular lymphocytes and decreased Mcl-1 expression. *J Clin Invest* 2001; 107(3):351–362.

107. Aoki Y, Feldman GM, Tosato G. Inhibition of STAT3 signaling induces apoptosis and decreases survivin expression in primary effusion lymphoma. *Blood* 2003; 101(4):1535–1542.

108. Klein B, Zhang XG, Lu ZY, Bataille R. Interleukin-6 in human multiple myeloma. *Blood* 1995; 85(4):863–872.

109. Hallek M, Bergsagel PL, Anderson KC. Multiple myeloma: increasing evidence for a multistep transformation process. *Blood* 1998; 91(1):3–21.

110. Catlett-Falcone R, Landowski TH, Oshiro MM, et al. Constitutive activation of Stat3 signaling confers resistance to apoptosis in human U266 myeloma cells. *Immunity* 1999; 10(1):105–115.

111. Quintanilla-Martinez L, Kremer M, Specht K, et al. Analysis of signal transducer and activator of transcription 3 (Stat 3) pathway in multiple myeloma: Stat 3 activation and cyclin D1 dysregulation are mutually exclusive events. *Am J Pathol* 2003; 162(5):1449–1461.

112. Rawat R, Rainey GJ, Thompson CD, Frazier-Jessen MR, Brown RT, Nordan RP. Constitutive activation of STAT3 is associated with the acquisition of an interleukin 6-independent phenotype by murine plasmacytomas and hybridomas. *Blood* 2000; 96(10):3514–3521.

113. Stahl N, Boulton TG, Farruggella T, et al. Association and activation of Jak-Tyk kinases by CNTF-LIF-OSM-IL-6 beta receptor components. *Science* 1994; 263(5143):92–95.

114. Kishimoto T, Akira S, Narazaki M, Taga T. Interleukin-6 family of cytokines and gp130. *Blood* 1995; 86(4):1243–1254.

115. Berger LC, Hawley TS, Lust JA, Goldman SJ, Hawley RG. Tyrosine phosphorylation of JAK-TYK kinases in malignant plasma cell lines growth-stimulated by interleukins 6 and 11. *Biochem Biophys Res Commun* 1994; 202(1):596–605.

116. Ogata A, Chauhan D, Teoh G, et al. IL-6 triggers cell growth via the Ras-dependent mitogen-activated protein kinase cascade. *J Immunol* 1997; 159(5):2212–2221.

117. Hideshima T, Chauhan D, Teoh G, et al. Characterization of signaling cascades triggered by human interleukin-6 versus Kaposi's sarcoma-associated herpes virus-encoded viral interleukin 6. *Clin Cancer Res* 2000; 6(3):1180–1189.

118. De Vos J, Jourdan M, Tarte K, Jasmin C, Klein B. JAK2 tyrosine kinase inhibitor tyrphostin AG490 downregulates the mitogen-activated protein kinase (MAPK) and signal transducer

and activator of transcription (STAT) pathways and induces apoptosis in myeloma cells. *Br J Haematol* 2000; 109(4):823–828.

119. Kopantzev Y, Heller M, Swaminathan N, Rudikoff S. IL-6 mediated activation of STAT3 bypasses Janus kinases in terminally differentiated B lineage cells. *Oncogene* 2002; 21(44):6791–6800.

120. Brenne AT, Baade Ro T, Waage A, Sundan A, Borset M, Hjorth-Hansen H. Interleukin-21 is a growth and survival factor for human myeloma cells. *Blood* 2002; 99(10):3756–3762.

121. Walters DK, Jelinek DF. A role for Janus kinases in crosstalk between ErbB3 and the interferon-alpha signaling complex in myeloma cells. *Oncogene* 2004; 23(6):1197–1205.

122. Hirano T, Ishihara K, Hibi M. Roles of STAT3 in mediating the cell growth, differentiation and survival signals relayed through the IL-6 family of cytokine receptors. *Oncogene* 2000; 19(21):2548–2556.

123. Puthier D, Bataille R, Amiot M. IL-6 up-regulates mcl-1 in human myeloma cells through JAK / STAT rather than ras / MAP kinase pathway. *Eur J Immunol* 1999; 29(12): 3945–3950.

124. Oshiro MM, Landowski TH, Catlett-Falcone R, et al. Inhibition of JAK kinase activity enhances Fas-mediated apoptosis but reduces cytotoxic activity of topoisomerase II inhibitors in U266 myeloma cells. *Clin Cancer Res* 2001; 7(12):4262–4271.

125. Alas S, Bonavida B. Inhibition of constitutive STAT3 activity sensitizes resistant non-Hodgkin's lymphoma and multiple myeloma to chemotherapeutic drug–mediated apoptosis. *Clin Cancer Res* 2003; 9(1):316–326.

126. Mitsiades CS, Mitsiades N, Poulaki V, et al. Activation of NF-kappaB and upregulation of intracellular anti-apoptotic proteins via the IGF-1/Akt signaling in human multiple myeloma cells: therapeutic implications. *Oncogene* 2002; 21(37):5673–5683.

127. Mahboubi K, Li F, Plescia J, et al. Interleukin-11 up-regulates survivin expression in endothelial cells through a signal transducer and activator of transcription-3 pathway. *Lab Invest* 2001; 81(3):327–334.

128. Chauhan D, Auclair D, Robinson EK, et al. Identification of genes regulated by dexamethasone in multiple myeloma cells using oligonucleotide arrays. *Oncogene* 2002; 21(9):1346–1358.

129. Brocke-Heidrich K, Kretzschmar AK, Pfeifer G, et al. Interleukin-6-dependent gene expression profiles in multiple myeloma INA-6 cells reveal a Bcl-2 family-independent survival pathway closely associated with Stat3 activation. *Blood* 2004; 103(1):242–251.

130. Hideshima T, Anderson KC. Molecular mechanisms of novel therapeutic approaches for multiple myeloma. *Nat Rev Cancer* 2002; 2(12):927–937.

131. Schaper F, Gendo C, Eck M, et al. Activation of the protein tyrosine phosphatase SHP2 via the interleukin-6 signal transducing receptor protein gp130 requires tyrosine kinase Jak1 and limits acute-phase protein expression. *Biochem J* 1998; 335(Pt 3):557–565.

132. Chauhan D, Pandey P, Hideshima T, et al. SHP2 mediates the protective effect of interleukin-6 against dexamethasone-induced apoptosis in multiple myeloma cells. *J Biol Chem* 2000; 275(36):27845–27850.

133. Tu Y, Gardner A, Lichtenstein A. The phosphatidylinositol 3-kinase/AKT kinase pathway in multiple myeloma plasma cells: roles in cytokine-dependent survival and proliferative responses. *Cancer Res* 2000; 60(23):6763–6770.

134. Hideshima T, Nakamura N, Chauhan D, Anderson KC. Biologic sequelae of interleukin-6 induced PI3-K/Akt signaling in multiple myeloma. *Oncogene* 2001; 20(42):5991–6000.

135. Shi Y, Hsu JH, Hu L, Gera J, Lichtenstein A. Signal pathways involved in activation of p70S6K and phosphorylation of 4E-BP1 following exposure of multiple myeloma tumor cells to interleukin-6. *J Biol Chem* 2002; 277(18):15712–15720.

136. Hsu JH, Shi Y, Hu L, Fisher M, Franke TF, Lichtenstein A. Role of the AKT kinase in expansion of multiple myeloma clones: effects on cytokine-dependent proliferative and survival responses. *Oncogene* 2002; 21(9):1391–1400.

137. Liu Y, Rohrschneider LR. The gift of Gab. *FEBS Lett* 2002; 515(1–3):1–7.

138. Levitzki A, Gazit A. Tyrosine kinase inhibition: an approach to drug development. *Science* 1995; 267(5205):1782–1788.

139. Blaskovich MA, Sun J, Cantor A, Turkson J, Jove R, Sebti SM. Discovery of JSI-124 (cucurbitacin I), a selective Janus kinase/signal transducer and activator of transcription 3 signaling pathway inhibitor with potent antitumor activity against human and murine cancer cells in mice. *Cancer Res* 2003; 63(6):1270–1279.

140. Sharfe N, Dadi HK, Roifman CM. JAK3 protein tyrosine kinase mediates interleukin-7-induced activation of phosphatidylinositol-3' kinase. *Blood* 1995; 86(6):2077–2085.

141. Wang LH, Kirken RA, Erwin RA, Yu CR, Farrar WL. JAK3, STAT, and MAPK signaling pathways as novel molecular targets for the tyrphostin AG-490 regulation of IL-2-mediated T cell response. *J Immunol* 1999; 162(7):3897–3904.

142. Kirken RA, Erwin-Cohen R, Behbod F, Wang M, Stepkowski SM, Kahan BD. Tyrphostin AG490 selectively inhibits activation of the JAK3/STAT5/MAPK pathway and rejection of rat heart allografts. *Transplant Proc* 2001; 33(1–2):95.

143. Gazit A, Osherov N, Posner I, et al. Tyrphostins. 2. Heterocyclic and alpha-substituted benzylidenemalononitrile tyrphostins as potent inhibitors of EGF receptor and ErbB2/neu tyrosine kinases. *J Med Chem* 1991; 34(6):1896–1907.

144. Simon HU, Yousefi S, Dibbert B, Levi-Schaffer F, Blaser K. Anti-apoptotic signals of granulocyte-macrophage colony-stimulating factor are transduced via Jak2 tyrosine kinase in eosinophils. *Eur J Immunol* 1997; 27(12):3536–3539.

145. Kirken RA, Erwin RA, Taub D, et al. Tyrphostin AG-490 inhibits cytokine-mediated JAK3/STAT5a/b signal transduction and cellular proliferation of antigen-activated human T cells. *J Leukoc Biol* 1999; 65(6):891–899.

146. Burger R, Bakker F, Guenther A, et al. Functional significance of novel neurotrophin-1/B cell-stimulating factor-3 (cardiotrophin-like cytokine) for human myeloma cell growth and survival. *Br J Haematol* 2003; 123(5):869–878.

147. Burdelya L, Catlett-Falcone R, Levitzki A, et al. Combination therapy with AG-490 and interleukin 12 achieves greater antitumor effects than either agent alone. *Mol Cancer Ther* 2002; 1(11):893–899.

148. Nielsen M, Kaestel CG, Eriksen KW, et al. Inhibition of constitutively activated Stat3 correlates with altered Bcl-2/Bax expression and induction of apoptosis in mycosis fungoides tumor cells. *Leukemia* 1999; 13(5):735–738.

149. Sun X, Layton JE, Elefanty A, Lieschke GJ. Comparison of effects of the tyrosine kinase inhibitors AG957, AG490, and STI571 on Bcr-Abl-expressing cells, demonstrating synergy between AG490 and STI571. *Blood* 2001; 97(7):2008–2015.

150. Marley SB, Davidson RJ, Goldman JM, Gordon MY. Effects of combinations of therapeutic agents on the proliferation of progenitor cells in chronic myeloid leukaemia. *Br J Haematol* 2002; 116(1):162–165.

151. Donato NJ LG, Wu JY, Estrov Z, Ford RJ, Levitzki A, Talpaz M. Targeting Jak2-dependent signaling pathways for therapeutic intervention in multiple myeloma. *Blood* 2002; 100(11):814a.

152. Kong L TM, Priebe W, Levitzki A, Aggarwal B, Fokt I, Szymanski S, Donato NJ. Novel tyrphostin analogues inhibit Stat3 activation and induce apoptosis in multiple hematological malignancies. *Blood* 2003; 102(11):(#4534).

153. Fiorucci G, Percario ZA, Marcolin C, Coccia EM, Affabris E, Romeo G. Inhibition of protein phosphorylation modulates expression of the Jak family protein tyrosine kinases. *J Virol* 1995; 69(9):5833–5837.

154. Elder RT, Xu X, Williams JW, Gong H, Finnegan A, Chong AS. The immunosuppressive metabolite of leflunomide, A77 1726, affects murine T cells through two biochemical mechanisms. *J Immunol* 1997; 159(1):22–27.

155. Siemasko K, Chong AS, Jack HM, Gong H, Williams JW, Finnegan A. Inhibition of JAK3 and STAT6 tyrosine phosphorylation by the immunosuppressive drug leflunomide leads to a block in IgG1 production. *J Immunol* 1998; 160(4):1581–1588.

156. Wasik MA, Nowak I, Zhang Q, Shaw LM. Suppression of proliferation and phosphorylation of Jak3 and STAT5 in malignant T-cell lymphoma cells by derivatives of octylamino-undecyl-dimethylxanthine. *Leuk Lymphoma* 1998; 28(5-6):551–560.

157. Ferrigni NR, McLaughlin JL, Powell RG, Smith CR, Jr. Use of potato disc and brine shrimp bioassays to detect activity and isolate piceatannol as the antileukemic principle from the seeds of Euphorbia lagascae. *J Nat Prod* 1984; 47(2):347–352.

158. Su L, David M. Distinct mechanisms of STAT phosphorylation via the interferon-alpha/beta receptor. Selective inhibition of STAT3 and STAT5 by piceatannol. *J Biol Chem* 2000; 275(17):12661–12666.

159. Aggarwal BB, Kumar A, Bharti AC. Anticancer potential of curcumin: preclinical and clinical studies. *Anticancer Res* 2003; 23(1A):363–398.

160. Natarajan C, Bright JJ. Curcumin inhibits experimental allergic encephalomyelitis by blocking IL-12 signaling through Janus kinase-STAT pathway in T lymphocytes. *J Immunol* 2002; 168(12):6506–6513.

161. Bharti AC, Donato N, Aggarwal BB. Curcumin (Diferuloylmethane) inhibits constitutive and IL-6-inducible STAT3 phosphorylation in human multiple myeloma cells. *J Immunol* 2003; 171(7):3863–3871.

162. Goodman PA, Niehoff LB, Uckun FM. Role of tyrosine kinases in induction of the c-jun proto-oncogene in irradiated B-lineage lymphoid cells. *J Biol Chem* 1998; 273(28):17742–17748.

163. Sudbeck EA, Liu XP, Narla RK, et al. Structure-based design of specific inhibitors of Janus kinase 3 as apoptosis-inducing antileukemic agents. *Clin Cancer Res* 1999; 5(6):1569–1582.

164. Uckun FM, Ek O, Liu XP, Chen CL. In vivo toxicity and pharmacokinetic features of the janus kinase 3 inhibitor WHI-P131 [4-(4'hydroxyphenyl)-amino-6,7- dimethoxyquinazo-line. *Clin Cancer Res* 1999; 5(10):2954–2962.

165. Brown GR, Bamford AM, Bowyer J, et al. Naphthyl ketones: a new class of Janus kinase 3 inhibitors. *Bioorg Med Chem Lett* 2000; 10(6):575–579.

166. Adams C, Aldous DJ, Amendola S, et al. Mapping the kinase domain of Janus Kinase 3. *Bioorg Med Chem Lett* 2003; 13(18):3105–3110.

167. Mortellaro A, Songia S, Gnocchi P, et al. New immunosuppressive drug PNU156804 blocks IL-2-dependent proliferation and NF-kappa B and AP-1 activation. *J Immunol* 1999; 162(12):7102–7109.

168. Stepkowski SM, Erwin-Cohen RA, Behbod F, et al. Selective inhibitor of Janus tyrosine kinase 3, PNU156804, prolongs allograft survival and acts synergistically with cyclosporine but additively with rapamycin. *Blood* 2002; 99(2):680–689.

169. Stepkowski SM, Nagy ZS, Wang ME, et al. PNU156804 inhibits Jak3 tyrosine kinase and rat heart allograft rejection. *Transplant Proc* 2001; 33(7–8):3272,3273.

170. Wang M, Kirken R, Behbod F, Erwin-Cohen R, Stepkowski SM, Kahan BD. Inhibition of Jak3 tyrosine kinase by PNU156804 blocks rat heart allograft rejection. *Transplant Proc* 2001; 33(1–2):201.

171. Changelian PS, Flanagan ME, Ball DJ, et al. Prevention of organ allograft rejection by a specific Janus kinase 3 inhibitor. *Science* 2003; 302(5646):875–878.

172. Kudlacz E, Perry B, Sawyer P, et al. The novel JAK-3 inhibitor CP-690550 is a potent immunosuppressive agent in various murine models. *Am J Transplant* 2004; 4(1):51–57.

173. Grunberger T, Demin P, Rounova O, et al. Inhibition of acute lymphoblastic and myeloid leukemias by a novel kinase inhibitor. *Blood* 2003; 102(12):4153–4158.

174. Thompson JE, Cubbon RM, Cummings RT, et al. Photochemical preparation of a pyridone containing tetracycle: a Jak protein kinase inhibitor. *Bioorg Med Chem Lett* 2002; 12(8):1219–1223.

175. Levitzki A. Tyrosine kinases as targets for cancer therapy. *Eur J Cancer* 2002; 38(Suppl 5): S11–S18.
176. Ravandi F, Talpaz M, Estrov Z. Modulation of cellular signaling pathways: prospects for targeted therapy in hematological malignancies. *Clin Cancer Res* 2003; 9(2):535–550.
177. Ito T, May WS. Drug development train gathering steam. *Nat Med* 1996; 2(4):403–404.
178. Lindauer K, Loerting T, Liedl KR, Kroemer RT. Prediction of the structure of human Janus kinase 2 (JAK2) comprising the two carboxy-terminal domains reveals a mechanism for autoregulation. *Protein Eng* 2001; 14(1):27–37.
179. Giordanetto F, Kroemer RT. Prediction of the structure of human Janus kinase 2 (JAK2) comprising JAK homology domains 1 through 7. *Protein Eng* 2002; 15(9):727–737.
180. Seidel HM, Lamb P, Rosen J. Pharmaceutical intervention in the JAK/STAT signaling pathway. *Oncogene* 2000; 19(21):2645–2656.

7 Glivec® (Gleevec®, Imatinib, STI571)

A Targeted Therapy for Chronic Myelogenous Leukemia

Elisabeth Buchdunger, PhD and Renaud Capdeville, MD

CONTENTS

CONCEPT AND TARGET SELECTION: BCR-ABL
MEDICINAL CHEMISTRY: DEVELOPMENT OF AN ABL
    TYROSINE KINASE INHIBITOR
PHARMACOLOGICAL PROFILE OF GLIVEC
CLINICAL DEVELOPMENT IN CML
MANAGING CLINICAL RESISTANCE TO GLIVEC
CONCLUSION
REFERENCES

1. CONCEPT AND TARGET SELECTION: BCR-ABL

Chronic myelogenous leukemia (CML) is a clonal hematological disorder characterized by a reciprocal translocation between chromosomes 9 and 22 (1,2) known as the Philadelphia (Ph) chromosome. The molecular consequence of this interchromosomal exchange is the creation of the *bcr-abl* gene coding for a protein with elevated tyrosine kinase activity. The demonstration that the expression of Bcr-Abl is both necessary and sufficient to cause a CML-like syndrome in murine bone marrow transplantation models (3–5) and the finding that the tyrosine kinase activity of Bcr-Abl is crucial for its transforming activity (6), has established the enzymatic activity of this deregulated protein as an attractive drug target addressing Bcr-Abl-positive leukemias. For the first time,

From: *Cancer Drug Discovery and Development:*
Protein Tyrosine Kinases: From Inhibitors to Useful Drugs
Edited by: D. Fabbro and F. McCormick © Humana Press Inc., Totowa, NJ

a drug target was identified that very clearly differed in its activity between normal and leukemic cells. It was conceivable that this enzyme could be approached with classical tools of pharmacology since its activity, the transfer of phosphate from adenosine triphosphate (ATP) to tyrosine residues of protein substrates, could clearly be described and measured in biochemical as well as cellular assays. Furthermore, cell lines were available that were derived from human leukemic cells that had the same chromosomal abnormality. Such cell lines were instrumental for in vitro and animal studies that laid the groundwork for the clinical trials. So, the essential tools were assembled to go forward aiming at identifying potent and selective inhibitors of the Abl tyrosine kinase.

2. MEDICINAL CHEMISTRY: DEVELOPMENT OF AN ABL TYROSINE KINASE INHIBITOR

The starting point for the medicinal chemistry project that led to the synthesis of Glivec was the identification of a lead compound from a screen for inhibitors of protein kinase C (PKC). This compound, a phenyl-amino pyrimidine derivative, had very promising "lead-like" properties *(7,8)* and had a high potential for diversity, allowing simple chemistry to be applied to produce compounds with more potent activity or selectivity. A high cellular PKC inhibitory activity was obtained with derivatives bearing a 3'-pyridyl group at the 3 position of the pyrimidine (Fig. 1A). During the optimization of this structural class, it was observed that the presence of an amide group on the phenyl ring provided inhibitory activity against tyrosine kinases, such as the Bcr-Abl kinase (Fig. 1B). At this point a key observation from analysis of structure–activity relationship (SAR) was that a substitution at position 6 of the diamino phenyl ring abolished PKC inhibitory activity completely. Indeed, the introduction of a simple "flag-methyl" led to loss of activity against PKC, whereas the activity against protein-tyrosin kinases was retained or even enhanced (Fig. 1C). However, the first series of selective inhibitors originally prepared showed poor oral bioavailability and low solubility in water. The attachment of a highly polar side chain (an *N*-methylpiperazine) was found to dramatically improve both solubility and oral bioavailability. To avoid the mutagenic potential of aniline moieties, a spacer was introduced between the phenyl ring and the nitrogen atom. The best compound from this series was the methyl piperazine derivative originally named STI571 (imatinib, now known as Glivec® or Gleevec®), which was selected as the most promising candidate for clinical development (Fig. 1D) *(9,10)*.

Docking studies *(11)* and X-ray crystallography *(12,13)* showed that binding of Glivec occurs at the ATP binding site. Analysis of the crystal structure showed that Glivec inhibits the Abl kinase by binding with high specificity to an inactive form of the kinase. The need for the kinase to adopt this unusual

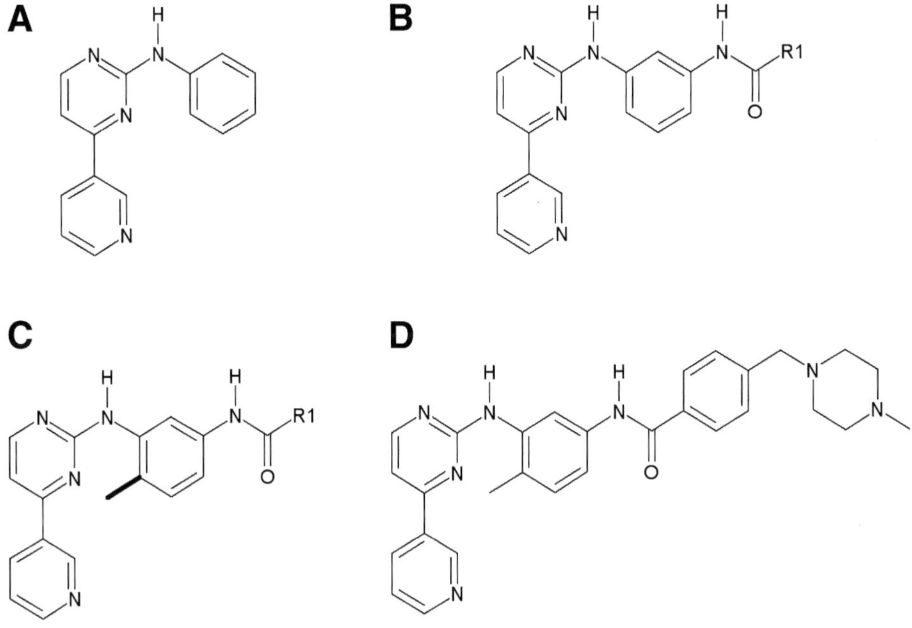

A

B

C

D

Fig. 1. Summary of the chemical optimization.

conformation favoring binding may contribute to the high selectivity of the compound. Unexpectedly, these analyses indicated that the *N*-methylpiperazine group (added to increase drug solubility) was also interacting strongly with Abl via hydrogen bonds to the backbone carbonyl of Ile360 and His 361.

3. PHARMACOLOGICAL PROFILE OF GLIVEC

3.1. In Vitro/Cellular Activity

In vitro studies using purified enzymes expressed as bacterial fusion proteins or immunoprecipitations of intact proteins showed that Glivec potently inhibited all of the *Abl* tyrosine kinases, including cellular Abl (c-Abl), v-Abl, the oncogenic form contained in the Abelson murine leukemia virus, and Bcr-Abl (Table 1) (14–16). STI571 is an ATP-competitive inhibitor of Abl with a K_i value of 85 nm (17). Extended profiling against various serine/threonine and tyrosine kinases revealed that the compound was also an inhibitor of the platelet-derived growth factor (PDGF) receptor and c-KIT tyrosine kinases, devoid of activity against most other kinases. The selective inhibitory activity of Glivec was also demonstrated at the cellular level (18–20; Table 1). The compound inhibited the constitutively activated fusion forms of Abl, such as the p210Bcr-Abl (14), p185Bcr-Abl (15,16), and Tel-Abl (15) tyrosine kinases with IC_{50} values between 0.1 and 0.35 μ*M*. The inhibition of autophosphorylation of

Table 1
Inhibition of Protein Kinases by Glivec

Enzyme	In vitro substrate phosphorylation IC_{50} [μM]	Cellular tyrosine phosphorylation IC_{50} [μM]
c-Abl	**0.17 ± 0.023; 0.025[a]**	
p210Bcr-Abl	**0.025[a]**	**0.25**
p185Bcr-Abl	**0.025[a]**	**0.25**
TEL-Abl		**0.35**
PDGFRβ	**0.87 ± 0.012**	**0.1**
TEL-PDGF-Rβ		**0.15**
c-Kit	**0.56 ± 0.092**	**0.1**
FGF-R1	>10	
c-Fms and v-Fms		>10
VEGFR1 (Flt-1)	>10	
VEGFR2 (Kdr)	>10	>10
Flt-3	>10	>10
Flt-4	5.7 ± 1.1	
EGFR (HER1)	>100	>100
ErbB2 (HER2)	>10	>10
ErbB4 (HER4)	>10	
IGF-IR	>10	>100
Insulin receptor	>10	>100
c-Met	>10	
Tie-2 (Tek)	>10	
Jak-2	>100[a]	>100
c-Fgr	>100	
Lck	9.0	
c-Lyn	>100	
Syk (TPK-IIB)	>100	
c-Src	>10	
Akt (PKB)	>10	
Cdk1/cyclin B	>10	
Jnk2	>10	
p38 MAPK	>10	
PDK1	>10	
PKA	>10	
PKCα, β1, β2, γ, δ, ε, ζ, or η	>10	
PPK	>10	
Protein kinase CK-1, CK-2	>10	
c-Raf-1	0.97 ± 0.16	>10

[a]IC_{50} was determined in immunocomplex assays. Data represent the mean ± SEM drug concentrations causing a 50% reduction in kinase activity (IC_{50} value; μM). PDGFR, platelet-derived growth factor receptor; FGFR1, fibroblast growth factor receptor 1; VEGFR, vascular endothelial

Bcr-Abl was closely related to the antiproliferative activity of Glivec. Incubation with submicromolar concentrations of Glivec selectively induced apoptosis in Bcr-Abl-positive cell lines and also induced cell killing in primary leukemia cells from Ph chromosome-positive CML and acute lymphoblastic leukemia patients, whereas Ph chromosome-negative cells were not affected *(14,16,21–24)*. The IC$_{50}$ values for inhibition of the KU812 and MC3 Bcr-Abl-positive CML blast crisis cell lines were 0.1–0.3 µ*M (24)*. Selective inhibition of CML colony formation by Glivec has been demonstrated. At concentrations of 1 µ*M*, the compound selectively inhibited colony formation from peripheral blood and bone marrow from Ph-positive CML patients, with a 92–98% decrease in Bcr-Abl-positive colonies but little effect on normal hematopoiesis *(14,22)*. The findings were confirmed by assessing the effects of Glivec on proliferation of peripheral blood progenitors under stroma-dependent long-term culture (LTC) conditions *(25)*.

Fundamental phenotypic features in Bcr-Abl-positive cells involve resistance to apoptosis, enhanced proliferation, and altered adhesion properties. The impact of Glivec on some known downstream signaling molecules of Bcr-Abl has been examined. A link between constitutive activation of signal transducer and activator of transcription (STAT) 5 and enhanced viability of Bcr-Abl-transformed cells has been demonstrated *(26,27)*. Glivec had a profound inhibitory effect on STAT5 activation in vitro and in vivo *(26–28)*. Furthermore, inhibition of the Bcr-Abl kinase activity by Glivec in Bcr-Abl-expressing cell lines and fresh leukemic cells from CML patients induced apoptosis by suppressing the capacity of STAT5 to activate the expression of the antiapoptotic protein Bcl-x$_L$ *(27)*. The adapter molecule CrkL is a prominent target of Bcr-Abl, and its tyrosine phosphorylation has been a useful marker of Bcr-Abl kinase activity *(29)*. As expected, a decrease in tyrosine phosphorylation of CrkL has been observed in Glivec-treated cell lines and has also served as an indicator of Bcr-Abl kinase activity in patients (*see* Section 4).

There is increasing evidence that cell cycle regulation is disturbed in Bcr-Abl-positive cells; however, the underlying molecular mechanisms are poorly understood. Recently, Bcr-Abl has been shown to promote cell cycle progression and activate cyclin-dependent kinases by interfering with the regulation of the cell cycle inhibitory protein p27 *(30)*. Glivec prevented downregulation of p27 levels in Bcr-Abl-expressing cells *(30,31)*.

The effects of Glivec on cytoskeletal changes and adhesion have been investigated using Bcr-Abl-transfected fibroblasts *(32)*. Glivec was shown to restore normal architecture and to increase adhesion in this model of Bcr-Abl expression.

Table 1 (*Continued*) growth factor receptor; EGFR, epidermal growth factor receptor; ErbB, oncogene B of Avian Erythroblastosis virus; HER, human EGF receptor family; IGF-IR, insulin-like growth factor receptor I; TPK, tyrosine-protein kinase; PKB, protein kinase B; Jnk2, c-Jun N-terminal kinase 2; MAPK, mitogen-activated protein kinase; PDK1, 3-phosphoinositide-dependent protein kinase-1; PKA, cAMP-dependent protein kinase; PKC, protein kinase C; PPK, phosphorylase kinase; CK, casein kinase.

3.2. Activity in Animal Models

The antiproliferative activity of Glivec has been confirmed in animal models. Once-daily intraperitoneal treatment with 2.5 to 50 mg/kg of Glivec starting 1 wk after injection of Bcr-Abl-transformed 32D cells in syngeneic mice caused dose-dependent inhibition of tumor growth *(14)*. In contrast, Glivec showed no anti-tumor activity against tumors derived from v-src-transformed 32D cells, in line with the lack of inhibition of Src kinase by the compound. The in vivo activity of Glivec against Bcr-Abl-driven tumors was confirmed using the KU812 cell line derived from a CML patient in blast crisis injected into nude mice. Oral treatment with 160 mg/kg daily in three divided doses for 11 consecutive days was associated with continuous blockage of $p210^{bcr-abl}$ tyrosine phosphorylation and resulted in tumor-free survival of the animals *(24)*. These data suggested that continuous exposure to Glivec would be important for optimal anti-leukemic effects. The anti-tumor effect of Glivec was specific for Bcr-Abl-expressing cells as no growth inhibition occurred in mice given injections of U937, a Bcr-Abl-negative myeloid cell line. Glivec has also been shown to have oral activity in a murine model of CML based on retro-viral $p210^{bcr-abl}$ transduction of transplanted bone marrow, where survival of animals was significantly prolonged, together with a marked improvement in peripheral white blood counts and splenomegaly *(28)*.

4. CLINICAL DEVELOPMENT IN CML

Clinically, CML is a chronic disease evolving through three successive stages from the chronic phase to the end stage of blast crisis that resembles acute leukemia. Overall, the median survival of patients with newly diagnosed CML is approx 5–6 yr with interferon (IFN)-based treatment regimen. The first trial with Glivec was a phase I study in patients with chronic phase and subsequently also with blast phase CML. In this trial, patients were treated at doses ranging from 25 to 1000 mg daily, and no maximal tolerated dose was identified despite a trend for a higher frequency of grade 3–4 adverse events at doses of 750 mg or higher. On the other hand, a clear dose–response relationship with respect to efficacy was described in patients with chronic-phase CML. At doses of 300 mg or higher, 98% of the patients achieved a complete hematological response, and trough serum levels were above the concentrations required for in vitro activity *(33,34)* (Fig. 2). In addition, effective inhibition of the Bcr-Abl kinase was documented in patient samples by the inhibition of the phosphorylation status of the downstream target CrkL *(34)*. From this study, taking into account a large interpatient variability in pharmacokinetic parameters, doses ranging from 400 mg (for chronic-phase patients) to 600 mg (for advanced-phase CML) were recommended for subsequent studies.

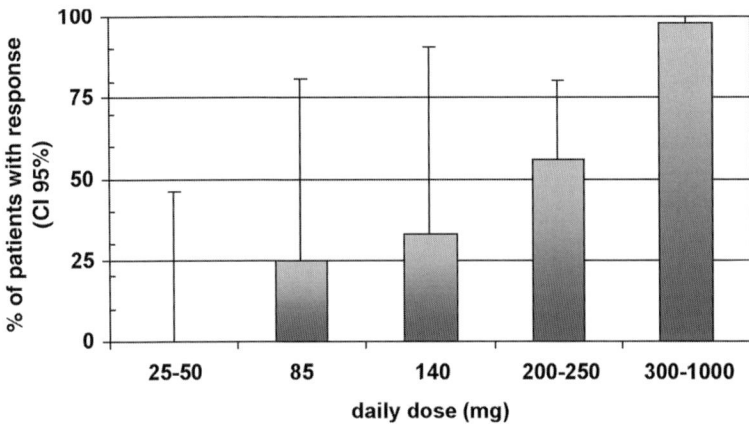

Fig. 2. Dose–response relationship of Glivec in chronic-phase chronic myeloid leukemia (CML) (phase I study): percentage of patients with a hematological response as a function of dose cohort. Hematological response was defined as complete hematological response (CHR) to be confirmed after at least 4 wk. CHR was defined as WBC $<10 \times 10^9/L$, platelet $<450 \times 10^9/L$, myelocytes + metamyelocytes $<5\%$ in blood, no blasts and promyelocytes in blood, basophils $<20\%$, no extramedullary involvement.

The clinical development program then focused on two main objectives: to better define the role of Glivec in the management of various stages of CML by characterizing efficacy and safety in large phase II and III trials; and to explore the role and signaling activity of the KIT and PDGF receptors in other malignancies as well as the potential therapeutic use of Glivec in these indications. Three large multinational studies have been performed in 532 patients with late-chronic-phase CML failing prior IFN therapy *(35)*, in 235 patients with accelerated-phase CML *(36)*, and in 260 patients with myeloid blast crisis *(37)*. Treatment was given at a dose of 400 mg in the chronic-phase trial and 400–600 mg in the two other studies. The results of these three studies indicated that the rate of both hematological and cytogenetic response increased as the treatment was started earlier in the course of the disease (Fig. 3). Importantly, the achievement of a hematological and/or cytogenic response was associated with an improved survival and progression-free survival *(35–37)*. In the chronic-phase study where patients started treatment within a median of 32 mo after their initial diagnosis, the estimated probability of being free of progression at 18 mo was 89.2% *(35)*. The most frequently reported adverse events were mild nausea, vomiting, edema, and muscle cramps. However, rare but serious adverse events such as liver toxicity or fluid retention syndromes were also reported. Neutropenias and thrombopenias were more common in patients with advanced disease, suggesting that hematological toxicity may be related more to an underlying compromised bone marrow reserve rather than a toxicity of the drug itself through inhibition of c-KIT-driven hematopoiesis.

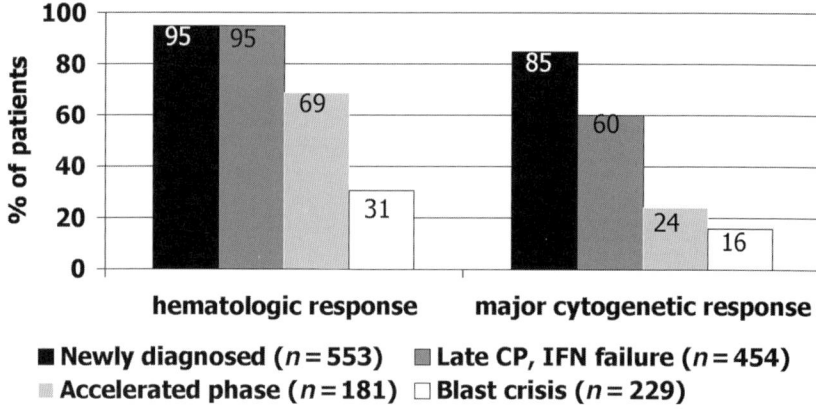

Fig. 3. Hematological and cytogenetic response in CML. In all studies, results are expressed as the percentage of responding patients among the patients for whom the diagnosis of the correct phase of chronic myeloid leukemia (CML) was confirmed on a central review of data. A major cytogenetic response combines both complete (0% Ph + metaphases) and partial responses (1–35%). Hematologic response was defined as complete hematological response (CHR) in the chronic phase study, and as either a CHR, a marrow response, or a return to chronic (RTC) phase in the advanced-phase studies, all to be confirmed after at least 4 wk. In the chronic phase study, CHR was defined as WBC $<10 \times 10^9$/L, platelet $<450 \times 10^9$/L, myelocytes + metamyelocytes $<5\%$ in blood, no blasts and promyelocytes in blood, basophils $<20\%$, no extramedullary involvement. In advanced-phase studies, CHR was defined as neutrophils $\geq 1.5 \times 10^9$/L, platelets $\geq 100 \times 10^9$/L, no blood blasts, marrow blasts $<5\%$ and no extramedullary disease. A marrow response was defined with the same criteria as for CHR but with neutrophils $\geq 1 \times 10^9$/L and platelets $\geq 20 \times 10^9$/L. An RTC phase was defined as $<15\%$ blasts in marrow and blood, $<30\%$ blasts + promyelocytes in marrow and blood, $<20\%$ basophils in blood, and no extramedullary disease.

   The activity of Glivec for patients with newly diagnosed CML has been investigated in a large randomized phase III study comparing first-line therapy with Glivec against standard interferon in combination with low-dose cytarabine. This study, known as the "IRIS" study (International Randomized study of Interferon vs STI571), has enrolled 1106 patients and is currently ongoing. The results of an interim analysis with a median follow-up of 19 mo indicated a better tolerability and a superior efficacy of first-line Glivec as compared to interferon and low-dose cytarabine, in terms of estimated rates of complete cytogenetic response (76.2% vs 14.5%) and hematologic response (96.8% vs 69%). More importantly, the estimated rate of freedom from progression to accelerated-phase or blast-crisis CML at 18 mo was improved from 91.5% with IFN + cytarabine, to 96.7% with Glivec (38). Importantly, in comparison with IFN + cytarabine, Glivec therapy induced a faster and more profound depletion of residual Bcr-Abl transcripts in patients with a complete cytogenetic response as measured by quantitative polymerase chain reaction *(39)*. Taken together, keeping in mind the limited follow-up available to date, these findings suggest that Glivec now represents a superior first-line drug therapy for patients with newly diagnosed Ph + CML. In this context, the therapeutic decisions related to the use of bone marrow transplantation are becoming increasingly complex.

5. MANAGING CLINICAL RESISTANCE TO GLIVEC

In CML advanced phase, even though the rate of hematologic responses with Glivec is high, these responses are usually short-lived and the majority of patients will ultimately develop resistance and undergo disease progression. In chronic-phase CML, a small proportion of patients either do not achieve a satisfactory hematological or cytogenetic response or lose a previous response to Glivec therapy *(40–45)*. In these patients experiencing resistance to Glivec, a variety of mechanisms of resistance have been reported, including amplification of the Bcr-Abl gene, the emergence of leukemic clones with mutations in the Abl kinase domain, and the development of new chromosomal abnormalities.

Different strategies may be conceivable to overcome clinical resistance and optimize the therapy with Glivec, by increasing the antileukemic activity and the durability of response: the use of higher doses of Glivec, the combination of Glivec with either established antileukemic agents or new investigational agents affecting signaling pathways downstream of the Bcr-Abl protein, and the development of new potent inhibitors of the Bcr-Abl kinase active against Glivec-resistant mutants. Clinical data have indicated that increasing the daily dose of Glivec from 400 to 600 or 800 mg can improve the hematological or cytogenetic response in selected patients *(46)*. Multiple combination studies with other agents have been performed preclinically in various laboratories. A large number of conventional chemotherapeutic drugs have been tested for synergy, additivity, and antagonism with Glivec using proliferation assays with Bcr-Abl-transfected cell lines and human CML cell lines and colony-forming assays with primary CML patient cells (recently reviewed in refs. *47* and *48*). However, because various experimental designs and different analytical paradigms have been used, results should be interpreted with caution. The data for the various combinations of Glivec with cytotoxic agents and γ-irradiation are summarized in Table 2. A synergistic or additive antiproliferative effect has been documented for a variety of agents including cytarabine (Ara-C), IFN-α, and homoharringtonine. In addition to standard chemotherapeutic agents, several groups have studied in vitro combinations of Glivec with novel antileukemic drugs and signal transduction inhibitors. Targeting downstream signaling pathways, the inhibition of Mek (PD184352) *(49)*, phosphatidylinositol-3 kinase (wortmannin and LY294002) *(50)*, and farnesyltransferase (SCH66336; L-744832) *(51,52)* enhanced the inhibitory effects of Glivec. Moreover, SCH66336

Table 2
**In Vitro Combination Studies of Glivec With Chemotherapeutic Agents
and Irradiation**

Combination partner	Synergy	Additivity	Antagonism	Reference
Ara-C	+	+		61–68
Busulfan		+		69
Carboplatin		+		65
Cladribine		+		65
Daunorubicin	+	+		61,62,68,70
Decitabine	+			54
Doxorubicin		+		63
Etoposide	+	+		63–65
Gemcitabine		+		65
Homoharringtonine	+	+		63,66,68
Hydroxyurea	+	+	+	62–65,68
4-Hydroperoxycyclo-phosphamide		+		63
Interferon-α	+	+		61–63
Mafosfamide		+		64
Methotrexate		+	+	63,65
Mitoxantrone	+			65
Nimustine (ACNU)		+		65
Taxotere		+		65
Thiotepa		+		65
Topotecan			+	65
Treosulfan		+		69
Vincristine	+	+		63
γ-Irradiation	+			69

has been shown to also sensitize Glivec-resistant cells to Glivec-induced apoptosis (52). Alternative strategies included agents that decrease Bcr-Abl levels such as geldanamycin, allylamino-17-demethoxygeldanamycin (17-AAG), or arsenic trioxide (47,51,53,54). Additive to synergistic effects were observed when these agents were combined with Glivec in Bcr-Abl-positive cells. Table 3 summarizes the new experimental approaches that have been tested in combination with Glivec.

Based on the promising preclinical data, clinical trials combining Glivec with standard antileukemic agents or investigational drugs have been initiated. In phase I/II trials, the combination of Glivec appears feasible with either pegylated-interferons (55,56) or Ara-C (57), although at the cost of an increased toxicity, particularly hematologic toxicity. Phase III trials are under preparation. Another phase I study is investigating the combination of Glivec

Table 3
In Vitro Combination Studies Of Glivec With Novel Antileukemic Agents

Investigational drug	Synergy	Additivity	Antagonism	Reference
17-AAG (Hsp90 chaperone inhibitor)		+		51
AG490 (JAK-2 inhibitor)	+	+		71,72
Arsenic trioxide	+			53,54
Flavopiridol (cyclin-dependent kinase inhibitor)	+			73
LY294002, Wortmannin (PI-3K inhibitors)	+			50,61
PD184352 (Mek inhibitor)	+			49
PS-341 (proteasome inhibitor)	+	+	+	74
SCH66336, L-744832 (farnesyl transferase inhibitors)	+			51,52,75
Suberoylanilide hydroxamic acid (histone deacetylase inhibitor)		+		76
Telomestatin (telomerase inhibitor)		+		77
TRAIL (apoptosis inducer)		+		78–80

PI-3K, phosphatidylinositiol-3 kinase; TRAIL, tumor necrosis factor–related apoptosis-inducing ligand.

and arsenic trioxide (58). As an alternative approach to the management of resistance, other studies are evaluating the effects of high doses of 800 mg of Glivec (59,60).

6. CONCLUSION

The research program has clearly shown that it is possible to define in vitro and animal models with high predictive quality since the results of the subsequent clinical studies have largely corroborated the preclinical findings. The predictive quality was achieved in this particular case by using models with the identical genetic abnormalities as those found in man. The case of Glivec also shows that compounds that affect not only one but two or more targets (which is frequently the case) can be beneficial in allowing several diseases to be addressed with differing molecular abnormalities, without paying too high a price in terms of toxicity.

The extensive clinical data available in CML in phase I to phase III studies indicate that the inhibition of Bcr-Abl can be achieved with Glivec in humans and translate into clinically meaningful patient benefit, as evidenced by a much higher rate of cytogenetic response and a lower rate of progression to advanced phases of the disease. These data validate the initial hypothesis of this program, and underscore the importance of rationally selecting the target diseases to be considered in the early phases of development of a molecule such as Glivec.

ACKNOWLEDGMENTS

We would like to thank Brian Druker for his critical input and a fruitful collaboration throughout this program, Nick Lydon for his contribution in the early phase of the program, Jürg Zimmermann who synthesized the compound, and Thomas Meyer, Helmut Mett, and Terence O'Reilly for the pharmacological profiling. We also gratefully acknowledge the members of the Glivec international project team for their critical contribution to the success of this program.

REFERENCES

1. Rowley JD. A new consistent abnormality in chronic myelogenous leukaemia identified by quinacrine fluorescence and giemsa staining. *Nature* 1973; 243:290–293.
2. Nowell PC, Hungerford DA. A minute chromosome in human chronic granulocytic leukemia. *Science* 1960; 132:1497.
3. Daley GQ, Van Etten RA, Baltimore D. Induction of chronic myelogenous leukemia in mice by the p210$^{bcr/abl}$ gene of the Philadelphia chromosome. *Science* 1990; 247:824–830.
4. Kelliher MA, McLaughlin J, Witte ON, et al. Induction of chronic myelogenous leukemia in mice by the v-abl and BCR/ABL. *Proc Nat Acad Sci USA* 1990; 87:6649–6653.
5. Heisterkamp N, Jenster G, ten Hoeve, et al. Acute leukaemia in bcr/abl transgenic mice. *Nature* 1990; 344:251–253.
6. Lugo TG, Pendergast AM, Muller AJ, et al. Tyrosine kinase activity and transformation potency of *bcr-abl* oncogene products. *Science* 1990; 247:1079–1082 .
7. Lipinsky CA. Drug-like properties and the causes of poor solubility and poor permeability. *J Pharmacol Toxicol Methods* 2001; 44:235–249.
8. Teague S, Davis A, Leeson P, et al. The design of leadlike combinatorial libraries. *Angew Chem Int Ed* 1999; 38:3743–3748.
9. Zimmermann J, Buchdunger E, Mett H, et al. (Phenylamino)pyrimidine (PAP) derivatives: a new class of potent and highly selective PDGF-receptor autophosphorylation inhibitors. *Bioorg Med Chem Lett* 1996; 6:1221–1226.
10. Zimmermann J, Buchdunger E, Mett H, Meyer T, Lydon NB. Potent and selective inhibitors of the ABL-kinase: phenylaminopyrimidine (PAP) derivatives. *Bioorg Med Chem Lett* 1997; 7:187–192.
11. Zimmermann J, Furet, Buchdunger E. STI571. A new treatment modality for CML. In: Ojima I, Vite G, Altmann K, eds. Anticancer Agents: Frontiers in Cancer Chemotherapy. *ACS Symposium Series* 796, American Chemical Society, Washington, DC, 2001:245–259.
12. Schindler T, Bornmann W, Pellicena P et al. Structural mechanism for STI571 inhibition of Abelson tyrosine kinase. *Science* 2000; 89:1938–1942.
13. Nagar B, Bornmann WG, et al. Crystal structures of the kinase domain of c-Abl in complex with the small molecule inhibitors PD173955 and imatinib (STI-571). *Cancer Res* 2002; 62:4236–4243.
14. Druker BJ, Tamura S, Buchdunger E, et al. Effects of a selective inhibitor of the Abl tyrosine kinase on the growth of Bcr-Abl positive cells. *Nat Med* 1996; 2:561–566.

15. Carroll M, Ohno-Jones S, Tamura S, et al. CGP 57148, a tyrosine kinase inhibitor, inhibits the growth of cells expressing BCR-ABL, TEL-ABL, and TEL-PDGFR fusion proteins. *Blood* 1997; 90:4947–4952.

16. Beran M, Cao X, Estrov Z, et al. Selective inhibition of cell proliferation and BCR-ABL phosphorylation in acute lymphoblastic leukemia cells expressing M_r 190,000 BCR-ABL protein by a tyrosine kinase inhibitor (CGP 57148). *Clin Cancer Res* 1998; 4:1661–1672.

17. Fabbro D, Furet P, Buchdunger E, et al. Structure–activity studies supporting a postulated binding mode of STI571 to Abl kinase. Molecular targets and cancer therapeutics: discovery, biology and clinical application. *NCI-EORTC International Conferences* 2001:abstract #593.

18. Buchdunger E, Cioffi CL, Law N, et al. The Abl protein-tyrosine kinase inhibitor, STI571, inhibits in vitro signal transduction mediated by c-Kit and PDGF receptors. *J Pharmacol Exp Ther* 2000; 295:139–145.

19. Heinrich MC, Griffith DJ, Druker BJ, et al. Inhibition of c-kit receptor tyrosine kinase activity by STI571, a selective tyrosine kinase inhibitor. *Blood* 2000; 96:925–932.

20. Buchdunger E, Zimmermann J, Mett H, et al. Effects of a selective inhibitor of the Abl tyrosine-kinase in vitro and in vivo by a 2-phenylaminopyrimidine derivative. *Cancer Res* 1996; 56:100–104.

21. Gambacorti-Passerini C, Le Coutre P, Mologni L, et al. Inhibition of the ABL kinase activity blocks the proliferation of BCR/ABL+ leukemic cells and induces apoptosis. *Blood Cells Mol Dis* 1997; 23:380–394.

22. Deininger M, Goldman J, Lydon N, et al. The tyrosine kinase inhibitor CGP57148B selectively inhibits the growth of BCR-ABL-positive cells. *Blood* 1997; 90:3691–3698.

23. Dan S, Naito M, Tsuruo T. Selective induction of apoptosis in Philadelphia chromosome-positive chronic myelogenous leukemia cells by an inhibitor of BCR-ABL tyrosine kinase, CGP 57148B. *Cell Death Differ* 1998; 5:710–715.

24. Le Coutre P, Mologni L, Cleris L, et al. In vivo eradication of human BCR/ABL-positive leukemia cells with an ABL kinase inhibitor. *J Natl Cancer Inst* 1999; 91:163–168.

25. Kasper B, Fruehauf S, Schiedlmeier B, Buchdunger E, Ho AD, Zeller WJ. Favourable therapeutic index of a p210[BCR-ABL]-specific tyrosine kinase inhibitor activity on lineage-commited and primitive chronic myelogenous leukemia progenitors. *Cancer Chemother Pharmacol* 1999; 44:433–438.

26. Sillaber C, Gesbert F, Frank DA, et al. STAT5 activation contributes to growth and viability in Bcr/Abl-transformed cells. *Blood* 2000; 95:2118–2125.

27. Horita M, Andreu EJ, Benito A, et al. Blockade of the Bcr-Abl kinase activity induces apoptosis of chronic myeloid leukemia cells by suppressing signal transducer and activator of transcription 5-dependent expression of Bcl-x_L. *J Exp Med* 2000; 191:977–984.

28. Wolff NC, Ilaria RL. Establishment of a murine model for therapy-treated chronic myelogenous leukemia using the tyrosine kinase inhibitor STI571. *Blood* 2001; 98:2808–2816.

29. Oda T, Heaney C, Hagopian JR, et al. Crkl is the major tyrosine-phosphorylated protein in neutrophils from patients with chronic myelogenous leukemia. *J Bio Chem* 1994; 269:22925–22928.

30. Jonuleit T, van der Kuip H, Miething C, et al. Bcr-Abl kinase down-regulates cyclin-dependent kinase inhibitor p27 in human and murine cell lines. *Blood* 2000; 96:1933–1939.

31. Gesbert F, Sellers WR, Signoretti S, et al. BCR/ABL regulates expression of the cyclin dependent kinase inhibitor p27Kip1 through the PI3K/AKT pathway. *J Biol Chem* 2000; 50:39223–39230.

32. Gaston I, Stenberg PE, Bhat A, et al. Abl kinase but not PI3-kinase links to the cytoskeletal defects in Bcr-Abl transformed cells. *Exp Hematol* 2000; 28:77–86.

33. Druker BJ, Sawyers CL, Kantarjian H, et al. Activity of a specific inhibitor of the BCR-ABL tyrosine kinase in the blast crisis of chronic myeloid leukemia and acute lymphoblastic leukemia with the Philadelphia chromosome. *N Engl J Med* 2001; 344:1038–1042.

34. Druker BJ, Talpaz M, Resta DJ, et al. Efficacy and safety of a specific inhibitor of the BCR-ABL tyrosine kinase in chronic myeloid leukemia. *N Engl J Med* 2001; 344:1031–1037.
35. Kantarjian H, Sawyers C, Hochhaus A, et al. Hematologic and cytogenetic responses to imatinib mesylate in chronic myelogenous leukemia. *N Engl J Med* 2002; 346:645–652.
36. Talpaz M, Silver RT, Druker B, et al. Glivec™ (imatinib mesylate) induces durable hematologic and cytogenetic responses in patients with accelerated phase chronic myeloid leukemia: results of a phase 2 study. *Blood* 2002; 99:1928–1937.
37. Sawyers C, Hochhaus A, Feldman E, et al. Imatinib induces hematologic and cytogenetic responses in patients with chronic myeloid leukemia in myeloid blast crisis: results of a Phase II study. *Blood* 2002; 99:3530–3539.
38. O'Brien SG, Guilhot F, Larson RA, et al. The IRIS study: international randomized study of interferon and low-dose Ara-C versus STI571 (imatinib) in patients with newly diagnosed chronic phase chronic myeloid leukemia. *N Engl J Med* 2003; 348:994–1004.
39. Hughes T, Kaeda J, Branford S et al. Molecular response to imatinib (STI571) or interferon + Ara-C as initial therapy for CML: results in the IRIS study. *Blood* 2002; 100(Suppl 11):93a (abstract #345).
40. Gorre ME, Mohammed M, Ellwood K, et al. Clinical resistance to STI-571 cancer therapy caused by BCR-ABL gene mutation or amplification. *Science* 2001; 263:876–880.
41. Branford S, Rudzki Z, Walsh S, et al. High frequency of point muatations clustered within the adenosine triphosphate-binding region of Bcr/Abl in patients with chronic leukemia or Ph-positive acute lymphoblastic leukemia who developp imatinib (STI571) resistance. *Blood* 2002; 99:3472–3475.
42. Shah NP, Nicoll JM, Nagar B, et al. Multiple Bcr-Abl kinase domain mutations confer polyclonal resistance to the tyrosine kinase inhibitor imatinib (STI571) in chronic phase and blast crisis chronic myeloid leukemia. *Cancer Cell* 2002; 2:117–125.
43. Hochhaus A, Kreil S, Corbin AS, et al. Molecular and chromosomal mechanisms of resistance to imatinib (STI571) therapy. *Leukemia* 2002; 16:2190–2196.
44. O'Dwyer ME, Mauro MJ, Kurilik G, et al. The impact of clonal evolution on response to imatinib mesylate (STI571) in accelerated phase CML. *Blood* 2002; 100:1628–1633.
45. Bumm T, Müller C, Al-Ali HK, et al. Emergence of clonal cytogenetic abmnormalities in PH-negative cells in some CML patients in cytogenetic remission to imatinib but resoration of polyclonal hematopoiesis in the majority. *Blood* 2003; 101:1941–1949.
46. Kantarjian HM, Talpaz M, O'Brien S, et al. Dose escalation of imatinib mesylate can overcome resistance to standard dose therapy in patients with chronic myelogenous leukemia. *Blood* 2003; 101: 473–475
47. La Rosée P, O'Dwyer ME, Druker BJ. Insights from pre-clinical studies for new combination treatment regimens with the Bcr-Abl kinase inhibtor imatinib mesylate (Gleevec/Glivec) in chronic myelogenous leukemia: a translational perspective. *Leukemia* 2002; 16:1213–1219.
48. Topaly J, Zeller WJ, Fruehauf S. Combination therapy with imatinib mesylate (STI571): synopsis of in in vitro studies. *Br J Hemat* 2002; 119:3–14.
49. Yu C, Krystal G, Varticovski L, et al. Pharmacologic mitogen-activated protein/extracellular signal-regulated kinase/mitogen-activated protein kinase inhibitors interact synergistically with STI571 to induce apoptosis in Bcr/Abl-expressing human leukemia cell. *Cancer Res* 2002; 62:188–199.
50. Klejman A, Rushen L, Morrione A, Slupianek A, Skorski T. Phosphatidylinositol-3 kinase inhibitors enhance the anti-leukemia effect of STI571. *Oncogene* 2002; 21:5868–5876.
51. Topaly J, Schad M, Zeller WJ, Ho AD, Fruehauf S. Strong synergism of different signal transduction inhihibitors in chronic myelogenous leukemia. *Blood* 2001; 98(Suppl. 11): 617a (abstract #2587).

52. Hoover R, Mahon F X, Melo JV, Daley GQ. Overcoming STI571 resistance with the farnesyl transferase inhibitor SCH66336. *Blood* 2002; 100:1068–1071.
53. Porosnicu M, Nimmanapalli R, Nguyen D, Worthington E, Perkins C, Bhalla KN. Co-treatment with As2O3 enhances selective cytotoxic effects of STI-571 against Brc-Abl-positive acute leukemia cells. *Leukemia* 2001; 15:772–778.
54. La Rosée P, Johnson K, Moseson EM, O'Dwyer M, Druker BJ. Preclinical evaluation of the efficacy of STI571 in combination with a variety of novel anticancer agents. *Blood* 2001; 98(Suppl 11):839a (abstract #3488).
55. Hochhaus A, Fischer T, Brümmendorf TH, et al. Imatinib (Glivec) and pegylated interferon-alpha 2a (Pegasys) phase I/II combination study in chronic phase chronic myelogenous leukemia (CML). *Blood* 2002; 100(Suppl 11):164a (abstract #616).
56. Baccarani M, Trabacchi E, Bassi S, et al. Results of a phase II trial testing the combination of imatinib and pegylated alpha-2b interferon in Ph+ chronic myeloid leukemia in early chronic phase. The early cytogenetic response is significantly risk related. By the Italian Cooperative Study Group on CML. *Blood* 2002; 100(Suppl 11):94a (abstract #348).
57. Gardembas M, Rousselot P, Tulliez M, et al. Imatinib (Gleevec) and cytarabine (Ara-C) is an effective regimen in philadelphia-positive chronic myelogenous leukemia chronic phase patients. *Blood* 2002; 100(Suppl 11):95a (abstract #351).
58. Mauro M, Deininger MW, O'Dwyer ME, et al. Phase I/II study of arsenic trioxide (Trisenox) in combination with imatinib mesylate (Gleevec, STI571) in patients with gleevec-resistant chronic myelogenous leukemia in chronic phase. *Blood* 2002; 100(Suppl 11):781a (abstract #3090).
59. Cortes JE, Talpaz M, Giles F, et al. High dose imatinib mesylate (STI571, Gleevec) in patients with chronic myeloid leukemia resistant or intolerant to interferon alpha induces molecular remission. *Blood* 2002; 100(Suppl 11):164a (abstract #615).
60. Kantarjian H, Talpaz M, O'Brien S, et al. Dose escalation of imatinib mesylate can overcome resistance to standard dose therapy in patients with chronic myelogenous leukemia. *Blood* 2003; 101:473–475.
61. Tipping AJ, Zalfirides G, Mahon FX, Goldman JM, Melo JV. Response of STI571-resistant cells to other chemotherapeutic drugs and signal transduction inhibititors. *Blood* 2000; 98(Suppl 11):98a (abstract #420).
62. Thiesing JT, Ohno-Jones S, Kolibaba KS, Druker BJ. Efficacy of STI571, an Abl tyrosine kinase inhibitor, in conjunction with other antileukemic agents against Bcr-Abl-positive cells. *Blood* 2000; 96:3195–3199.
63. Kano Y, Akutsu M, Tsuoda S, et al. In vitro cytotoxic effects of a tyrosine kinase inhibitor STI571 in combination with commonly used antileukemic agents. *Blood* 2001; 97: 1999–2007.
64. Topaly J, Zeller, WJ, Fruehauf S. Synergistic activity of the new ABL-specific tyrosine kinase inhibitor STI571 and chemotherapeutic drugs on BCR-ABL-positive chronic myelogenous leukemia cells. *Leukemia* 2001; 15:342–347.
65. Frühauf S, Topaly J, Hochhaus A, Zeller WJ, Ho AD. Synergistic activity of STI571 with chemotherapeutic drugs. *Onkologie* 2001; 24(Suppl. 6): 186a.
66. Scappini B, Onida F, Kantarjian HM, Dong L, Verstovsek S, Keating MJ, Beran M. In vitro effects of ST571-containing drug combinations on the growth of Philadelphia-positive chronic myelogenous leukemia cells. *Cancer* 2002; 94:2653–2662.
67. Liu WM, Stimson LA, Joel SP. The in vitro activity of the tyrosine kinase inhibitor STI571 in Bcr-Abl positive chronic myeoloid leukemia cells: synergistic interactions with antileukaemic agents. *Br J Cancer* 2002; 86:1742–1448.
68. Tipping AJ, Mahon FX, Zafirides G, Lagarde V, Goldman JM, Melo JV. Drug responses of imatinib-mesylate-resistant cells: synergism of imatinib with other chemotherapeutic drugs. *Leukemia* 2002;16:2349–2357.

69. Topaly J, Ho A.D. Fruehauf S, Zeller W. Rationale for combination therapy of chronic myelogenous leukemia with imatinib and irradiation or alkylating agents: implications for pretransplant conditioning. *Br J Cancer* 2002; 86:1487–1493.

70. Tabrizi R, Mahon FX, Makhoul P, et al. Resistance to daunorubicin-induced apoptosis is not completely reversed in CML blast cells by STI571. *Leukemia* 2002; 16:1154–1159.

71. Sun X, Layton JE, Elefanty A, Lieschke, GJ. Comparison of the effects of the tyrosine kinase inhibitors AG957, AG490, and STI571 on BCR-ABL-expressing cells, demonstrating synergy between AG490 and STI571. *Blood* 2001; 97:2008–2015.

72. Marley SB, Davidson RJ, Goldman JM, Gordon MY. Effects of combinations of therapeutic agents on the proliferation of progenitor cells in chronic myeloid leukemia. *Br J Haematol* 2002; 116:162–165.

73. Yu C, Dai Y, Dent P, Krystal G, Grant S. The cyclin-dependent kinase inhibitor flavopiridol interacts synergistically with Bcr-Abl kinase inhibitor STI571 to induce mitochondrial damage and apoptosis in Bcr-Abl+ human leukemia cells (K562 and LAMA-84). *Blood* 2001; 98(Suppl 11):146a (abstract #615).

74. Gatto SR, Scappini B, Verstovsek S, et al. In vitro effects of PS-341 alone and in combination with STI571 in Bcr-Abl positive cell liones both sensitive and resistant to STI571. *Blood* 2001; 98(Suppl 11):101a (abstract #424).

75. Nakajima A, Tauchi T, Sumi M, Bishop WR, Ohyashiki K. Efficacy of SCH66336, the farnesyl transferase inhibitor, in conjunction with other antileukemic agents against Glivec-resistant BCR-ABL-positive cells. *Blood* 2001; 98(Suppl 11):575a (abstract #2409).

76. Nimmanapalli R, Fuino L, Stobaugh C, Richon VM, Bhalla K. Co-treatment with the histone deacetylase inhibitor suberoylanilide hydroxamic acid (SAHA) enhances Gleevec-induced apoptosis of Bcr-Abl positive human acute leukemia cells. *Blood* 2002; [prepublished online November 21, 2002].

77. Tauchi T, Sumi M, Nakajima A, et al. Activity of a novel telomerase inhbitor, telomestatin, against Glivec-resistant Bcr-Abl-positive cells. *Blood* 2001; 98(Suppl 11):616a (abstract #2581).

78. Barteneva N, Kantarjjian H, Somasundaram B, Estrov Z, Donato N, Talpaz M. Interferon-a (INFa) augments the apoptotic effect of STI571 in Ph+ blast crisis cell lines by inducing TRAIL and FAS (CD95/APO1). *Blood* 2000; 96(Suppl 11):345a (abstract #1489).

79. O'Dwyer W, Rathbun K, Druker BJ, Bagby GC. STI571 induces apoptosis of Bcr-Abl positive cells in vitro; the role of caspase 3, 9 and TRAIL. *Blood* 2000; 96(Suppl 11):343a (abstract #1481).

80. Nimmanapalli R, Porosnicu M, Nguyen D, et al. Cotreatment with STI-571 enhances tumor necrosis factor alpha-related apoptosis-inducing ligand (TRAIL or apo-2L)-induced apoptosis of Bcr-Abl-positive human acute leukemia cells. *Clin Cancer Res* 2001; 7:350–357.

8 Platelet-Derived Growth Factor
Normal Function, Role in Disease, and Application of PDGF Antagonists

Tobias Sjöblom, PhD, Kristian Pietras, PhD, Arne Östman, PhD, and Carl-Henrik Heldin, PhD

CONTENTS

INTRODUCTION
PDGF ISOFORMS AND RECEPTORS
DEVELOPMENTAL AND PHYSIOLOGICAL ROLES OF PDGF
PDGF ANTAGONISTS
PDGF RECEPTORS IN DISEASE
PERSPECTIVES
REFERENCES

1. INTRODUCTION

Platelet-derived growth factor (PDGF) is a family of isoforms that stimulate the growth, survival, and motility of fibroblasts, smooth muscle cells, and other cell types (reviewed in ref. *1*) (Fig. 1). PDGF was originally identified in human platelets and purified from this source; however, subsequent studies have shown that PDGF is synthesized by a number of different cell types.

PDGF has important roles in the regulation of growth and differentiation of various mesenchymal cell types during embryonal development. In the adult, PDGF stimulates wound healing and also regulates the homeostasis of the connective tissue compartment.

Overactivity of PDGF or constitutive activation of PDGF receptors has been implicated in several disorders, including malignancies, atherosclerosis, and fibrotic conditions. Therefore, much effort has recently been devoted to the

From: *Cancer Drug Discovery and Development:*
Protein Tyrosine Kinases: From Inhibitors to Useful Drugs
Edited by: D. Fabbro and F. McCormick © Humana Press Inc., Totowa, NJ

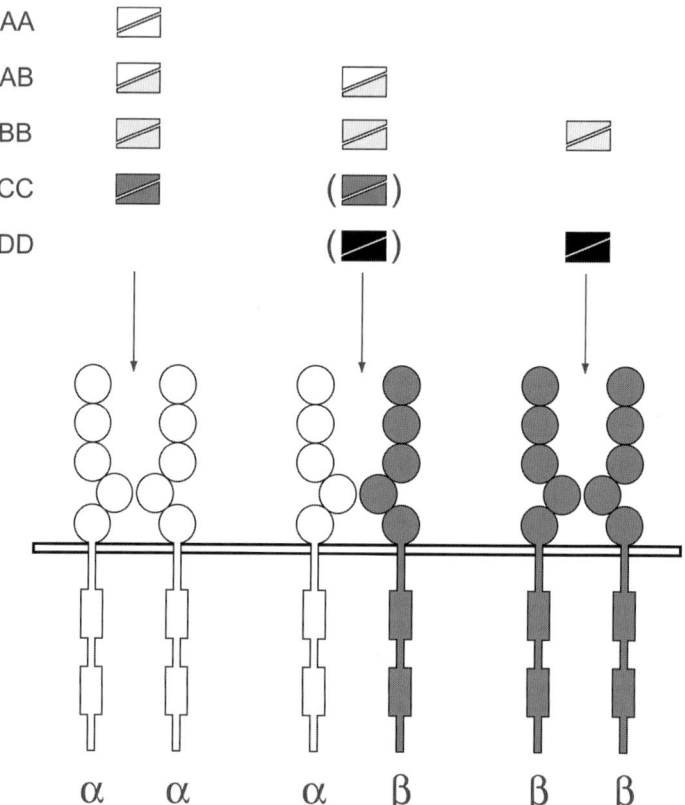

Fig. 1. Schematic illustration of platelet-derived growth factor (PDGF) ligands and their interactions with PDGF receptors. Homodimeric PDGF-α-receptor complexes are induced upon binding of PDGF-AA, -AB, -BB, and -CC, whereas homodimeric PDGF-β-receptor complexes are induced by PDGF-BB or -DD stimulation. Heterodimers, composed of one PDGF-α-receptor and one PDGF-β-receptor, are induced by PDGF-AB and -BB. Activation of both PDGF receptor types after PDGF-CC and -DD stimulation can occur in cells expressing both receptor types, but direct binding of these ligands to heterodimeric receptor complexes has not yet been demonstrated.

development of specific and efficient PDGF antagonists. This review will focus on the validation of PDGF antagonists in animal models and on the initial clinical studies in which PDGF antagonists have been used to treat patients.

2. PDGF ISOFORMS AND RECEPTORS

Five PDGF isoforms are known, i.e., homodimeric forms of the related A-, B-, C-, and D-polypeptide chains, and a PDGF-AB heterodimer (reviewed in ref. *1*). All PDGF isoforms are synthesized as precursor molecules, which undergo proteolytic processing. Processing of the A- and B-chains occur in conjunction with their secretion from the producer cell by removal of N-, and in the case of the B-chain, C-terminal sequences. In contrast, PDGF-CC and -DD are secreted and stored extracellularly as precursors that are kept inactive through the presence of N-terminal CUB domains; thus, for these isoforms an important regulatory step in vivo is the activation of the proteases involved in the cleavage of their precursors. The mature parts of the PDGF chains are about 100 amino acid residues long, and contain a perfect conservation of eight cysteine residues; the same spacing of cysteine residues are seen in members of the vascular endothelial growth factor (VEGF) family. The two subunits of the PDGF dimers are arranged in an antiparallel manner, and both subunits in the dimer contribute to each of the two receptor-binding epitopes (Fig. 1). Two of the conserved cysteine residues are involved in interchain disulfide bonds, whereas the remaining six form a characteristic cystine knot structure in which one disulfide bond passes through the hole formed by two other disulfide bonds and intervening peptide sequences *(2)*.

The PDGF isoforms exert their cellular effects via binding to structurally related α- and β-tyrosine kinase receptors (reviewed in ref. *1*). Each receptor contains an extracellular part of five immunoglobulin (Ig)-like domains, and an intracellular tyrosine kinase domain that is divided into two parts by a characteristic inserted sequence without homology to kinases (Fig. 1). Ligand binding, which occurs mainly to Ig domains 2 and 3, induces dimerization of the receptors; the receptor dimers are further stabilized by direct receptor–receptor interactions involving Ig domain 4 (Fig. 2). The composition of the dimeric receptor complex is determined by the stimulating PDGF isoform and by which of the receptor types the target cell expresses. The α-receptor binds the A-, B-, and C-chains of PDGF, whereas the β-receptor binds the B- and D-chains. Thus, PDGF-AA and -CC induce α-α receptor homodimers, PDGF-DD β-β receptor homodimers, PDGF-AB α-α homodimers, and α-β heterodimers, and PDGF-BB can induce all types of receptor dimers (Fig. 1). In addition, there are indications that PDGF-CC and PDGF-DD also can activate the β- and α-receptors, respectively, on cells expressing both receptor types, although direct binding has not been observed.

Ligand-induced dimerization of PDGF receptors leads to autophosphorylation in trans between in the intracellular parts of the receptors. One autophosphorylation site is localized in the activation loop of the kinase; autophosphorylation of this tyrosine residue leads to an increase of the catalytic activity of the kinase. Other autophosphorylated tyrosine residues are localized outside the kinase domain, and serve as docking sites for signaling molecules containing Src homology (SH)2 domains (Fig. 2).

About 10 different types of SH2 domain-containing molecules bind to and are activated by PDGF receptors (reviewed in ref. *1*). These include the tyrosine kinase Src, phosphatidylinositol-3′-kinase (PI3-K), phospholipase C-γ1 (PLC-γ1), the Grb2/Sos1 complex that activates Ras and the Erk mitogen-activated

A
Ligand binding
Receptor dimerization
Kinase activation and recruitment of signaling molecules
c-Src
PI-3-K
PLC-γ
Survival
Proliferation
Migration
B
Inhibitors of ligand binding
Inhibitors of receptor dimerization
Kinase inhibitors
ligand Abs
receptor Abs
soluble receptors
PDGF aptamers
Ig-domain 4 Abs
soluble Ig-domain 4
e.g. STI571/Glivec
AG1296
PKC412
SU11248

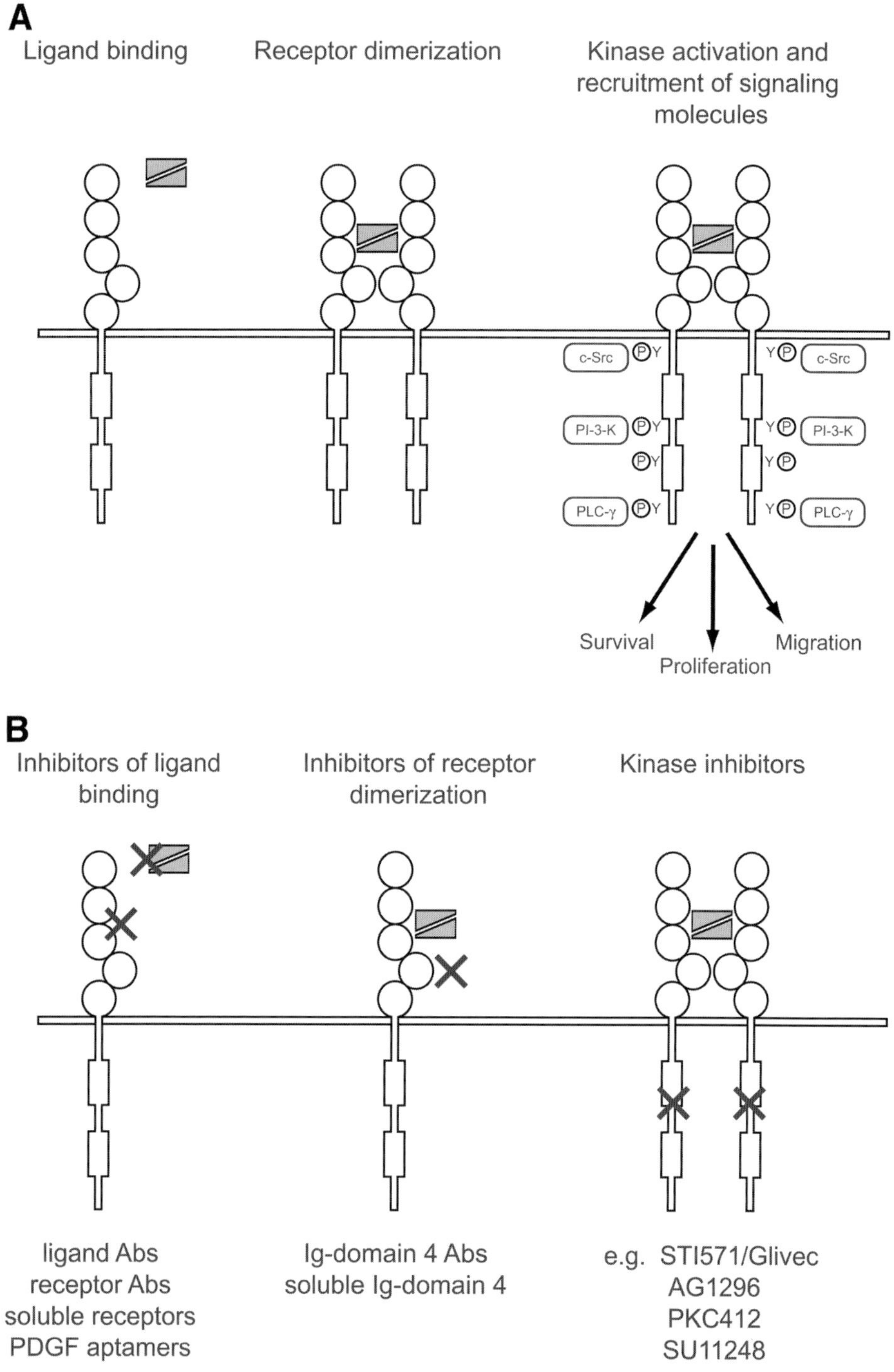

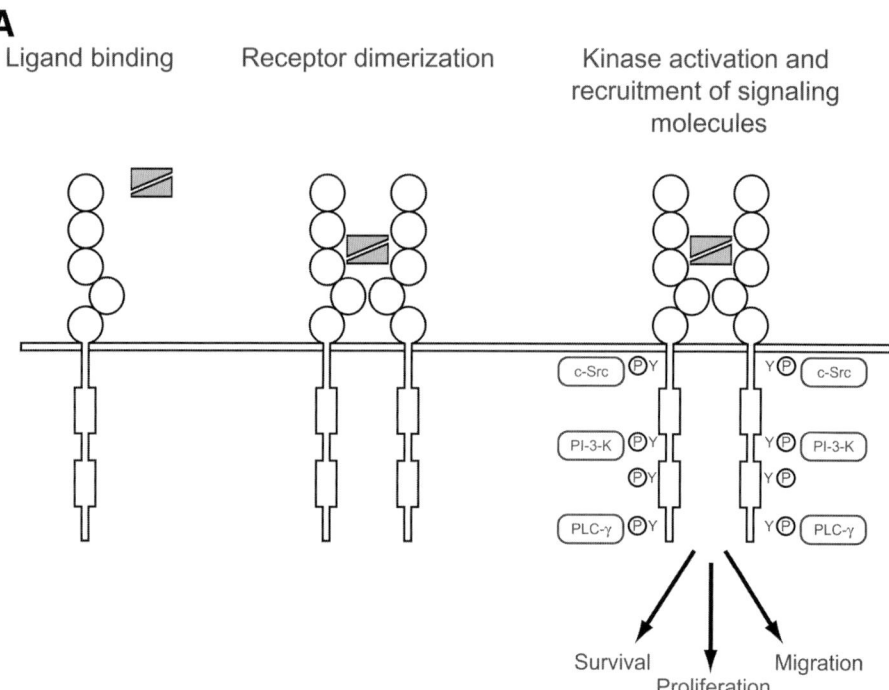

protein kinase (MAPK) pathway, guanosine triphosphatase activating protein for Ras (Ras-GAP), the tyrosine phosphatase SHP-2, and transcription factors of the signal transducers and activators of transcription family. Many of these signaling molecules interact with both α-α and β-β receptor dimers, but there are certain differences. Moreover, there is evidence that α-β heterodimers have distinct properties compared to the two homodimeric receptor complexes. The three different PDGF receptor combinations thus have overlapping but distinct signaling capacities.

Attempts to link the various PDGF-induced signaling pathways to the effects of PDGF on cell behavior, have revealed that activation of PI3-kinase and PLC-γ1 are of particular importance for stimulation of actin reorganization and chemotaxis, whereas activation of the Erk MAPK pathway and Src are important for the mitogenic effect of PDGF. However, there is extensive cross-talk between different signaling pathways. Thus, intracellular signaling occurs by activation of a signaling network rather than by activation of individual parallel pathways, and the roles of individual signaling components can vary between different cell types.

Another interesting observation is that stimulatory and inhibitory signals are often induced in parallel, which thus modulates the strength of signaling. One important example is chemotaxis, where the PDGF β-receptor mediates a powerful stimulatory effect, whereas activation of the α-receptor inhibits chemotaxis. The molecular mechanism behind this difference is not known. There are also examples of modulatory mechanisms affecting the activation of individual signaling pathways. One example is Ras, which is converted to its active GTP-bound form by the exchange factor Sos1, and converted back to its inactive guanosine diphosphate-bound form by Ras-GAP. Both Grb2/Sos1 and Ras-GAP bind to PDGF receptors, but to different phosphotyrosines. Thus, the net effect on Ras activation is determined by the balance in recruitment of Grb2/Sos1 vs Ras-GAP, which in turn is dependent on the stoichiometry of phosphorylation of individual tyrosines. Another example is the tyrosine phosphatase SHP-2, which after activation by PDGF receptors can dephosphorylate the receptors and some of their substrates, thus modulating the strength in signaling.

Fig. 2. Platelet-derived growth factor (PDGF) receptor activation and mechanisms of action of PDGF antagonists. (**A**) Ligand binding entails PDGF receptor dimerization and upregulation of the tyrosine kinase activity. Signaling molecules are recruited to active PDGF receptors through SH2 domain interactions with specific phosphotyrosine residues, and convey signals to induce growth, survival, and migration. (**B**) Macromolecular antagonists have been used to block ligand binding or receptor dimerization, whereas low-molecular-weight compounds have been used to inhibit the PDGF receptor tyrosine kinases.

3. DEVELOPMENTAL AND PHYSIOLOGICAL ROLES OF PDGF

Gene-targeting studies have assigned roles for PDGFs in the ontogeny of specialized smooth muscle cell lineages, such as alveolar smooth muscle cells in the case of PDGF-A and α-receptor *(3,4)*, and pericytes and mesangial cells in the case of PDGF-B and β-receptor *(5,6)*. Knockouts of the C- and D-chains of PDGF have not yet been reported.

3.1. Developmental Roles of the PDGF A-Chain and α-Receptor

Animals deficient for the A-chain have a lethal, emphysema-like phenotype, with deficient migration of alveolar smooth muscle cell progenitors during lung development, and death at 3 wk of age *(3,7)*. This phenotype, together with an insufficient development of intestinal villi, hair follicles, and Leydig cells of the testis, stems from decreased recruitment and proliferation of mesenchymal cell progenitors owing to disturbed epithelial–mesenchymal interactions *(8,9)*. In addition, PDGF-A-chain-deficient mice have impaired proliferation of oligo-dendrocyte precursor cells, resulting in severe hypomyelination of the central nervous system *(10,11)*.

Deletion of the PDGF-α-receptor results in a more severe phenotype than deletion of the PDGF-A-chain. This suggests that the α-receptor mediates signaling from other PDGF ligands, such as PDGF-BB, -CC, or -DD, that is of critical importance for proper embryonic development. Knockout of the α-receptor led to embryonic lethality between E8 and E16, with abnormal somite patterning, cleft face, spina bifida, subepidermal blebbing, hemorrhaging, and skeletal malformations *(4,12,13)*.

3.2. Developmental Roles of the PDGF B-Chain and β-Receptor

PDGF-B-chain and β-receptor deficient mice show highly similar pheno-types, with abnormal placenta, anemia, thrombocytopenia, impaired kidney development, defective blood vessel maturation, edema, and perinatal death owing to rupture of microvascular aneurysms *(5,6)*. The high phenotypic simi-larity of PDGF-B-chain and β-receptor knockout animals implies that PDGF-BB is the major ligand for the PDGF-β-receptor during development.

In microvessels of developing and newborn mouse embryos, PDGF-B-chain expression is restricted to immature arterial endothelial cells and PDGF-β-receptor expression is confined to smooth muscle cells and pericytes *(14)*. Loss of capillary pericytes in B-chain and β-receptor knockout mice is the likely cause of tissue-restricted vascular malformation and edema in these ani-mals. Mural cell investment was not affected in major vascular plexa, but reduced around capillary vessels of the brain, heart, lung, and gastrointestinal villi *(14)*. Furthermore, the presence of capillary aneurysms in the brain, where blood vessels are formed by angiogenesis during late embryonic development, was correlated to insufficient pericyte proliferation and recruitment to sprouting blood vessels *(15)*. Interestingly, the microvessel density in the brain is similar in B-chain or β-receptor deficient and wild-type animals, suggesting that PDGF signaling is not essential for the migration and tube formation of endothelial cells *per se* during developmental angiogenesis *(16)*. This is supported by the contribution of PDGF-β-receptor negative cells to the endothelial cell lineage but not to any muscle cell lineage, including smooth muscle cells and pericytes, during the development of mouse chimeras *(17)*.

Several studies support a role for PDGF-B and β-receptor signaling in pericyte recruitment to neovessels, and subsequent control of endothelial cell growth by the pericyte. Endothelial cell hyperplasia occurs in brain capillaries of B-chain and β-receptor knockout mice, suggesting a role of the pericyte in controlling endothelial cell proliferation *(16)*. In retinas of adult mice with endothelium-specific ablation of PDGF-B, regions of low pericyte density showed capillary malformations and regression, whereas a dense and chaotic vessel network formed in regions with high pericyte density *(18)*. Also, ectopic PDGF-BB disrupted pericyte–endothelial cell interactions and induced regression of immature retinal vessels *(19)*.

In the placenta of B-chain and β-receptor knockout animals, dilation of embryonic vessels alongside with a reduced density of pericytes and cytotrophoblasts caused an altered ratio of maternal to embryonic blood vessel surface area *(20)*. Insufficient nutrient exchange in the malformed placenta, together with vascular and cardiac abnormalities, is the likely explanation of hematological abnormalities in PDGF-B-chain and β-receptor deficient animals because *PDGFB –/–* and *PDGFRB –/–* bone marrow can fully reconstitute hematopoiesis *(21)*.

The impaired kidney development in B-chain and β-receptor deleted embryos stems from the absence of mesangial cells in the kidney glomerulus. Loss of mesangial cells and replacement of capillary tufts by a few dilated capillary loops resulted in decreased renal filtration in knockout animals *(6,22)*.

3.3. Pathway-Specific and Isoform-Specific Biological Responses

In order to elucidate the specific developmental contributions from different signaling pathways downstream of PDGF receptors, different knock-in mouse models have been generated.

PDGF-α-receptor knock-in mice, lacking a functional Src binding site, showed deficient oligodendrocyte recruitment, whereas mutation of the phosphoinositide 3-kinase (PI3-K) binding sites also led to lung emphysema and abnormal skeletal development *(23)*. However, mutation of the PI3-K binding site of the PDGF-β-receptor, alone or together with mutation of the PLC-γ binding site, did not result in abnormal development *(24,25)*. Replacement of the

PDGF-α-receptor intracellular part with the corresponding part from Torso, a tyrosine kinase receptor from *Drosophila*, which can recruit and activate similar signal transduction pathways as the PDGF-α-receptor but with different kinetics, resulted in rescue of the neural crest phenotype but not of the skeletal defects *(26)*. This partial rescue was associated with normal level of MAP kinase activation, but decreased activation of PI3-K. Thus, timely activation of specific signal transduction pathways is of crucial importance in the biological response to PDGF.

Evidence for intrinsic differences in signaling capacity between the PDGF α- and β-receptor intracellular domains was gained by analysis of knock-in mice, where the intracellular part of one receptor was exchanged for the intracellular part of the other receptor *(27)*. The β-receptor intracellular domain was able to substitute for the corresponding α-receptor part, resulting in normal development. However, the α-receptor intracellular domain could not fully substitute for the corresponding β-receptor part, resulting in partial phenocopy of the B-chain/β-receptor knockouts with lack of mesangial cells and deficient pericyte recruitment.

3.4. Functions of PDGF in the Adult

Although significant progress has been made in understanding the developmental role of PDGF, its physiological role in the adult is less clear. In animal models of wound healing, exogenous PDGF-BB has been shown to enhance granulation tissue formation *(28–30)*. Furthermore, PDGF-BB has been shown to promote wound healing in patients with decubitus ulcers *(31)*. Interestingly, though PDGF-β-receptor +/– and –/– cells take part in formation of the endothelial cell lineage in developing mouse chimeras *(17)*, they fail to contribute to endothelium and fibroblasts during wound healing in adult chimeras *(32)*. In hematopoietic chimeras created by grafting of PDGF-B-chain –/– bone marrow to wild-type recipients, the absence of PDGF-AB/BB of hematopoietic origin had no effect on fibroblast-rich granulation tissue formation, but significantly increased the blood vessel content in a sponge assay of wound healing *(33)*. These studies thus indicate that endogenous PDGF is not required for all aspects of wound healing, and that the role of PDGF in wound healing remains to be fully elucidated. A positive role in regulation of interstitial fluid pressure, mediated through activation of PI3-K, has also been assigned to the PDGF β-receptor *(24,34)*.

4. PDGF ANTAGONISTS

The involvement of PDGF in several serious disorders makes clinically useful PDGF antagonists highly desirable. Inhibitors acting at the different steps of PDGF receptor activation have been developed, including macromolecules interfering with ligand:receptor interactions or receptor dimerization and low-molecular-weight kinase inhibitors (Fig. 2). Isoform-specificity has only been achieved with agents targeting ligand–receptor interactions. A general feature distinguishing extracellularly acting agents is that they are not affected by drug resistance mechanisms involving cellular excretion.

4.1. Antagonists Targeting Ligand–Receptor Interactions or Receptor Dimerization

Ligand- or receptor-targeting antibodies constitute the prototypic agents interfering with ligand binding. A number of monoclonal antibodies against PDGF or the receptors have been described (reviewed in ref. *35*). In addition, SELEX aptamers, which bind PDGF-AB/BB with high affinity and inhibit their action, have been developed *(36)*. SELEX aptamers are oligonucletides, which have been modified and polyethylene glycol-conjugated to improve pharmaco-kinetics *(37–40)*. A third type of agents blocking ligand–receptor interactions are soluble PDGF receptor domains. Whereas native extracellular domains block binding at 10- to 100-nM concentrations, predimerized variants block PDGF action in vitro at low or subnanomolar concentrations *(41,42)*. Soluble receptors have shown in vivo efficacy after gene transfer, e.g., in models of lung fibrosis *(43)*.

Attempts have also been made to target the fourth Ig-like domain of PDGF receptors, which contributes to receptor dimerization through direct receptor–receptor interactions *(44)*. Inhibitory effects of soluble Ig-like domain 4, and of antibodies against this domain, have been demonstrated *(44–46)*. However, as of now, none of these agents have shown effects in models of PDGF-driven disease.

4.2. PDGF Receptor Kinase Inhibitors

STI571/Glivec is the best-characterized and most widely used PDGF kinase inhibitor at present (*see* Chapter 7 for characteristics of this compound). However, other PDGF receptor kinase inhibitors also have shown efficacy in animal models, including AG1296, CT53518, PKC412, SU11248, RPR101511A, CT52923, and SU9518 *(47–53)*. Glivec inhibits, in addition to the α- and β-receptors for PDGF, at least three other kinases, i.e., the stem cell factor receptor (c-kit), Abl, and Arg. Likewise, none of the other compounds display complete selectivity for the PDGF α- or β-receptor. The compounds vary with regard to target spectrum, suggesting variations in their mode of binding to the PDGF receptors (*see* table in ref. *54*). This notion is further supported by the recent demonstration of activity of Protein Kinase C (PKC) 412 against a Glivec-resistant activated form of the PDGF α-receptor *(50)*.

5. PDGF RECEPTORS IN DISEASE

5.1. PDGF and Vasculoproliferative Disease

Atherosclerosis, as well as vessel obstruction following percutaneous transluminal angioplasty, bypass grafting, or transplantation, is characterized by smooth muscle cell proliferation. Upregulation of PDGF receptors has been observed in smooth muscle of samples derived from atherosclerotic vessels, as well as from neointima lesions (55,56). Similarly, production of PDGF at these sites has been documented (57,58). A causative role for PDGF in these pathological processes was also suggested by induction of smooth muscle proliferation after local delivery of PDGF in injured rat carotid arteries or porcine arteries (59,60). These findings have led to large series of studies exploring the effects of PDGF antagonists in animal models of vasculoproliferative diseases.

5.1.1. RESTENOSIS

In 1991, beneficial effects were obtained with neutralizing antibodies to PDGF in a rat model of restenosis (61). These initial observations have subsequently been confirmed in similar models with other types of inhibitors (38,52,62,63). In addition, partial inhibition of lesion formation has been obtained with PDGF antagonists in angioplasty- or stent-induced swine models of restenosis (48,64,65). Finally, two studies using PDGF receptor antibodies in primate models of restenosis in femoral or saphenous arteries, reported approx 40% lesion inhibition (66,67). These studies thus suggest that PDGF antagonists can be used clinically to prevent restenosis. However, some concerns have been raised about the efficacy and persistence of the beneficial effects. A possible way to achieve stronger and more long-lasting effects of PDGF inhibitors is to combine them with other agents. In line with this, combination of PDGF and FGF antibodies resulted in superior results, as compared to monotreatments in a rat model of restenosis (68). Also, persistent reduction in neointima formation was observed in a rabbit model where smooth muscle cell-targeting PDGF antagonists were combined with VEGF-C gene transfer and an associated enhancement of endothelialization (69).

5.2.2. ATHEROSCLEROSIS

The availability of atherosclerosis-prone Apo-E-deficient mice has allowed studies on the role of PDGF receptor signaling in atherosclerosis. Six weeks of treatment with PDGF-β-receptor monoclonal antibodies resulted in reduction of atherosclerotic lesion size and number with 67 and 80%, respectively (70). Targeting of the PDGF-α-receptor failed to yield any effects. PDGF-β-receptor-expressing smooth muscle cells were localized in the intima, whereas α-receptor positive cells occurred mainly in the media. In a separate study, 50 wk of treatment with a PDGF receptor tyrosine kinase inhibitor delayed, but failed to prevent, fibrous cap formation *(71)*. Delayed development of lesions was also observed in mice where PDGF receptor signaling was attenuated by grafting with bone marrow derived from PDGF-B-chain –/– mice. Therapeutic effects of PDGF receptor kinase inhibitors have also been obtained in models of allograft atherosclerosis *(72,73)*.

5.2. PDGF and Fibrotic Disease

The biological activities of PDGF, i.e., stimulation of fibroblast proliferation and matrix production, are compatible with a causal role of PDGF in development of different fibrotic conditions. Among the fibrotic conditions, glomerolunephritis and lung fibrosis have been studied in most detail with regard to possible involvement of PDGF receptor signaling.

5.2.1. GLOMERULONEPHRITIS

A therapeutic potential for PDGF antagonists in treatment of renal disease has been suggested by findings that PDGF and its receptors are upregulated in clinical samples and by the effects of PDGF antagonists in animal models (reviewed in ref. *74*). Some recent studies have further emphasized this notion.

The discovery of novel PDGF isoforms has prompted studies on their expression in association with kidney disease. Both PDGF-C and -D are expressed in the adult kidney *(75,76)*. PDGF-D expression occurred predominantly in visceral epithelial cells, and upregulation of PDGF-C was observed in rats after anti-Thy-1.1-induced nephritis *(77)*. Also, studies on clinical samples of IgA nephropathy, membranous nephropathy, and transplant glomerulopathy indicated upregulation, in podocytes, of PDGF-C in a disease-associated manner *(78)*.

Recently, previous studies with PDGF aptamers and PDGF antibodies in models of glomerulonephritis have been consolidated. Glivec treatment of Thy-1.1 glomerulonephritis led to significant reduction in mesangial cell proliferation and collagen deposition *(79)*. Also, short-term treatment with PDGF aptamers prevented chronic effects of renal scarring *(80)*.

5.2.2. LUNG FIBROSIS

Fibroblast proliferation and matrix accumulation are hallmarks of lung fibrosis. In support of a causal role for PDGF in lung fibrosis, induction of fibrosis was observed after intratracheal injection of PDGF-BB and after lung-specific transgenic overexpression of PDGF-BB *(81,82)*. A limited number of animal studies have tested the effects of PDGF antagonists in models of lung fibrosis.

The PDGF receptor kinase inhibitor AG1296, administered intraperitoneally, has been tested in a rat model of vanadium pentoxide pulmonary fibrosis *(83)*. Reduced cell proliferation and collagen synthesis was observed. Gene transfer of a construct encoding a soluble form of the PDGF-β-receptor also reduced lung fibrosis induced by bleomycin *(43)*.

5.3. PDGF in Malignant Disease

Initial evidence for the involvement of PDGF in tumorigenesis came with the discovery of homology between the PDGF-B gene and the simian sarcoma viral oncogene v-*sis (84,85)*. The transforming activity of the *sis* gene product occurs by autocrine stimulation of PDGF receptors (reviewed in ref. *86)*. Autocrine stimulation by PDGF-A, -C, and -D also transforms PDGF receptor-expressing cells *(87,88)*. In addition, there are mutational events that cause constitutive activation of PDGF receptors (Fig. 3). Moreover, paracrine action of PDGF ligands, produced by tumor cells and acting on the vascular and fibroblastic compartments of the tumor stroma, has recently been implicated in tumor angiogenesis, desmoplasia, and tumor physiology (reviewed in ref. *89)* (Fig. 3).

5.3.1. Autocrine PDGF Stimulation in Glioblastoma and Sarcomas

In human glioma, PDGF-α-receptor and PDGF-A-chain are co-expressed in tumor cells with increasing frequency in high-grade as compared to low-grade tumors, suggesting a role in autocrine stimulation of tumor growth. In particular, PDGF-α-receptor expression was observed in tumors that did not overexpress the epidermal growth factor (EGF) receptor *(90)*. Moreover, expression of PDGF-C and PDGF-D was recently demonstrated in human glioma cells *(91,92)*. A minority of glioblastoma multiforme tumors has an amplified PDGF α-receptor gene, with subsequent receptor overexpression, and an oncogenic form of the PDGF-α-receptor has been isolated from human glioblastoma *(93–95)*. Further evidence for the oncogenic potential of autocrine PDGF signaling stems from retroviral transduction of the PDGF-B-chain into primate or murine brain, which induced glioblastoma-like brain tumors *(96,97)*. Experimental glioblastoma with autocrine PDGF-A-chain/PDGF-α-receptor loop has been inhibited in vitro by dominant-negative PDGF or PDGF-β-receptor mutants, and in vivo by the PDGF receptor antagonist Glivec *(98–100)*. These observations provide a rationale for ongoing clinical studies with Glivec in treatment of glioblastoma.

Co-expression of the PDGF-B-chain and the PDGF-β-receptor has also been demonstrated in various soft tissue tumors and sarcomas *(101)*. In Ewing family sarcomas, which harbor EWS/ETS fusion transcription factors, the expression of PDGF-C is upregulated *(102)*. Moreover, dominant-negative PDGF-C, as well as the PDGF receptor kinase inhibitor AG1296, inhibited growth of Ewing sarcoma cells in vitro *(103)*. The usefulness of PDGF antagonists in the treatment of sarcoma is addressed in ongoing clinical trials. Recently, a subset of gastrointestinal stroma tumors (GISTs) was found to have activating mutations in the PDGF α-receptor, and responded to treatment with Glivec *(104,105)*.

Dysregulation of PDGF receptors in tumor cells

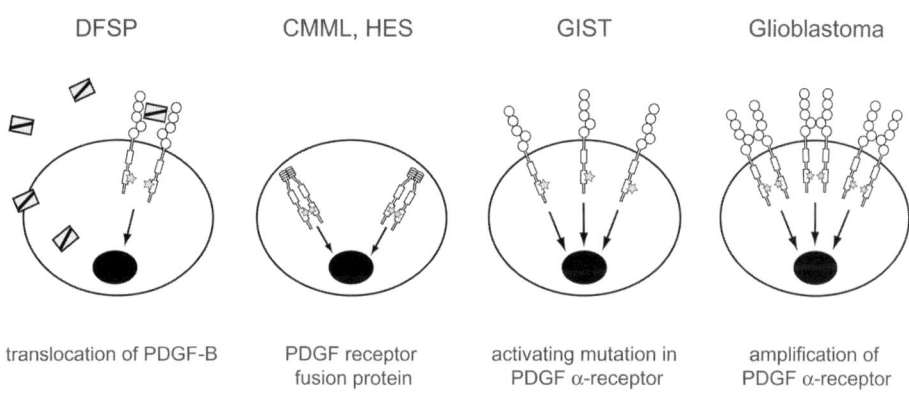

DFSP   CMML, HES   GIST   Glioblastoma

translocation of PDGF-B   PDGF receptor fusion protein   activating mutation in PDGF α-receptor   amplification of PDGF α-receptor

Paracrine PDGF stimulation of normal cell types in tumors

recruitment of stromal fibroblasts   stimulation of angiogenesis

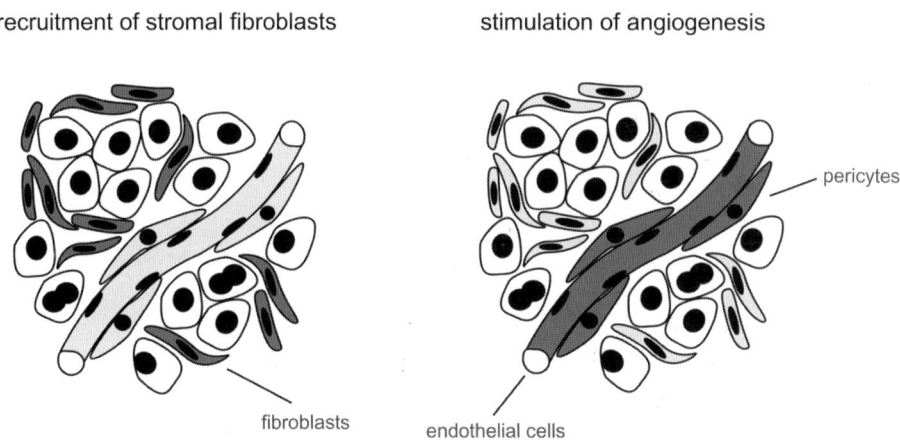

pericytes

fibroblasts   endothelial cells

Fig. 3. Autocrine and paracrine functions of platelet-derived growth factor (PDGF) in malignant disease. **(A)** Dysregulation of PDGF receptor activity can occur by overexpression of PDGF ligand following translocation to a strong promoter, as is the case in dermatofibrosarcoma protuberans. Fusion proteins, composed of a dimerizing domain attached to the intracellular part of a PDGF receptor, have constitutive kinase activity and sustain transformation in some hematopoietic malignancies, including chronic myelomonocytic leukemia and hypereosinophilic syndrome. Point mutations, leading to constitutive activation of the PDGF receptor tyrosine kinase, are found in some cases of gastrointestinal stroma tumors. Moreover, autoactivation of PDGF receptors, as a consequence of gene amplification and protein overexpression, is found in a subset of glioblastomas. **(B)** PDGF isoforms produced by tumor cells can induce formation of a fibroblast-rich tumor stroma, and enhance angiogenesis by recruitment of pericytes and endothelial cells. PDGF receptor expressing cells in light and dark gray. DFSP, dermatofibrosarcoma protuberans; CMML, chronic myelomonocytic leukemia; HES, hypereosinophilic syndrome; GIST, gastrointestinal stroma tumors.

Dermatofibrosarcoma protuberans (DFSP) is a locally invasive soft-tissue tumor of the dermis that is characterized by translocations involving chromosomes 17 and 22 *(106–109)*. Sequencing of t(17;22)(q22;q13) break point regions in DFSP demonstrated fusion of the collagen type Iα1-chain gene *(COL1A1)* with the PDGF-B-chain gene *(PDGFB)* in all tumors examined, alongside with expression of *COL1A1-PDGFB* fusion mRNA (110). The *COL1A1-PDGFB* translocation has also been detected in some superficial adult fibrosarcomas *(111)*, but not in malignant fibrous histiocytoma or dermatofibroma *(112)*.

Expression of *COL1A1-PDGFB* in fibroblasts resulted in production of a dimeric, disulfide-linked precursor molecule containing procollagen Iα1 and PDGF-B sequences *(113,114)*. The precursor was processed to mature PDGF-BB, and induced autocrine PDGF-β-receptor phosphorylation, morphological transformation, and in vivo tumor growth. In primary cultures of human DFSP, increased tyrosine phosphorylation of the PDGF-β-receptor was detected, which was decreased by treatment with Glivec *(115)*. Moreover, tumor growth of primary DFSP was strongly reduced by Glivec treatment.

In small-scale clinical trials, Glivec reduced the tumor volume of a paravertebral metastasis of DFSP by 75%, thereby enabling surgery *(116)*. Tumor regression, as monitored by magnetic resonance imaging, was paralleled by reduced contrast enhancement, and no viable tumor tissue was detected in the excised tumor. Out of two patients with lung metastases of DFSP treated with Glivec, one patient had a transient response with subsequent tumor progression and death, whereas the other experienced tumor regression, which continued at 6 mo of therapy *(117)*. Tumor regression by Glivec treatment was also observed in two patients with primary DFSP (M. Heinrich, personal communication).

5.3.2. PDGF RECEPTOR FUSION PROTEINS IN MYELOID DISORDERS

Chronic myelomonocytic leukemia (CMML) is a subtype of the myelodysplastic syndrome, a group of hematopoietic disorders with clonal, dysplastic growth of progenitor cells and progression to acute myeloid leukemia *(118)*. As a consequence of t(5;12), the *PDGFRB* gene is fused to the *ets*-like transcription factor *TEL* *(ETV6)*. The t(5;12) is an early mutation in the pathogenesis of CMML, as evidenced by subsequent progression to acute myeloid leukemia *(119)*. To date, at least 35 patients with CMML and a reciprocal t(5;12)(q33;p13) chromosomal translocation have been reported. Apart from *TEL/ETV6-PDGFRB*, fusions have been reported between *H4/D10S170* *(120,121)*, *HIP1 (122,123)*, and *RAB5* and *PDGFRB (124)*. The break point in *PDGFRB* is conserved, and the fusion proteins all contained the transmembrane and intracellular domains of the PDGF-β-receptor. Despite divergent biological functions, the fusion partners all encode putative oligomerization domains, such as coiled-coil or leucine zipper motifs. In general, leukemia driven by

PDGF β-receptor mutations has presented as a myeloproliferative disorder characterized by eosinophilia, such as eosinophilic leukemia or CMML, with a strong male predominance *(125)*.

The *TEL-PDGFRB* fusion transcript encodes the amino-terminal part of Tel, including a Pointed domain, and the transmembrane and intracellular parts of the PDGF-β-receptor. Pointed domain interactions lead to oligomerization, resulting in constitutive kinase activity and autophosphorylation *(113,126,127)*. The signaling molecules downstream of Tel-PDGF-β-receptor are essentially those recruited by the stimulated PDGF-β-receptor *(126,128)*. Transformation by PDGF receptor fusion proteins requires the tyrosine kinase activity, and kinase inhibition blocks the growth of cells transformed by these oncogenes *(129,130)*. Retroviral transduction of *TEL-PDGFRB* into whole bone marrow of recipient mice induced a fatal myeloproliferative disorder, and Glivec inhibited leukemia induced by transgenic expression of *TEL-PDGFRB (131,132)*.

Currently, Glivec treatment of leukemia with *PDGFRB* translocation is evaluated in clinical trials. In four CMML cases with *TEL-PDGFRB* translocation, Glivec administration yielded prompt responses, with resolution of eosinophilia and normalized peripheral blood count within 1 wk, and continued remission after 9 mo of treatment *(133,134)*. In one case of *RAB5-PDGFRB*-positive leukemia, the disease responded rapidly to Glivec with no detectable fusion transcript after 6 wk of therapy, and continued molecular remission at 6 mo of treatment *(135)*.

Recently, deregulated PDGF-α-receptor signaling was linked to the pathogenesis of hypereosinophilic syndrome (HES), which is characterized by persistent eosinophilia and subsequent damage to various organ systems *(136)*. Initial studies showed effective control of HES by Glivec treatment, with resolution of eosinophilia after 1 wk of therapy *(137)*. Half of HES patients with lasting responses to Glivec therapy were found to have a *FIP1L1-PDGFRA* fusion oncogene, yielding a constitutively active tyrosine kinase with transforming properties *(138,139)*. Because *FIP1L1-PDGFRA*-negative HES patients responded to Glivec, other genetic aberrations resulting in activation of Glivec-sensitive kinases are likely to be identified in this disease. The *FIP1L1-PDGFRA* fusion was also found in patients with Glivec-responsive systemic mast cell disease *(140)*.

5.3.3. PARACRINE PDGF EFFECTS ON TUMOR STROMA

Perivascular and stromal expression of the PDGF-β-receptor has been found in more than 90 and 50%, respectively, of common human solid tumors (T. Sjöblom, et al., unpublished observation). Furthermore, several lines of evidence suggest that paracrine PDGF stimulation has a role in tumor formation, making PDGF-β-receptors candidate targets for anti-angiogenic and antistromal therapy.

Whereas expression of PDGF ligands is frequently detected in epithelial tumor cells, PDGF receptors are mainly found on mesenchymal cells of the tumor stroma *(141,142)*. The vascular PDGF β-receptor staining pattern observed in human tumors has been attributed to pericytes, as well as to endothelial cells *(143–145)*. However, PDGF β-receptor expression was found to colocalize with the pericyte antigen high-molecular-weight-melanoma-associated antigen (HMW-MAA), and not with the endothelial cell antigen on Willebrand factor (vWF) *(146)*. In experimental fibrosarcoma, PDGF-B was expressed in endothelial cells whereas vessel-associated mural cells expressed the PDGF β-receptor *(147)*.

Pro-angiogenic effects of PDGF-BB, and to some extent of PDGF-AA and -AB, have been demonstrated in the chick chorioallantoic membrane assay *(148,149)*. Interestingly, neovessels induced by PDGF-BB showed extensive branching and recruitment of smooth muscle α-actin expressing mural cells, in contrast to the brush-like vascular formations induced by VEGF-A *(150,151)*. Recently, PDGF-CC and -DD were also shown to induce angiogenesis in the mouse corneal pocket assay *(152)*. In experimental models of glioblastoma and murine melanoma, PDGF-BB and -DD production enhanced pericyte recruitment and thereby accelerated tumor growth *(153,153a)*. Simultaneous inhibition of pericyte recruitment, using PDGF antagonists, was also shown to potentiate the anti-angiogenic effect of VEGF inhibition *(154)*. Anti-angiogenic effects of Glivec, in combination with paclitaxel, were also obtained in a bone metastasis model of prostate cancer that had activated PDGF-β-receptors in the endothelium *(155)*.

Observations from in vivo models where the tumor cells produce PDGF, but lack PDGF receptors, are consistent with a role of PDGF signaling in formation of the tumor stroma. PDGF-BB production accelerated the tumor take in a mouse model of human melanoma, concomitantly with the induction of a well-vascularized, fibroblast-rich tumor stroma and a decrease in tumor necrosis *(156)*. Expression of PDGF-B in nontumorigenic epithelial HaCaT cells induced formation of progressively enlarging cysts, with a high blood vessel density and high proliferative activity in stromal fibroblasts *(157)*. In contrast, control-transfected HaCaT cells did not form cysts, implying that stroma cells activated by PDGF-BB can stimulate the growth of benign tumor cells and tumor progression. Moreover, PDGF-A induced desmoplasia in a breast carcinoma model *(158)*. Thus, PDGF signaling contributes to stroma formation in experimental tumors, thereby accelerating tumor growth rate and progression.

In addition to its role in connective tissue formation, PDGF stimulation was shown to contribute to the increased interstitial fluid pressure in experimental tumors *(39)*. PDGF antagonists reduced the intratumoral fluid pressure in two rodent models of cancer. Interestingly, PDGF antagonists increased uptake and therapeutic efficacy of cytotoxic drugs in experimental tumors *(40,159)*. The putative ability of Glivec to lower interstitial fluid pressure of human solid tumors is currently investigated in clinical trials.

6. PERSPECTIVES

The explorative studies on PDGF receptors as potential drug targets, which have now been ongoing for three decades, have in the last years been rewarded with early clinical findings indicating therapeutic effects of PDGF receptor inhibition. Additional challenges for continued translational successes lie ahead.

Accumulating data from therapies in which specific signaling molecules are targeted emphasize the need for selection of proper patient cohorts for optimal design of early clinical trials. Thus, improved tools, like activation-specific antibodies, for determination of PDGF receptor status are highly warranted. Emerging evidence for molecular resistance in CML and GIST also point toward the utility of multiple PDGF receptor kinase inhibitors with distinct and well-characterized inhibitory profiles. The recent discoveries of PDGF receptor activation through point mutations or translocations also illustrate the need for novel methods allowing screening for occurrence of mutations in PDGF receptor genes.

A series of experimental therapy studies in models of cancer and vasculoproliferative disease have also indicated the potential of rational combinations involving PDGF receptor inhibitors. These should be continued and are likely to present yet unexploited clinical opportunities.

Concerning clinical cancer studies, the reported results from malignancies characterized by autocrine PDGF receptor stimulation are encouraging. These should be extended to other tumor types predicted to involve autocrine PDGF signaling, like glioblastomas. The results from such studies will definitely be more informative if they include efforts to find molecular markers for identification of the expected responding and nonresponding subpopulations of patients. Recent implications of additional, possibly general, roles of PDGF receptors in control of tumor drug uptake and angiogenesis should also stimulate to clinical studies with PDGF inhibitors, in combination with other drugs, for treatment of the common solid tumors. Furthermore, the results with PDGF antagonists in animal models of atherosclerosis, restenosis, and fibrotic disease appear to be sufficiently promising to motivate clinical studies on the efficacy of PDGF antagonists also in these settings.

In summary, the initial preclinical and clinical studies with PDGF antagonists have given encouraging results. Continued efforts to develop clinically useful inhibitors, and to explore their usefulness in the treatment of various diseases, are highly warranted.

REFERENCES

1. Heldin C-H, Eriksson U, Östman A. New members of the platelet-derived growth factor family of mitogens. Arch Biochem Biophys 2002; 398:284–290.

2. Oefner C, D'Arcy A, Winkler FK, Eggimann B, Hosang M. Crystal structure of human platelet-derived growth factor BB. *EMBO J* 1992; 11:3921–3926.

3. Boström H, Willetts K, Pekny M, et al. PDGF-A signaling is a critical event in lung alveolar myofibroblast development and alveogenesis. *Cell* 1996; 85:863–873.

4. Soriano P. The PDGFα receptor is required for neural crest cell development and for normal patterning of the somites. *Development* 1997; 124:2691–2700.

5. Levéen P, Pekny M, Gebre-Medhin S, Swolin B, Larsson E, Betsholtz C. Mice deficient for PDGF B show renal, cardiovascular, and hematological abnormalities. *Genes Dev* 1994; 8:1875–1887.

6. Soriano P. Abnormal kidney development and hematological disorders in PDGF β-receptor mutant mice. *Genes Dev* 1994; 8:1888–1896.

7. Lindahl P, Karlsson L, Hellström M, et al. Alveogenesis failure in PDGF-A-deficient mice is coupled to lack of distal spreading of alveolar smooth muscle cell progenitors during lung development. *Development* 1997; 124:3943–3953.

8. Karlsson L, Lindahl P, Heath JK, Betsholtz C. Abnormal gastrointestinal development in PDGF-A and PDGFR-α deficient mice implicates a novel mesenchymal structure with putative instructive properties in villus morphogenesis. *Development* 2000; 127:3457–3466.

9. Gnessi L, Basciani S, Mariani S, et al. Leydig cell loss and spermatogenic arrest in platelet-derived growth factor (PDGF)-A-deficient mice. *J Cell Biol* 2000; 149:1019–1026.

10. Calver AR, Hall AC, Yu WP, et al. Oligodendrocyte population dynamics and the role of PDGF in vivo. *Neuron* 1998; 20:869–882.

11. Fruttiger M, Karlsson L, Hall AC, et al. Defective oligodendrocyte development and severe hypomyelination in PDGF-A knockout mice. *Development* 1999; 126:457–467.

12. Tallquist MD, Weismann KE, Hellstrom M, Soriano P. Early myotome specification regulates PDGFA expression and axial skeleton development. *Development* 2000; 127:5059–5070.

13. Tallquist MD, Soriano P. Cell autonomous requirement for PDGFRα in populations of cranial and cardiac neural crest cells. *Development* 2003; 3:507-518.

14. Hellström M, Kalén M, Lindahl P, Abramsson A, Betsholtz C. Role of PDGF-B and PDGFR-β in recruitment of vascular smooth muscle cells and pericytes during embryonic blood vessel formation in the mouse. *Development* 1999; 126:3047–3055.

15. Lindahl P, Johansson BR, Levéen P, Betsholtz C. Pericyte loss and microaneurysm formation in PDGF-B-deficient mice. *Science* 1997; 277:242–245.

16. Hellström M, Gerhardt H, Kalén M, et al. Lack of pericytes leads to endothelial hyperplasia and abnormal vascular morphogenesis. *J Cell Biol* 2001; 153:543–553.

17. Crosby JR, Seifert RA, Soriano P, Bowen-Pope DF. Chimaeric analysis reveals role of PDGF receptors in all muscle lineages. *Nat Genet* 1998; 18:385–388.

18. Enge M, Bjarnegard M, et al. Endothelium-specific platelet-derived growth factor-B ablation mimics diabetic retinopathy. *EMBO J* 2002; 21:4307–4316.

19. Benjamin LE, Hemo I, Keshet E. A plasticity window for blood vessel remodelling is defined by pericyte coverage of the preformed endothelial network and is regulated by PDGF-B and VEGF. *Development* 1998; 125:1591–1598.

20. Ohlsson R, Falck P, Hellström M, et al. PDGFB regulates the development of the labyrinthine layer of the mouse fetal placenta. *Dev Biol* 1999; 212:124–136.

21. Kaminski WE, Lindahl P, Lin NL, et al. Basis of hematopoietic defects in platelet-derived growth factor (PDGF)-B and PDGF β-receptor null mice. *Blood* 2001; 97:1990–1998.

22. Lindahl P, Hellström M, Kalén M, et al. Paracrine PDGF-B/PDGF-Rβ signaling controls mesangial cell development in kidney glomeruli. *Development* 1998;125:3313–3322.

23. Klinghoffer RA, Hamilton TG, Hoch R, Soriano P. An allelic series at the PDGFaR locus indicates unequal contributions of distinct signaling pathways during development. *Dev Cell* 2002; 2:103–113.

24. Heuchel R, Berg A, Tallquist M, et al. Platelet-derived growth factor β receptor regulates interstitial fluid homeostasis through phosphatidylinositol-3' kinase signaling. *Proc Natl Acad Sci USA* 1999; 96:11410–11415.

25. Tallquist MD, Klinghoffer RA, Heuchel R, et al. Retention of PDGFR-β function in mice in the absence of phosphatidylinositol-3' kinase and phospholipase Cγ signaling pathways. *Genes Dev* 2000; 14:3179–3190.

26. Hamilton TG, Klinghoffer RA, Corrin PD, Soriano P. Evolutionary divergence of platelet-derived growth factor α receptor signaling mechanisms. *Mol Cell Biol* 2003; 11:4013–4025.

27. Klinghoffer RA, Mueting-Nelsen PF, Faerman A, Shani M, Soriano P. The two PDGF receptors maintain conserved signaling in vivo despite divergent embryological functions. *Mol Cell* 2001; 7:343–354.

28. Grotendorst GR, Martin GR, Pancev D, Sodek J, Harvey AK. Stimulation of granulation tissue formation by platelet-derived growth factor in normal and diabetic rats. *J Clin Invest* 1985; 76:2323–2339.

29. Lepistö J, Laato M, Niinikoski J, Lundberg C, Gerdin B, Heldin C-H. Effects of homodimeric isoforms of platelet-derived growth factor (PDGF-AA and PDGF-BB) on wound healing in rat. *J Surgi Res* 1992; 53:596–601.

30. Sprugel KH, McPherson JM, Clowes AW, Ross R. Effects of growth factors in vivo. I. Cell ingrowth into porous subcutaneous chambers. *Am J Pathol* 1987; 129:601–613.

31. Robson MC, Phillips LG, Thomason A, Robson LE, Pierce GF. Platelet-derived growth factor BB for the treatment of chronic pressure ulcers. *Lancet* 1992; 339:23–25.

32. Crosby JR, Tappan KA, Seifert RA, Bowen-Pope DF. Chimera analysis reveals that fibroblasts and endothelial cells require platelet-derived growth factor receptor β expression for participation in reactive connective tissue formation in adults but not during development. *Am J Pathol* 1999; 154:1315–1321.

33. Buetow BS, Crosby JR, Kaminski WE, et al. Platelet-derived growth factor B-chain of hematopoietic origin is not necessary for granulation tissue formation and its absence enhances vascularization. *Am J Pathol* 2001; 159:1869–1876.

34. Rodt SÅ, Åhlén K, Berg A, Rubin K, Reed RK. A novel physiological function for platelet-derived growth factor-BB in rat dermis. *J Physiol* 1996; 495:193–200.

35. Östman A, Heldin C-H. Involvement of platelet-derived growth factor in disease: development of specific antagonists. *Adv Cancer Res* 2001; 80:1–38.

36. Green LS, Jellinek D, Jenison R, Östman A, Heldin C-H, Janjic N. Inhibitory DNA ligands to platelet-derived growth factor B-chain. *Biochem* 1996; 35:14413–14424.

37. Floege J, Ostendorf T, Janssen U, et al. Novel approach to specific growth factor inhibition in vivo . Antagonism of platelet-derived growth factor in glomerulonephritis by aptamers. *Am J Pathol* 1999; 154:169–179.

38. Leppänen O, Janjic N, Carlsson M-A, et al. Intimal hyperplasia recurs after removal of PDGF-AB and -BB inhibition in the rat carotid artery injury model. *Arterioscler Thromb Vasc Biol* 2000; 20:E89–E95.

39. Pietras K, Östman A, Sjöquist M, et al. Inhibition of platelet-derived growth factor receptors reduces interstitial hypertension and increases transcapillary transport in tumors. *Cancer Res* 2001; 61:2929–2934.

40. Pietras K, Rubin K, Sjöblom T, et al. Inhibition of PDGF receptor signaling in tumor stroma enhances anti-tumor effect of chemotherapy. *Cancer Res* 2002; 62:5476–5484.

41. Heidaran MA, Mahadevan D, Larochelle WJ. β PDGFR-IgG chimera demonstrates that human β PDGFR Ig-like domains 1 to 3 are sufficient for high affinity PDGF BB binding. *FASEB J* 1995; 9:140–145.

42. Leppänen O, Miyazawa K, Bäckström G, et al. Predimerization of recombinant platelet-derived growth factor receptor extracellular domains increases antagonistic potency. *Biochemistry* 2000; 39:2370–2375.

43. Yoshida M, Sakuma-Mochizuki J, Abe Ky, et al. In vivo gene transfer of an extracellular domain of platelet-derived growth factor β receptor by the HVJ-liposome method ameliorates bleomycin-induced pulmonary fibrosis. *Biochem Biophys Res Commun* 1999; 265:503–508.

44. Omura T, Heldin C-H, Östman A. Immunoglobulin-like domain 4-mediated receptor-receptor interactions contribute to platelet-derived growth factor–induced receptor dimerization. *J Biol Chem* 1997; 272:12676–12682.

45. Lokker NA, O'Hare JP, Barsoumian A, et al. Functional importance of platelet-derived growth factor (PDGF) receptor extracellular immunoglobulin-like domains. Identification of PDGF binding site and neutralizing monoclonal antibodies. *J Biol Chem* 1997; 272:33037–33044.

46. Shulman T, Sauer FG, Jackman RM, Chang CN, Landolfi NF. An antibody reactive with domain 4 of the platelet-derived growth factor β receptor allows BB binding while inhibiting proliferation by impairing receptor dimerization. *J Biol Chem* 1997; 272:17400–17404.

47. Kovalenko M, Gazit A, Böhmer A, et al. Selective platelet-derived growth factor receptor kinase blockers reverse *sis*-transformation. *Cancer Res* 1994; 54:6106–6114.

48. Bilder G, Wentz T, Leadley R, et al. Persons P, Page K, Perrone M, Dunwiddie C. Restenosis following angioplasty in the swine coronary artery is inhibited by an orally active PDGF-receptor tyrosine kinase inhibitor, RPR101511A. *Circulation* 1999; 99:3292–3299.

49. Pandey A, Volkots DL, Seroogy JM, et al. Identification of orally active, potent, and selective 4-piperazinylquinazolines as antagonists of the platelet-derived growth factor receptor tyrosine kinase family. *J Med Chem* 2002; 45:3772–3793.

50. Cools J, Stover EH, Boulton CL, et al. PKC412 overcomes resistance to imatinib in a murine model of FIP1L1-PDGFRα-induced myeloproliferative disease. *Cancer Cell* 2003; 5:459–469.

51. Mendel DB, Laird AD, Xin X, et al. In vivo antitumor activity of SU11248, a novel tyrosine kinase inhibitor targeting vascular endothelial growth factor and platelet-derived growth factor receptors: determination of a pharmacokinetic/pharmacodynamic relationship. *Clin Cancer Res* 2003; 1:327–337.

52. Yamasaki Y, Miyoshi K, Oda N, et al. Weekly dosing with the platelet-derived growth factor receptor tyrosine kinase inhibitor SU9518 significantly inhibits arterial stenosis. *Circ Res* 2001; 88:630–636.

53. Yu JC, Lokker NA, Hollenbach S, et al. Efficacy of the novel selective platelet-derived growth factor receptor antagonist CT52923 on cellular proliferation, migration, and suppression of neointima following vascular injury. *J Pharmacol Exp Ther* 2001; 298:1172–1178.

54. Sawyers CL. Finding the next Gleevec: FLT3 targeted kinase inhibitor therapy for acute myeloid leukemia. *Cancer Cell* 2002; 1:413–415.

55. Rubin K, Tingström A, Hansson GK, et al. Induction of B-type receptors for platelet-derived growth factor in vascular inflammation. possible implications for development of vascular proliferative lesions. *Lancet* 1988; 1:1353–1356.

56. Tanizawa S, Ueda M, Van der Loos CM, Van der Wal AC, Becker AE. Expression of platelet derived growth factor B chain and β receptor in human coronary arteries after percutaneous transluminal coronary angioplasty. An immunohistochemical study. *Br Heart J* 1996; 75:549–556.

57. Ross R, Masuda J, Raines EW, et al. Localization of PDGF-B protein in macrophages in all phases of atherogenesis. *Science* 1990; 248:1009–1012.

58. Ueda M, Becker AE, Kasayuki N, Kojima A, Morita Y, Tanaka S. In situ detection of platelet-derived growth factor-A and -B chain mRNA in human coronary arteries after percutaneous transluminal coronary angioplasty. *Am J Pathol* 1996; 149:831–843.

59. Jawien A, Bowen-Pope DF, Lindner V, Schwartz SM, Clowes AW. Platelet-derived growth factor promotes smooth muscle migration and intimal thickening in a rat model of balloon angioplasty. *J Clin Invest* 1992; 89:507–511.

60. Nabel EG, Yang Z, Liptay S, et al. Recombinant platelet-derived growth factor B gene expression in porcine arteries induce intimal hyperplasia in vivo. *J Clin Invest* 1993; 91:1822–1829.

61. Ferns GA, Raines EW, Sprugel KH, Motani AS, Reidy MA, Ross R: Inhibition of neointimal smooth muscle accumulation after angioplasty by an antibody to PDGF. *Science* 1991; 253:1129–1132.

62. Sirois MG, Simons M, Edelman ER. Antisense oligonucleotide inhibition of PDGFR- β receptor subunit expression directs suppression of intimal thickening. *Circulation* 1997; 95:669–676.

63. Myllarniemi M, Calderon L, Lemström K, Buchdunger E, Hayry P. Inhibition of platelet-derived growth factor receptor tyrosine kinase inhibits vascular smooth muscle cell migration and proliferation. *FASEB J* 1997; 11:1119–1126.

64. Banai S, Wolf Y, Golomb G, et al. PDGF-receptor tyrosine kinase blocker AG1295 selectively attenuates smooth muscle cell growth in vitro and reduces neointimal formation after balloon angioplasty in swine. *Circulation* 1998; 97:1960–1969.

65. Bilder G, Amin D, Morgan L, et al. Stent-induced restenosis in the swine coronary artery is inhibited by a platelet-derived growth factor receptor tyrosine kinase inhibitor, TKI963. *J Cardiovasc Pharmacol* 2003; 41:817–829.

66. Giese NA, Marijianowski MMH, McCook O, et al. The role of α and β platelet-derived growth factor receptor in the vascular response to injury in nonhuman primates. Arterioscler. *Thromb Vasc Biol* 1999; 19:900–909.

67. Hart CE, Kraiss LW, Vergel S, et al. PDGFβ receptor blockade inhibits intimal hyperplasia in the baboon. *Circulation* 1999; 99:564–569.

68. Rutherford C, Martin W, Salame M, Carrier M, Änggård E, Ferns G. Substantial inhibition of neo-intimal response to balloon injury in the rat carotid artery using a combination of antibodies to platelet-derived growth factor-BB and basic fibroblast growth factor. *Atherosclerosis* 1997; 130:45–51.

69. Leppänen O, Rutanen J, Hiltunen MO, et al. Oral imatinib mesylate (STI571/Gleevec) improves the efficacy of local intravasular endothetial growth factor-c gene transfer in reducing neointimal growth in hypercholesterolemic rabbits. *Circulation* 2004; 109:1140–1146.

70. Sano H, Sudo T, Yokode M, et al. Functional blockade of platelet-derived growth factor receptor-β but not of receptor-α prevents vascular smooth muscle cell accumulation in fibrous cap lesions in apolipoprotein E-deficient mice. *Circulation* 2001; 103: 2955–2960.

71. Kozaki K, Kaminski WE, Tang J, et al. Blockade of platelet-derived growth factor or its receptors transiently delays but does not prevent fibrous cap formation in ApoE null mice. *Am J Pathol* 2002; 161:1395–1407.

72. Sihvola R, Koskinen P, Myllärniemi M, et al. Prevention of cardiac allograft arteriosclerosis by protein tyrosine kinase inhibitor selective for platelet-derived growth factor receptor. *Circulation* 1999; 99:2295–2301.

73. Sihvola RK, Tikkanen JM, Krebs R, et al. Platelet-derived growth factor receptor inhibition reduces allograft arteriosclerosis of heart and aorta in cholesterol-fed rabbits. *Transplantation* 2003; 75:334–339.

74. Floege J, Ostendorf T. Platelet-derived growth factor: a new clinical target on the horizon. *Kidney Int* 2001; 59:1592–1593.

75. Aase K, Abramsson A, Karlsson L, Betsholtz C, Eriksson U. Expression analysis of PDGF-C in adult and developing mouse tissues. *Mech Dev* 2002; 110:187–191.

76. Changsirikulchai S, Hudkins KL, Goodpaster TA, et al. Platelet-derived growth factor-D expression in developing and mature human kidneys. *Kidney Int* 2002; 62:2043–2054.

77. Eitner F, Ostendorf T, Van Roeyen C, et al. Expression of a novel PDGF isoform, PDGF-C, in normal and diseased rat kidney. *J Am Soc Nephrol* 2002; 13:910–917.

78. Eitner F, Ostendorf T, Kretzler M, et al. PDGF-C expression in the developing and normal adult human kidney and in glomerular diseases. *J Am Soc Nephrol* 2003; 14:1145–1153.

79. Gilbert RE, Kelly DJ, McKay T, et al. PDGF signal transduction inhibition ameliorates experimental mesangial proliferative glomerulonephritis. *Kidney Int* 2001; 59:1324–1332.

80. Ostendorf T, Kunter U, Grone HJ, et al. Specific antagonism of PDGF prevents renal scarring in experimental glomerulonephritis. *J Am Soc Nephrol* 2001; 12:909–918.

81. Yi ES, Lee H, Yin S, et al. Platelet-derived growth factor causes pulmonary cell proliferation and collagen deposition in vivo. *Am J Pathol* 1996; 149:539–548.

82. Gurujeyalakshmi G, Hollinger MA, Giri SN. Pirfenidone inhibits PDGF isoforms in bleomycin hamster model of lung fibrosis at the translational level. *Am J Physiol Lung Cell Mol Physiol* 1999; 276:L311–L318.

83. Rice PL, Porter SE, Koski KM, Ramakrishna G, Chen A, Schrump D, Kazlauskas A, Malkinson AM. Reduced receptor expression for platelet-derived growth factor and epidermal growth factor in dividing mouse lung epithelial cells. *Mol Carcinog* 1999; 25:285–294.

84. Doolittle RF, Hunkapiller MW, Hood LE, et al. Simian sarcoma virus *onc* gene, v-*sis*, is derived from the gene (or genes) encoding a platelet-derived growth factor. *Science* 1983; 221:275–277.

85. Waterfield MD, Scrace GT, Whittle N, et al. Platelet-derived growth factor is structurally related to the putative transforming protein p28^{sis} of simian sarcoma virus. *Nature (London)* 1983; 304:35–39.

86. Westermark B, Johnsson A, Betsholtz C, Heldin C-H. Biological properties of simian sarcoma virus and its oncogene product. *Contr. Oncol.,* vol. 24. Basel. Karger, 1987:51–61.

87. Beckmann MP, Betsholtz C, Heldin C-H, et al. Comparison of biological properties and transforming potential of human PDGF-A and PDGF-B chains. *Science* 1988; 241:1346–1349.

88. Li H, Fredriksson L, Li X, Eriksson U. PDGF-D is a potent transforming and angiogenic growth factor. *Oncogene* 2003; 22:1501–1510.

89. Pietras K, Sjöblom T, Rubin K, Heldin C, Östman A. PDGF receptors as cancer drug targets. *Cancer Cell* 2003; 3:439–443.

90. Hermanson M, Funa K, Koopmann J, et al. Association of loss of heterozygosity on chromosome 17p with high platelet-derived growth factor a receptor expression in human malignant gliomas. *Cancer Res* 1996; 56:164–171.

91. LaRochelle WJ, Jeffers M, Corvalan JR, et al. Platelet-derived growth factor D. tumorigenicity in mice and dysregulated expression in human cancer. *Cancer Res* 2002; 62:2468–2473.

92. Lokker NA, Sullivan CM, Hollenbach SJ, Israel MA, Giese NA. Platelet-derived growth factor (PDGF) autocrine signaling regulates survival and mitogenic pathways in glioblastoma cells: evidence that the novel PDGF-C and PDGF-D ligands may play a role in the development of brain tumors. *Cancer Res* 2002; 62:3729–3735.

93. Fleming TP, Matsui T, Heidaran MA, Molloy CJ, Artrip J, Aaronson SA. Demonstration of an activated platelet-derived growth factor autocrine pathway and its role in human tumor cell proliferation in vitro. *Oncogene* 1992; 7:1355–1359.

94. Kumabe T, Sohma Y, Kayama T, Yoshimoto T, Yamamoto T. Amplification of α-platelet-derived growth factor receptor gene lacking an exon coding for a portion of the extracellular region in a primary brain tumor of glial origin. *Oncogene* 1992; 7:627–633.

95. Clarke ID, Dirks PB. A human brain tumor-derived PDGFR-α deletion mutant is transforming. *Oncogene* 2003; 22:722–733.

96. Deinhardt F. The biology of primate retrovirus. In: Klein, G, ed. *Viral Oncology*, Vol. New York: Raven Press, 1980:359–398.

97. Uhrbom L, Hesselager G, Nistér M, Westermark B. Induction of brain tumors in mice using a recombinant platelet-derived growth factor B-chain retrovirus. *Cancer Res* 1998; 58:5275–5279.

98. Shamah SM, Stiles CD, Guha A. Dominant-negative mutants of platelet-derived growth factor revert the transformed phenotype of human astrocytoma cells. *Mol Cell Biol* 1993; 13:7203–7212.

99. Strawn LM, Mann E, Elliger SS, et al. Inhibition of glioma cell growth by a truncated platelet-derived growth factor-β receptor. *J Biol Chem* 1994; 269:21215–21222.

100. Kilic T, Alberta JA, Zdunek PR, et al. Intracranial inhibition of platelet-derived growth factor–mediated glioblastoma cell growth by an orally active kinase inhibitor of the 2-phenylaminopyrimidine class. *Cancer Res* 2000; 60:5143–5150.

101. Smits A, Funa K, Vassbotn FS, et al. Expression of platelet-derived growth factor and its receptors in proliferative disorders of fibroblastic origin. *Am J Pathol* 1992; 140:639–648.

102. Zwerner JP, May WA. PDGF-C is an EWS/FLI induced transforming growth factor in Ewing family tumors. *Oncogene* 2001; 20:626–633.

103. Zwerner JP, May WA. Dominant negative PDGF-C inhibits growth of Ewing family tumor cell lines. *Oncogene* 2002; 21:3847–3854.

104. Heinrich MC, Corless CL, Duensing A, et al. PDGFRA activating mutations in gastrointestinal stromal tumors. *Science* 2003; 299:708–710.

105. Heinrich MC, Corless CL, von Mehren M, et al. PDGFRA and KIT mutations correlate with the clinical responses to imatinib mesylate in patients with advanced gastrointestinal stromal tumors (GIST). In *Proceedings of ASCO, 2003*.

106. Mandahl N, Heim S, Willen H, Rydholm A, Mitelman F. Supernumerary ring chromosome as the sole cytogenetic abnormality in a dermatofibrosarcoma protuberans. *Cancer Genet Cytogenet* 1990; 49:273–275.

107. Naeem R, Lux ML, Huang SF, Naber SP, Corson JM, Fletcher JA. Ring chromosomes in dermatofibrosarcoma protuberans are composed of interspersed sequences from chromosomes 17 and 22. *Am J Pathol* 1995; 147:1553–1558.

108. Craver RD, Correa H, Kao YS, Van Brunt T, Golladay ES. Aggressive giant cell fibroblastoma with a balanced 17; 22 translocation. *Cancer Genet Cytogenet* 1995; 80:20–22.

109. Pedeutour F, Simon MP, Minoletti F, et al. Translocation, t(17;22)(q22;q13), in dermatofibrosarcoma protuberans. a new tumor-associated chromosome rearrangement. *Cytogenet Cell Genet* 1996; 72:171–174.

110. Simon M-P, Pedeutour F, Sirvent N, et al. Deregulation of the platelet-derived growth factor B-chain gene via fusion with collagen gene *COL1A1* in dermatofibrosarcoma protuberans and giant-cell fibroblastoma. *Nature Genet* 1997; 15:95–98.

111. Sheng WQ, Hashimoto H, Okamoto S, et al. Expression of COL1A1-PDGFB fusion transcripts in superficial adult fibrosarcoma suggests a close relationship to dermatofibrosarcoma protuberans. *J Pathol* 2001; 194:88–94.

112. Wang J, Hisaoka M, Shimajiri S, Morimitsu Y, Hashimoto H. Detection of COL1A1-PDGFB fusion transcripts in dermatofibrosarcoma protuberans by reverse transcription-polymerase chain reaction using archival formalin-fixed, paraffin-embedded tissues. *Diagn Mol Pathol* 1999; 8:113–119.

113. Shimizu A, O'Brien KP, Sjöblom T, et al. The dermatofibrosarcoma protuberans-associated collagen type Iα1/platelet-derived growth factor (PDGF) B-chain fusion gene generates a transforming protein that is processed to functional PDGF-BB. *Cancer Res* 1999; 59:3719–3723.

114. Greco A, Fusetti L, Villa R, et al. Transforming activity of the chimeric sequence formed by the fusion of collagen gene COL1A1 and the platelet derived growth factor b-chain gene in dermatofibrosarcoma protuberans. *Oncogene* 1998; 17:1313–1319.

115. Sjöblom T, Shimizu A, O'Brien KP, et al. Growth inhibition of dermatofibrosarcoma protuberans tumors by the platelet-derived growth factor receptor antagonist STI571 through induction of apoptosis. *Cancer Res* 2001; 61:5778–5783.

116. Rubin BP, Schuetze SM, Eary JF, et al. Molecular targeting of platelet-derived growth factor B by imatinib mesylate in a patient with metastatic dermatofibrosarcoma protuberans. *J Clin Oncol* 2002; 20:3586–3591.

117. Maki RG, Awan RA, Dixon RH, Jhanwar S, Antonescu CR. Differential sensitivity to imatinib of 2 patients with metastatic sarcoma arising from dermatofibrosarcoma protuberans. *Int J Cancer* 2002; 100:623–626.

117a. McArthur GA, Demetri GD, van Oosterom A, et al. Molecular and clinical analysis of locally advanced dermatofibrosarcoma protuberans treated with imatinib: Imatinib Target Exploration Consortium Study B2225. *J Clin Oncol* 2005; 23: 866–873.

118. Ganser A, Hoelzer D. Clinical course of myelodysplastic syndromes. *Hematol Oncol Clin North Am* 1992; 6:607–618.

119. Golub TR, Barker GF, Lovett M, Gilliland DG. Fusion of PDGF receptor β to a novel *ets*-like gene, *tel*, in chronic myelomonocytic leukemia with t(5;12) chromosomal translocation. *Cell* 1994; 77:307–316.

120. Kulkarni S, Heath C, Parker S, et al. Fusion of H4/D10S170 to the platelet-derived growth factor receptor β in BCR-ABL-negative myeloproliferative disorders with a t(5;10)(q33;q21). *Cancer Res* 2000; 60:3592–3598.

121. Schwaller J, Anastasiadou E, Cain D, et al. H4(D10S170), a gene frequently rearranged in papillary thyroid carcinoma, is fused to the platelet-derived growth factor receptor β gene in atypical chronic myeloid leukemia with t(5;10)(q33;q22). *Blood* 2001; 97:3910–3918.

122. Ross TS, Bernard OA, Berger R, Gilliland DG. Fusion of Huntingtin Interacting protein 1 to platelet-derived growth factor β receptor (PDGFβR) in chronic myelomonocytic leukemia with t(5;7)(q33;q11.2). *Blood* 1998; 91:4419–4426.

123. Ross TS, Gilliland DG. Transforming properties of the Huntingtin interacting protein 1/platelet-derived growth factor β receptor fusion protein. *J Biol Chem* 1999; 274:22328–22336.

124. Magnusson MK, Meade KE, Brown KE, et al. Rabaptin-5 is a novel fusion partner to platelet-derived growth factor β receptor in chronic myelomonocytic leukemia. *Blood* 2001; 98:2518–2525.

125. Steer EJ, Cross NC. Myeloproliferative disorders with translocations of chromosome 5q31-35: role of the platelet-derived growth factor receptor β. *Acta Haematol* 2002; 107:113–122.

126. Carroll M, Tomasson MH, Barker GF, Golub TR, Gilliland DG. The TEL/platelet-derived growth factor β receptor (PDGFβR) fusion in chronic myelomonocytic leukemia is a

transforming protein that self-associates and activates PDGFβR kinase-dependent signaling pathways. *Proc Natl Acad Sci USA* 1996; 93:14845–14850.

127. Jousset C, Carron C, Boureux A, et al. A domain of TEL conserved in a subset of ETS proteins defines a specific oligomerization interface essential to the mitogenic properties of the TEL-PDGFRβ oncoprotein. *EMBO J* 1997; 16:69–82.

128. Sternberg DW, Tomasson MH, Carroll M, et al. The TEL/PDGFβR fusion in chronic myelomonocytic leukemia signals through STAT5-dependent and STAT5-independent pathways. *Blood* 2001; 98:3390–3397.

129. Carroll M, Ohno-Jones S, Tamura S, et al. CGP 57148, a tyrosine kinase inhibitor, inhibits the growth of cells expressing BCR-ABL, TEL-ABL, and TEL-PDGFR fusion proteins. *Blood* 1997; 90:4947–4952.

130. Sjöblom T, Boureux A, Rönnstrand L, Heldin C-H, Ghysdael J, Östman A. Characterization of the chronic myelomonocytic leukemia associated TEL-PDGFβR fusion protein. *Oncogene* 1999; 18:7055–7062.

131. Tomasson MH, Williams IR, Hasserjian R, et al. TEL/PDGFβR induces hematologic malignancies in mice that respond to a specific tyrosine kinase inhibitor. *Blood* 1999; 93:1707–1714.

132. Tomasson MH, Sternberg DW, Williams IR, et al. Fatal myeloproliferation, induced in mice by TEL/PDGFβR expression, depends on PDGFβR tyrosines 579/581. *J Clin Invest* 2000; 105:423–432.

133. Apperley JF, Gardembas M, Melo JV, et al. Response to imatinib mesylate in patients with chronic myeloproliferative diseases with rearrangements of the platelet-derived growth factor receptor β. *N Engl J Med* 2002; 347:481–487.

134. Pitini V, Arrigo C, Teti D, Barresi G, Righi M, Alo G. Response to STI571 in chronic myelomonocytic leukemia with platelet derived growth factor β receptor involvement. a new case report. *Haematologica* 2003; 88:ECR18.

135. Magnusson MK, Meade KE, Nakamura R, Barrett J, Dunbar CE. Activity of STI571 in chronic myelomonocytic leukemia with a platelet-derived growth factor β receptor fusion oncogene. *Blood* 2002; 100:1088–1091.

136. Weller PF, Bubley GJ. The idiopathic hypereosinophilic syndrome. *Blood* 1994; 83:2759–2779.

137. Gleich GJ, Leiferman KM, Pardanani A, Tefferi A, Butterfield JH. Treatment of hypereosinophilic syndrome with imatinib mesylate. *Lancet* 2002; 359:1577–1578.

138. Cools J, DeAngelo DJ, Gotlib J, et al. A tyrosine kinase created by fusion of the PDGFRA and FIP1L1 genes as a therapeutic target of imatinib in idiopathic hypereosinophilic syndrome. *N Engl J Med* 2003; 348:1201–1214.

139. Griffin JH, Leung J, Bruner RJ, Caligiuri MA, Briesewitz R. Discovery of a fusion kinase in EOL-1 cells and idiopathic hypereosinophilic syndrome. *Proc Natl Acad Sci USA* 2003; 100:7830–7835.

140. Pardanani A, Ketterling RP, Brockman SR, et al. CHIC2 deletion, a surrogate for FIP1L1-PDGFRA fusion, occurs in systemic mastocytosis associated with eosinophilia and predicts response to imatinib therapy. *Blood* 2003; 102:3093–3096.

141. Bhardwaj B, Klassen J, Cossette N, et al. Localization of platelet-derived growth factor β receptor expression in the periepithelial stroma of human breast carcinoma. *Clin Cancer Res* 1996; 2:773–782.

142. Sundberg C, Branting M, Gerdin B, Rubin K. Tumor cell and connective tissue cell interactions in human colorectal adenocarcinoma—Transfer of platelet-derived growth factor-AB/BB to stromal cells. *Am J Pathol* 1997; 151:479–492.

143. Franklin WA, Christison WH, Colley M, Montag AG, Stephens JK, Hart CE. In situ distribution of the β-subunit of platelet-derived growth factor receptor in nonneoplastic tissue and in soft tissue tumors. *Cancer Res* 1990; 50:6344–6348.

144. Hermanson M, Nistér M, Betsholtz C, Heldin C-H, Westermark B, Funa K. Endothelial cell hyperplasia in human glioblastoma: coexpression of mRNA for platelet-derived growh factor (PDGF) B chain and PDGF receptor suggests autocrine growth stimulation. *Proc Natl Acad Sci USA* 1988; 85:7748–7752.

145. Plate KH, Breier G, Farrell CL, Risau W. Platelet-derived growth factor receptor-β is induced during tumor development and upregulated during tumor progression in endothelial cells in human gliomas. *Lab Invest* 1992; 67:529–534.

146. Sundberg C, Ljungström M, Lindmark G, Gerdin B, Rubin K. Microvascular pericytes express platelet-derived growth factor-β receptors in human healing wounds and colorectal adenocarcinoma. *Am J Pathol* 1993; 143:1377–1388.

147. Abramsson A, Berlin O, Papayan H, Paulin D, Shani M, Betsholtz C. Analysis of mural cell recruitment to tumor vessels. *Circulation* 2002; 105:112–117.

148. Oikawa T, Onozawa C, Sakaguchi M, Morita I, Murota S. Three isoforms of platelet-derived growth factors all have the capability to induce angiogenesis in vivo. *Biol Pharm Bull* 1994; 17:1686–1688.

149. Risau W, Drexler H, Mironov V, et al. Platelet-derived growth factor is angiogenic in vivo. *Growth Factors* 1992; 7:261–266.

150. Oh SJ, Kurz H, Christ B, Wilting J. Platelet-derived growth factor-B induces transformation of fibrocytes into spindle-shaped myofibroblasts in vivo. *Histochem Cell Biol* 1998; 109:349–357.

151. Wilting J, Christ B, Bokeloh M, Weich HA. In vivo effects of vascular endothelial growth factor on the chicken chorioallantoic membrane. *Cell Tissue Res* 1993; 274:163–172.

152. Cao R, Bråkenhielm E, Li X, et al. Angiogenesis stimulated by PDGF-CC, a novel member in the PDGF family, involves activation of PDGFR-αα and -αβ receptors. *FASEB J* 2002; 16:1575–1583.

153. Guo P, Hu B, Gu W, Xu L, et al. Platelet-derived growth factor-B enhances glioma angiogenesis by stimulating vascular endothelial growth factor expression in tumor endothelia and by promoting pericyte recruitment. *Am J Pathol* 2003; 162:1083–1093.

153a. Furnhashi M, Sjöblom T, Abramsson A, et al. Platelet-derived growth factor production by B16 melanoma cells leads to increased pericyte abundance in tumors and an associated increase in tumor growth rate. *Cancer Res* 2004; 64:2725–2733.

154. Bergers G, Song S, Meyer-Morse N, Bergsland E, Hanahan D. Benefits of targeting both pericytes and endothelial cells in the tumor vasculature with kinase inhibitors. *J Clin Invest* 2003; 111:1287–1295.

155. Uehara HK, Karashima T, Shepherd DL, et al. Effects of blocking platelet-derived growth factor-receptor signaling in a mouse model of experimental prostate cancer bone metastases. *J Natl Cancer Inst* 2003; 95:458–470.

156. Forsberg K, Valyi-Nagy I, Heldin C-H, Herlyn M, Westermark B. Platelet-derived growth factor (PDGF) in oncogenesis: development of a vascular connective tissue stroma in xenotransplanted human melanoma producing PDGF-BB. *Proc Natl Acad Sci USA* 1993; 90:393–397.

157. Skobe M, Fusenig NE. Tumorigenic conversion of immortal human keratinocytes through stromal cell activation. *Proc Natl Acad Sci USA* 1998; 95:1050–1055.

158. Shao Z-M, Nguyen M, Barsky SH. Human breast carcinoma desmoplasia is PDGF initiated. *Oncogene* 2000; 19:4337–4345.

159. Pietras K, Hubert M, Buchdunger E, et al. STI571 enhances the therapeutic index of Epothilone B by a tumor-selective increase of drug uptake. *Clin Cancer Res* 2003; 3779–3787.

9   Structural Biology of Protein Tyrosine Kinases

Sandra W. Cowan-Jacob, PhD,
Paul Ramage, PhD, Wilhelm Stark, PhD,
Gabriele Fendrich, PhD,
and Wolfgang Jahnke, PhD

CONTENTS

INTRODUCTION
PROTEIN PREPARATION FOR TYROSINE KINASES
STRUCTURES OF PTKS
NMR AND PROTEIN KINASES
REFERENCES

1. INTRODUCTION

Structural understanding of protein tyrosine kinases (PTKs) has significantly progressed during the past 3 to 5 yr. Three-dimensional structures of several PTKs are now available, in both phosphorylated or nonphosphorylated forms, active or inactive states, unliganded or complexed to substrate analogs or inhibitors, and with only the catalytic domain present or in a multidomain construct including Src homology domain (SH)3 and SH2 domains. Detailed knowledge about the structural basis for kinase activity and regulation has been gained, and the structures are widely used for drug design.

The key for success in determining kinase structures lies in great part in the ability to produce milligram amounts of well-characterized, homogeneous, high-purity stable protein. Major advances have been achieved in this field, and the following section summarizes general guidelines and principles for the preparation of PTKs for structural studies. Methods for protein structure determination include X-ray crystallography and nuclear magnetic resonance

From: *Cancer Drug Discovery and Development:*
Protein Tyrosine Kinases: From Inhibitors to Useful Drugs
Edited by: D. Fabbro and F. McCormick © Humana Press Inc., Totowa, NJ

```
                                 β1           β2          β3              αC
                              *      *  *           *  *           *  #         * A
                                     A A           *              #  #*          #   *
                                     # # #*                                      *
ABL    1m52 223 ------GAM  DPSSPNYDKW  EMERTDITMK  HKLGGGQYGE  VYEGVWKK--  --YSLTV   AVKTLKED--  TMEVEEFLKE  AAVMKEIKHP
CSK    1byg 173 -----MGGSVA AQDEFYRSGW  ALNMKELKLL  QTIGKGEFGD  VMLGDYRG--  ---NKV    AVKCIKND--  -ATAQAFLAE  ASVMTQLRHS
FAK    1mp8 406 GAMGSSTRDY EIQRERIELG  RCIGEGQFGD  VHQGIYMS--  -PENPALAV AIKTCKNCTS  DSVREKFLQE  ALTMRQFDHP
HCK    2hck 241 KLSVPCMSSK PQKPWEKDAW  KKLGAGQFGE  VWMATYNK--  -HTKV     AVKTMKPG-   SMSVEAFLAE  ANVMKTLQHD
LCK    3lck 231 -KPWWEDEW  EVPRETLKLV  ERLGAGQFGE  VWMGYYNG--  -TTRV     AVKSLKQG-   SMSPDAFLAE  ANLMKQLQHQ
SRC    1fmk 241 LTTVCPTSKP EIPRESRLE   VKLGQGCFGE  VWMGTWNG--  -GQYDV    AIKTLKPG-   TMSPEAFLQE  AQVMKKLRHE
BTK    1k2p 397 -IDPKDLTFL  KELGTGQFGV  VKYGKWR--  AIKMIKEG-   SMSEDEFIEE  AKVMMNLSHE
ZAP70       331 ---F       LKRDNLLIAD  IELGCGNFGS  VRQGVYRM--  RKKQIDV AIKVLKQGTE  AQIMHQLDNP
```

```
                                  β4           β5            αD                      β3                  αC
                                  #  *         *      *        *          *  #                            **
EGFR   1M14 666 ---GSHMAS  GEAPNQALLR  ILKETEFKKI  KVLGSGAFGT  VYKGLWIP-  --EGEKVKIPV  AIKELREATS  PKANKEILDE  AYVMASVDNP
EPHB2  1jpa 603 IFIDPFTFED PNEAVREFAK  EIDISCVKIE  QVIGAGEFGE  VCSGHLKL--  -PGKREIFV   AIKTLKSGYT  EKQRRDFLSE  ASIMGQFDHP
FGFR1  1fgi 456 --MVAGVSE  YELPED-PRW  ELPRDRLVLG  KPLGEGAFGQ  VVLAEAIGLD  KDKPNRVTKV  AVMLKSDAT   EKDLSDLISE  MEMMKMIGKH
IRK    1p4o 943 MASVNPEYFS AADVYVPDEW  EVAREKITMS  RELQGSFGM   VYEGVAKG--  -VVKDEPETRV AIKTVNEAAS  MRERIEFLNE  ASVMKEFNCH
IGF1R  1jqh 979 --GSFS     AADVTVPDEW  EVAREKITMS  RELQGSFGM   VYEGVAKG--  -VVKDEPETRV AIKTVNEAAS  MRERIEFNCH  ASVMKEENCH
MET    1r0p 1052 HIDLSALNPE LVQAVQHVVI  GPSSLIVHFN  EVIGRGHFGC  VYHGTLL--  -DNDGKKIHC  AVKSLNRITD  IGEVSQFLTE  GIIMKDFSHP
MUSK   1luf 548 HPNPMYQRMP LILNPKLLSI  EYPRNNIEVV  RDIGEGAFGR  VFQARAPG--  LLPYEPFTMV  AVMLKKEAS   ADMQADFQRE  AALMAEFDNP
KIT    1pkg 563 INGNNYYID  PTQLPYDHKW  EFPRNRLSFG  KTLGAGAFGK  VVEATAYG-  LIKSDAAMTV  AVKMLKPSAH  LTEREAIMSE  LKVLSYLGNH
TIE2   1fvr 798 MKKHHHHHHG KNNPDPTIYP  VLDWNDIKFQ  DVIGEGNFGQ  VLKARIKK--  -DGLRMDA    AIKRMKEYAS  KODHRDFAGE  LEVLCKLGHH
VEGFR2 1vr2 808 PDELPLDEHC ERLPYDASKW  EFPRDRLNLG  KPLGRGAFGQ  VIEADAFG--  IDKTATCRTV  AVMLKEGAT   HSEHRALMSE  LKILIHIGHH
```

Kinase insertion domain

```
                                 β4           β5              αD                                   αE
                                 # *          *      *          *  #                                **
ABL    1m52 297 N-LVQLLGVC  TREP-PFYII  TEFMTYGNLL  DYLRECNRQE  V------    SAVLLYM    ATQISSAMEY
CSK    1byg 247 N-LVQLLGVI  VEEKGGLYIV  TEYMAKGSLV  DYLRSRGRSV  L------    -GGDCLLKF  SLDVCEAMEY
FAK    1mp8 482 H-IVKLIGVI  TEN -PVWII  MELCTLGELR  SFLQVRKY--  -SLDLASLILY AYQLSTALAY
HCK    2hck 321 K-LVKLHAVV  TKE-PIYII   TEFMAKGSLL  DFLKSDEGSK  Q------    --PLPKLIDF SAQIAEGMAF
LCK    3lck 299 R-LVRLYAVV  TQE-PIYII   TEYMENGSLV  DFLKTPSGI  ------      KLTINKLLDM AAQIAEGMAF
SRC    1fmk 321 K-LVQLYAVV  SEE-PIYIV   TEYMSKGSLL  DFIKGETGKY  LR-----    --LPQLVDM  AAQIASGMAY
BTK    1k2p 456 K-LVQLYGVC  TKQR-PIFII  TEYMANGCLL  NYLREMRHRF  ----       QTQQLLEM   CKDVCEAMEY
ZAP70       397 Y-IVRLIGVC  QAE-ALMLV   MEMAGGGPLH  KFIVGKREEI  ----       PVSNVAEI   LHQVSMGMKY
```

Fig. 1.

188

```
                                                                                    αEF              αF
                                                                                    __              __
EGFR     1M14  749  H-VCRLLGIC LTS--TVQLI TQLMPFGCLL DYVREHKDNI --------- --------- --------- --GSQYLLNW CVQIAKGMNY
EPHB2    1jpa  689  N-VIHLEGVV TKST-PVMII TEFMENGSLD SFLRQNDGQ- --------- --------- --------- --FTVIQLVGM LRGIAAGMKY
FGFR1    1fgi  542  KNIINLLGAC TQDG-PLXVI VEYASKGNLR EYLQARRPPG LEYSYNPSHN PEEQLS---- --------- ----SKDIVSC AYQVARGMEY
IRK      1p4o  1031 H-VVRLLGVV SQGQ-PTIVI MELMTRGDLK SYLRSLRPAM ANN------- --------- ----PVLA PPSLSKMIQM AGEIADGMAY
IGF1R    1jqh  1061 H-VVRLLGVV SQGQ-PTIVI MELMTRGDLK SYLRSLRPE- --------- --------- ----MENNPVLA PPSLSKMIQM AGEIADGMAY
MET      1r0p  1138 N-VLSLLGVC LPYMKHGDLR NFIRNETHN- --------- --------- --------- ----PTVKDLIGF GLQVAKGMKF
MUSK     1luf  636  N-IVKLLGVC AVGK-PMCLL FEIMAYGDLN EFLRSMSPHT VCSLSHSDLS --------- --ARVSSPGPP PLSCAEQLCI ARQVAAGMAY
Kit      1pkg  651  MNIVNLLGAC TIGG-PTIVI TEYCCYGDIL NFLRRKRDSF ICSKTS---- --//------ --PAIMEDDEL ALDLEDLLSF SYQVAKGMAF
TIE2     1fvr  883  PNIINLLGAC EHRG-YLYLA IEYAPHGNLL DFLRKSRVLE TDPA------ --//------ --FAIANSTAS TLSSQQLLHF AADVARGMDY
VEGFR2   1vr2  896  LNVVNLLGAC TKPGGPLMVI VEFCKFGNLS TYLRSKRNEF VPYKV----- --//------ --APEDLYKD FLTLEHLICY SFQVAKGMEF

             αE              β7          β8                          Activation loop        αEF                αF
             __              __          __                                                 __                __
             # ##            *        A  # A                       #        * *            #      *            ###### ###  *  #
ABL      1m52  354  LEKNFIHRD LAARNCLVGE NHILVKVADFG LSRLMTGDTY TAHA----GA KFPIKWTAPE SLAYNKFSIK SDVWAFGVLL WEIATYGMSP
CSK      1byg  305  LEGNNFVHRD LAARNVLVSE DNVAKVSDFG LTKEASSTQD T------G   KLPVKWTAPE ALREKFSTK  SDVWSFGILL WEIYSFGRVP
FAK      1mp8  537  LESKRFVHRD LAARNVLVSS NDCVKLGDFG LSRYMEDSTY YKAS----KG KLPIKWMAPE SINFRRFTSA SDVWMFGVCM WEILMHGVKP
HCK      2hck  377  IEQRNYIHRD LRAANILVSA SLVCKIADFG LARVIEDNEY TARE----GA KFPIKWTAPE AINFGSFTIK SDVWSFGILL MEIVTYGRIP
LCK      3lck  355  IEERNYIHRD LRAANILVSD TLSCKIADFG LARLIEDNEY TARE----GA KFPIKWTAPE AINYGTFTIK SDVWSFGILL TEIVTHGRIP
SRC      1fmk  377  VERMNYVHRD LRAANILVGE NLVCKVADFG LARLIEDNEY TARQ----GA KFPVKWTAPE AALYGRFTIK SDTWAFGVLM WEIYSLGKMP
BTK      1k2p  512  LESKQFLHRD LAARNCLVND QGVVKVSDFG LSRYVLDDEY TSSV----GS KFPVRWSPPE VLMYSKFSSK SDVWSYGVTM WEIYSIGKMP
ZAP70          452  LEEKNFVHRD LAARNILLVN RHYAKISDFG LSKALGADDS YTTA-RSAG  KWPLKWYAPE CINFRKFSSR SDVWSYGVTM WEALSYGQKP

             αE              β7          β8                          Activation loop        αEF                αF
EGFR     1M14  804  LEDRRLVHRD LAARNVLVKT PQHVKITDFG LAKLLGAEEK EYHA---EGG KVPIKWMALE SILHRIYTHQ SDVWSYGVTV WELMTFGSKP
EPHB2    1jpa  745  LADMNYVHRD LAARNILVNS NLVCKVSDFG LSRFLEDDTS DPTYTSALGG KIPIRWTAPE AIQYRKFTSA SDVWSYGIVM WEVMSYGERP
FGFR1    1fgi  614  LASKKCIHRD LAARNVLVTE DNVMKIADFG LARDIHHIDY YKKT---TNG RLPVKWMAPE ALFDRIYTHQ SDVWSFGVLL WEIFTLGGSP
IRK      1p4o  1096 LNANKFVHRD LAARNCMVAE DFTVKIGDFG MTRDIYETDY YRKG---GKG LLPVRWMSPE SLKDGVFTTY SDVWSFGVVL WEIATLAEQP
IGF1R    1jqh  1126 LNANKFVHRD LAARNCMVAE DFTVKIGDFG MTRDIYETDY YRKG---GKG LLPVRWMSPE SLKDGVFTTY SDVWSFGVVL WEIATLAEQP
MET      1r0p  1195 LASKKFVHRD LAARNCMLDE KFTVKVADFG LARDMYDKEF DSVHNKT-GA KLPVKWMALE SLQTQKFTTK SDVWSFGVLL WELMTRGAPP
MUSK     1luf  715  LSERKFVHRD LATRNCLVGE NMVKIADFG  LSRNISADY  YKAD---GNA AIPIRWMPPE SIFYNRYTTE SDVWAYGVVL WEIFSYGLQP
Kit      1pkg  783  LASKNCIHRD LAARNILLTH GRITKICDFG LARDIKNDSN YVVK---GNA RLPVKWMAPE SIFNCVYTFE SDVWSYGIFL WELFSLGSSP
TIE2     1fvr  955  LSQKQEIHRD LAARNILLSE NYVAKIADFG LSRGQEYVVK KT-----MG  RLPVRWMAIE SINYSVYTTN SDVWSYGVLL WEIVSLGGTP
VEGFR2   1vr2  1019 LASRKCIHRD LAARNILLSE KNVVKICDFG LARDIYKDPD YVRK---GDA RLPLKWMAPE TIFDRVVTIQ SDVWSFGVLL WEIFSLGASP
```

Fig. 1. (Continued)

189

```
                                 αG                         αH                    αI
                         **                  #            #   #     ##
ABL     1m52   440  YPGIDLSQVY ELLEKDYRME R-PEGCPEKV YELMRACWQW NPSDRPSFAE IHQAFETMFQ ESSISDEVEK ELGKRGT
CSK     1byg   387  YPRIPLKDVV PRVEKGYKMD A-PDGCPPAV YEVMKNCWHL DAAMRPSFLQ LREQLEHIKT HELHL
FAK     1mp8   623  FQGVKNNDVI GRIENGERLP M-PPNCPPTL YSLMTKCWAY DPSRRPRFTE LKAQLSTILE EEKAQ
HCK     2hck   463  YPGMSNPEVI RALERGYRMP R-PENCPEEL YNIMMRCWKN RPEERPTFEY IQSVLDDFYT ATESQXQQQP
LCK     31ck   441  YPGMTNPEVI QNLERGYRMV R-PDNCPEEL YQLMRLCWKE RPEDRPTFDY LRSVLEDFFT AT
SRC     1fmk   463  YPGMVNREVL DQVERGYRMP C-PPECPESL HDLMCQCWRK EPEERPTFEY LQAFLEDYFT STEPQXQPGE NL
BTK     1k2p   598  YERFTNSETA EHIAQGLRLY R-PHLASEKV YTIMYSCWHE KADERPTFKI LLSNILDVMD EES
ZAP70          540  YKKMKGPEVM AFIEQGKRME C-PPECPPEL YALMSDCWIY KWEDRPDFLT VEQRMRACYY SLASK

EGFR    1M14   891  YDGIPASEIS SILEKGERLP Q-PPICTIDV YMIMVKCWMI DADSRPKFRE LIIEFSKMAR DPQRYLVIQG DERMHLPSPT
EPHB2   1jpa   835  YWDMTNQDVI NAIEQDYRLP P-PMDCPSAL HQLMLDCWQK DRNHRPKFGQ IVNTLDKMIR NPNSLKAMAP LSS
FGFR1   1fgi   701  YPGVPVEELF KLLKEGHRMD K-PSNCTNEL YMMMRDCWHA VPSQRPTFKQ LVEDLDRIVA LTSNQE
IRK     1p4o  1183  YQGLSNEQVL RFVMEGGLLD K-PDNCPDML FELMRMCWQY NPKMRPSFLE IISSIKEEME PGFREVSFYY SEENKLPEPE
IGF1R   1jqh  1213  YQGLSNEQVL RFVMEGGLLD K-PDNCPDML FELMRMCWQY NPKMRPSFLE IISSIKEEME PGFREVSFYY SEENK
MET     1r0p  1284  YPDVNTFDIT VYLLQGRRLL Q-PEYCPDPL YEVMLKCWHP KAEMRPSFSE LVSRISAIFS TFIGEHYVHV NATYVNVK
MUSK    1luf   802  YYGMAHEEVI YYVRDGNILA C-PENCPLEL YNLMRLCWSK LPADRPSFCS IHRILQRMCE RAEGTVGV
Kit     1pkg   870  YPGMPVDSKF YKMIKEGFRM LSPEHAPAEM YDIMKTCWDA DPLKRPTFKQ IVQLIEKQIS ESTNHI
TIE2    1fvr  1039  YCGMTCAELY EKLPQGYRLE K-PLNCDDEV YDLMRQCWRE KPYERPSFAQ ILVSLNRMLE ERKTYVNTTL YEKFTYAGID CS
VEGFR2  1vr2  1106  YPGVKIDEEF CRRLKEGTRM RAPDYTTPEM YQTMLDCWHG EPSQRPTFSE LVEHLGNLLQ ANAQQD
```

Fig. 1. Structure based sequence alignment of 8 nonreceptor protein tyrosine kinases (PTKs) (top part of each panel) and 10 receptor PTKs (bottom), many of them belonging to different PTK families (compare Table 1 and 2). Secondary structural elements and residue numbering were taken mainly from the protein database PDB file and are colored in cyan for extended strands and green for helices. Yellow indicates residues that are disordered or missing in the structures but present in the constructs used to produce the protein for crystallization. The sequence may not be identical to that of the corresponding sequence database entries. For example, the kinase insertion domains of cKit and VEGFR2 are not present (indicated by // in the sequence). Phosphorylation sites in the sequence are indicated by Y. Some of the positions in the alignment are marked by a "#" or a "*" (100% identity, highly conserved). Active site residues are indicated by an A.

(NMR) spectroscopy. So far, all structures of kinase catalytic domains have been determined by crystallography, and a detailed description and comparison of PTK structural families can be found in Section 3. NMR has the potential to characterize solution structure and dynamics of PTKs, although PTKs have only recently been discovered as being amenable for NMR studies. NMR is also a valuable technique in fragment-based screening approaches to identify PTK inhibitors. Both applications are described in Section 4. Taken together, progress in protein preparation and structure determination have yielded the three-dimensional structures of representative members of 7 out of 12 nonreceptor PTK families, and 8 out of 19 receptor PTK families (Fig. 1) as of January 2004. This chapter is meant to provide an overview of these known PTK structures.

2. PROTEIN PREPARATION FOR TYROSINE KINASES

In 1994, Hubbard et al. published the 2.1Å structure of the human insulin receptor kinase (IRK) domain *(1)* and although there are now more than 60 protein kinase structures in the protein databank, most of them have been deposited during the last 5 yr (Table 1). Why has it taken so long for kinases to be considered feasible structural targets and what are the difficulties one is likely to experience along the route to crystallizing a kinase ?

Typically tyrosine kinases are rarely expressed as soluble proteins in *Escherichia coli (2)*, in contrast to some of the serine/threonine kinases *(3–5)*— instead scientists are forced to resort to the more cumbersome, labor-intensive baculovirus system. Expression yields tend to be lower and because the expressed kinase domains can be relatively unstable they are difficult to concentrate as well as exhibiting heterogeneous phosphorylation. The process of isolating structural-grade protein can be a difficult one. Once purified and concentrated, crystallization is not necessarily a simple task; Munshi et al., describe the purification of more than two dozen constructs of the insulin-like growth factor receptor kinase (Igf-1R), before one was found that crystallized *(6)*. The following subheading deals with some of the practical aspects to be taken into consideration when producing tyrosine kinases for both X-ray crystallography and NMR studies.

2.1. Construct Definition

Kinases are multidomain proteins (Fig. 2) and the definition of the boundaries of the domain to be expressed is of crucial importance and may influence expression levels, stability of the purified protein, as well as crystallisability *(6)*. Bioinformatics can be used to define kinase domains with a certain degree of accuracy. Limited proteolysis remains a favored technique for identifying optimal kinase domains. In the case of the IRK *(7)*, a 48 kDa soluble kinase-containing domain was first overexpressed *(8)* and subjected to trypsin and elastase treatment to define C- and N-termini respectively. Four subsequent N-terminal deletion mutants and two C-termini were expressed and tested for activity to identify an

Table 1
Known Structures of Nonreceptor Tyrosine Kinases

Kinase	PDB entry code	Source	Domains	Complex with	Reference
Abelson (Abl)	1FPU	Mouse	Kinase	Glivec™ variant	52
	1IEP	Mouse	Kinase	Glivec	53
	1M52	Mouse	Kinase	PD173955	53
	1OPL	Human	SH3, SH2, kinase D382N	PD166326 myristate	11
	1OPJ	Mouse	Kinase	Glivec myristate	11
	1OPK	Mouse	SH3, SH2, kinase	PD166326 myristate	11
Carboxy-terminal Src kinase (Csk)	1BYG	Human	Kinase	Staurosporine	13
	1K9A	Rat	SH3–SH2-kinase	—	35
Focal adhesion kinase (Fak)	1MP8	Human	Kinase	ATP	63
Hck	1AD5	Human	SH3, SH2, kinase	AMP-PNP	40
	2HCK	Human	SH3, SH2, kinase	Quercetin	40
	1QCF	Human	SH3, SH2, kinase	Ylamine	30
Lck	3LCK	Human	Kinase	—	45
	1QPC	Human	Kinase	AMP-PNP	102
	1QPD	Human	Kinase	Staurosporine	102
	1QPE	Human	Kinase	PP2	102
	1QPJ	Human	Kinase	Staurosporine	102
Src	1FMK	Human	SH3-SH2-kinase	—	34
	2PTK	Chicken	SH3-SH2-kinase	—	41
	2SRC	Human	SH3-SH2-kinase	AMP-PNP	42
	1KSW	Human	SH3-SH2-kinase Mutant T388G	N6-benzyl-ADP	103
Bruton's tyrosine Kinase (Btk)	1K2P	Human	Kinase	—	27
ZAP70	—	Human	Kinase	Staurosporine	n.p.

n.p. = not yet published; PDB, Protein databank

Table 2
Known Structures of Receptor Tyrosine Kinases

Kinase	PDB entry code	Source	Domains in structure	Complex with	Reference
EGFR	1M14	Human	Kinase	—	72
	1M17	Human	Kinase	Erlotinib	72
EphA2	1MQB	Human	Kinase	AMP-PNP	63
EphB2	1JPA	Mouse	Kinase + partial juxtamembrane	AMP-PNP	14
FGFR1	1AGW	Human	Kinase	SU4984	74
	1FGI	Human	Kinase	SU5402	74
	1FGK	Human	Kinase (L457V,C488A,C584S)	—	9
	2FGI	Human	Kinase (L457V,C488A,C584S)	PD173074	68
FGFR2	1GJO	Human	Kinase	—	n.p.
	1OEC	—	—	On hold	n.p.
Insulin receptor	1IRK	Human	Kinase (C981S, Y984F)	—	1
	1IR3	Human	Kinase (C981S, Y984F)	Substrate and AMP-PNP	64
	1GAG	Human	Kinase	Bisubstrate inhibitor	78
	1I44	Human	Kinase (activation loop mutant)	ACP	104
	1P14	Human	Kinase (catalytic loop mutant)	—	105
Insulin-like growth factor-1 receptor	1JQH	Human	Kinase	AMP-PNP	31
	1K3A	Human	Kinase	ACP	28
	1M7N	Human	Kinase	—	6
Met	1R0P	Human	Kinase	K-252A	10
	1R1W	Human	Kinase	—	10
MuSK	1LUF	Rat	Kinase	—	12
FLT3	—	Human	Kinase ΔKID	—	66
Kit	1PKG	Human	Kinase ΔKID	ADP	69
Tie2	1FVR	Human	Kinase	—	29
VEGFR2 (KDR)	1VR2	Human	Kinase ΔKID(940–989)	—	67

n.p.; not published.

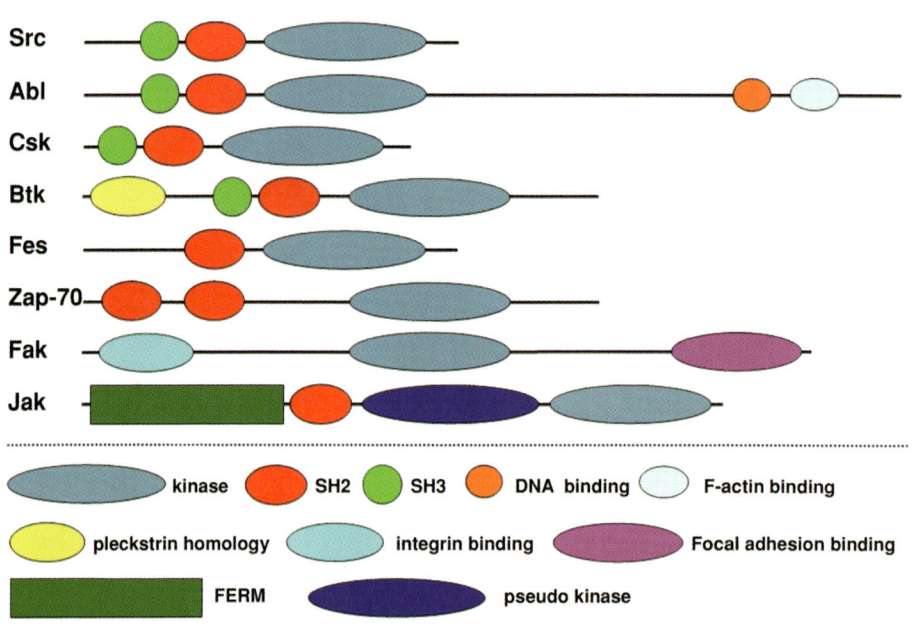

Fig. 2. Domain organization of nonreceptor tyrosine kinases, adapted from Hubbard and Till *(106)*. The size of the structural elements is not to scale.

active core region, a variant of which was subsequently crystallized. In our own experience with a multidomain nonreceptor kinase, the full-length protein was heterogeneously and unstably phosphorylated. Treatment of purified enzyme with again trypsin, thermolysin, and elastase were used to define the N-terminus of the protein, which was eventually expressed, purified, and crystallized. As well as defining the N- and C-termini, thought should also be given to the introduction of point mutations to improve the behavior of the protein. The replacement of cysteine residues with either Ala or Ser can prevent unwanted disulphide formation *(9)*. Unwanted autophosphorylation is also a problem, in effect reducing the yield of protein by producing multiple forms. In the case of cMet, two tyrosines (1194 and 1234) were replaced with phenyalanine and at the same time a constitutively active form of the kinase was produced by mutating tyrosine1235 to aspartate *(10)*. Autophosphorylation problems with c-Abelson kinase (Abl) were abolished by inactivating the kinase with an Asp382-Asn mutation *(11)*. Phosphorylation problems are sometimes a result of host cell kinase activity. MuSK, prior to alkaline phosphatase treatment, was phosphorylated at Ser-690 *(12)*.

2.2. Cloning and Expression

In the vast majority of published tyrosine kinase structures, the recombinant protein used was expressed in SF9 cells using the baculovirus system. This is because tyrosine kinases are usually expressed in the insoluble fraction in *E. coli* and are extremely difficult to refold. There are a few exceptions: carboxy-teminal Src kinase (Csk) and EphB2 were both crystallized from protein produced in *E. coli (13,14)*. Csk was produced as C-terminally His-tagged protein in DH5α cells, whereas EphB2 was grown in B834 cells as a glutathione *S*-transferase (GST) fusion protein. There are several other instances of catalytically active kinases being produced in *E. coli*; Csk (15–17), Txk (18), and v-Abl (2,19). Co-expression with bacterial chaperones has also been used successfully to produce soluble, active c-Src family kinases (20) and p50[csk] (21). In addition to *E. coli*, yeast has been used to express c-Src (22) and HEK293 cells to express c-Abl (23).

The baculovirus system, for the reasons mentioned in the above paragraph, is the expression system of choice for tyrosine kinases. However, it is a complicated system that is not very well suited to a high-throughput/multiparallel approach. As kinases are expressed intracellularly, and the baculovirus system is a lytic one, there is a fine balance to be made between the multiplicity of infection used and the length of time between infection and harvest. On the one hand, expression levels increase with time, up to a point where the cells start to lyse; on the other hand, for some kinases the solubility may be reduced by prolonged infection, presumably owing to the action of phosphatases (P. Ramage, unpublished results) or proteolytic degradation may become significant. When trying to maximize yields, it is useful to be able to carry out analytical affinity chromatography to optimize time of harvest. Similarly, the degree of autophosphorylation observed will often vary with the fermentation conditions.

New advances in fermentation technology, such as the Wave™ Bioreactors (Wave Biotech, Bridgewater, NJ) allow relatively large-scale fermentations to be carried out, with minimum manpower, owing to the disposable nature of the material. In-house results show such fermentations to be comparable in yield to those carried out in conventional bioreactors.

The production of isotopically labelled kinases for NMR poses particular problems; most kinases cannot be expressed in a soluble form in *E. coli* and labeling possibilities are limited using the baculovirus system. However, recent work by Strauss et al. has shown that it is possible to obtain high levels of incorporation of ^{15}N-amino acids in c-Abl and other proteins using amino acid–type selective labeling (24). This effectively means that kinases and other proteins that are reasonably well expressed in the baculovirus system (~10 mg/L), can be economically isotope labeled, permitting a variety of NMR studies to be carried out (*see* Section 4).

2.3. Purification

2.3.1. CELL LYSIS

At the time of harvest, insect cells such as Sf9 and Hi5 are very fragile and can even be lysed by passage through a syringe needle. Generally, mechanical techniques such as Dounce-homogenization *(6)*, sonication *(25)*, or high-pressure homogenization *(26)* are used. Particular care must be taken when sonicating cell suspensions, as many kinase catalytic domains are relatively unstable and can be driven into the insoluble fraction by localized overheating. Following lysis, centrifugation is used to generate a solution that can be filtered prior to chromatography. Insect cell lysates are actually quite difficult to clarify to a point that allows facile filtration. Typically spin times of 45 min to 1 h are used *(6,27)* with Rcf values of 30,000 g and higher. In our laboratories, following centrifugation, the clarified lysate is sometimes additionally filtered through a sintered glass filter funnel, prior to filtration through a membrane filter. This problem with clarification appears to be construct related, rather than cell line specific.

2.3.2. CHROMATOGRAPHY

In the rare instances where high expression levels were obtained, conventional chromatographic methods have been successfully used to purify kinases for crystallography. The method described by Hubbard et al. for the IRK involved anion exchange followed by size exclusion and a final high-resolution anion-exchange step *(1)*. We have also successfully used this combination for the purification of the IGF 1R kinase, as have Favelyukis et al. *(28)*. However, the vast majority of kinases produced for crystallography are purified with the aid of a purification tag. The most popular tag appears to be a hexa-histidine tag, normally at the N-terminus, although Csk was purified using a C-terminal tag *(13)*. Usually, although not always *(29)*, a proteolytic cleavage site is engineered into the construct to allow removal of the histidine tag. Viral proteases such as tobacco etch virus (TEV) or PreScission™ are the most popular *(27,30)* because of their unique specificity and high activity, although thrombin is still used *(31)*. His-tagged proteins are purified using chelating gels charged with either nickel or cobalt ions, although untagged Csk was purified using Zn-sepharose, based on its requirement for divalent cations *(17)*. Chelating-sepharose from Amersham Biosciences and Nitrilotriacetic acid (Ni-NTA) from Qiagen are both equally popular, elution being achieved by linear or stepwise gradients of imidazole. Kinases are very sensitive to oxidation and are normally kept in buffers containing low concentrations of dithiothreitol (DTT) or *tris* (2-carboxyethyl) phosphine hydrochloride. However, the metal ions on chelating gels are easily reduced by DTT, so normally β-mercaptoethanol is used at concentrations of up to 15 mM *(11)*, being replaced later by DTT. Often chelating gel eluates are either desalted into buffers containing both ethylene diamine tetracetic acid and reducing agents, or are simply added directly to prevent oxidation by metal ions leached from the column. Nagar et al. prepurified c-Abl over anion-exchange before using NTA-agarose *(11)*. Such an approach is useful when expression levels are low, because the enrichment that occurs on the ion-exchange column allows the chelating gel to function far more effectively. Additionally, a kinase such as c-Abl, without addition of inhibitor, is difficult to recover from ion exchange. However, in a crude lysate, such a separation can be carried out, facilitating removal of some phosphorylation states.

GST fusion proteins have also been widely used for the expression and purification of kinases *(10,19,31,32)*. The GST fusion partner can confer a degree of stability to the kinase, but can also keep misfolded proteins in solution. Purification over glutathione-agarose gives a high level of purity, if expression levels are high. This method is also compatible with reducing agents, required for kinase stability and activity. In the baculovirus system, when expression levels are lower, contaminating GSTs of insect origin may become problematic and can be quite difficult to remove, although hydrophobic interaction chromatography can be effective. The GST tag is removed by proteolysis and when the PreScission protease is used, both free GST and protease can be removed by rechromatography over glutathione peroxidase-agarose. This can also be done in one step because the kinase elutes and the PreScission protease remains bound to glutathione through its GST-tag.

Affinity chromatography using immobilized adenosine triphosphate (ATP) *(33)* has also been very successfully used for the purification of c-Src *(34)*. This method has the advantage that it uses a natural kinase ligand so that only functional kinase is bound to the matrix and therefore the purified protein will be of the highest quality. However, the affinity of binding is relatively low, which can result in a relatively dilute eluate. Also, certain kinases do not seem to bind well and high concentrations of NaCl will prevent binding. Otherwise this method is compatible with most buffers used in other chromatography steps. We successfully purified a full-length nonreceptor kinase for limited proteolysis studies from a relatively poorly expressing construct, first of all by using a C-terminal His-tag to produce an enriched protein fraction (~60% purity), then loading the NTA-eluate directly onto γ-aminophenyl ATP-sepharose, the eluate from which was pure enough to allow limited proteolytic definition of the catalytic domain.

Hydroxylapatite chromatography was used in the purification of the Tie2 kinase domain *(29)*. The technique was used in conjunction with anion exchange to try to increase the homogeneity in a kinase preparation that contained as many as six different phosphorylation states.

Ion-exchange chromatography is almost always used for the purification of kinases for crystallography because it is a powerful technique for separating different phosphorylation states. Most frequently anion-exchange chromatography is used; because such gels usually bind kinases at neutral to slightly basic pH values, under which conditions kinases are relatively stable. Cation-exchange columns on the other hand, tend to be run at lower pH, conditions under which many kinases are unstable. We have come across one exception for which we separated mono- from nonphosphorylated protein in a phosphate buffer at a pH of 7.2 on a Mono S cation-exchange column. However, this kinase domain had a relatively high isoelectric point, enabling the separation to occur above neutral pH. Interestingly, after the separation, the protein recovery was relatively low (~60%), but the recovered protein was much more stable, implying that this separation removed partially folded protein, which may have otherwise caused precipitation later. It is possible to obtain baseline or near-baseline separations of different phosphorylation states if the protein loading is kept very low and a shallow salt gradient used in the elution. Typically, one should not load more than 1 mg of kinase per milliliter pack bed and should use a gradient of 0–1 M NaCl over as many as 40 column volumes (PR unpublished).

As for most protein preparations for both NMR and X-ray crystallography, a size-exclusion step is a wise precaution although not essential. The Superdex 75 gel from Amersham Biosciences is almost ubiquitously used, although for unstable kinases it may be necessary to include up to 50% v/v glycerol in the buffers to minimize interactions with the matrix.

2.4. Formulation

Kinase domains are often difficult to concentrate and to handle through basic chromatographic steps. To alleviate these problems, a number of approaches have been taken. Ogawa et al. used the detergent β-octylglucoside to 0.5% w/v to aid with Csk stabilization (35), whereas Triton X-100 was used at 0.2% v/v for the IGF-1R kinase (28). One of the most effective ways to stabilize kinases is to add a specific inhibitor. In the case of c-Abl, the kinase domain is extremely difficult either to purify over anion exchange or to concentrate by ultrafiltration in the absence of inhibitor. Addition of a specific inhibitor to approx 3-M excess allowed the protein to be concentrated to 35 mg/mL (11). The degree to which inhibitors stabilize kinases, of course, varies considerably. Techniques such as differential scanning calorimetry can be used to quantify the extent of the stabilization and to screen buffer additives, pH, etc. to find the optimum conditions. The technique measures the denaturation point of a protein in a particular buffer. In our own laboratories, a series of inhibitors were tested on Bruton's tyrosine kinase (Btk). Increases in the melting point from 49°C to between 54 and 57°C were observed. More recently with c-Abl in complex with Gleevec™, a series of buffer formulations were compared to test the influence of salt concentration at different pH values. Stabilization with inhibitors can also have a large effect, and in such cases the inhibitor should be added as early as possible post-lysis.

Many kinases are stabilized by addition of glycerol to concentrations as high as 50% v/v. Although this may be useful during purification, concentrations over 10% v/v are problematic for later crystallization. In such cases, alternatives may need to be found. In one case we successfully exchanged a buffer containing 30% v/v glycerol for one containing 1% v/v ethylene glycol, enabling us to concentrate the kinase to between 30 and 40 mg/mL, whereas without glycerol it was impossible to concentrate the protein above approx 5 mg/mL.

3. STRUCTURES OF PTKs

PTKs have been studied extensively using structural biology over the past 15 yr, with the first atomic structure of a tyrosine kinase domain appearing in 1994 *(1)*. As of November 2003, there were 21 entries for nonreceptor PTKs and 24 entries for receptor PTKs in the protein data bank (PDB). In some cases, because of the high value of structures for drug design, structures have been published (some only in patents), but the coordinates are not yet publicly available.

The tyrosine kinase domain of PTKs closely resembles that of the Ser/Thr family of kinases, for which the structure of protein kinase A was the first kinase atomic structure determined *(36)*. The overall structure consists of an N-terminal lobe, containing a 5-stranded β-sheet and an α-helix (referred to as αC), and a larger, mainly α-helical C-terminal lobe (Fig. 3). ATP binds in the cleft that separates the two lobes, and forms hydrogen bonds with the portion of the structure that connects the two lobes, which is the so-called "hinge region" (after β5). The tyrosine-containing segment of the substrate protein binds mainly to the C-terminal lobe, where the platform of the binding site is partially formed by the activation loop (A-loop), the conformation of which is usually stabilized by phosphorylation of one or more tyrosines in the loop. This loop generally begins with a conserved sequence of three residues, Asp-Phe-Gly (DFG motif), and ends with Ala-Pro-Glu (Fig. 1). The aspartate residue in the DFG motif has an important role in coordinating magnesium ions that interact with the phosphate groups of ATP. Another flexible region of the protein is the glycine-rich or P-loop, which comes from the N-terminal lobe and is also involved in coordinating the phosphates of ATP. The active site is formed by residues from the P-loop, β3, and αC in the N-terminal lobe and the catalytic loop (between αE and β7) and the A-loop, from the C-terminal lobe. A characteristic of active kinases is a salt bridge in the active site between a glutamate residue from αC and a lysine from β3. This interaction also has an important role in correctly positioning the phosphates of ATP for catalysis. A useful description of the kinase active site can be found in a review by Johnson *(37)*.

The kinase domains of PTKs are rather flexible and can exist in more than one conformation *(38)*. The active conformation is common to all of them and is represented by the structure of IRK shown in Fig. 3. The inactive conformation is, however, unique to each kinase and probably depends more on the specific regulatory mechanism than anything else. The active conformation is sometimes referred to as a closed conformation because in most cases, the inactive state has a more open angle between the N- and C-terminal lobes. Receptor tyrosine kinases are generally activated by the binding of ligands to the extracellular domain, which induce and stabilize an oligomeric association

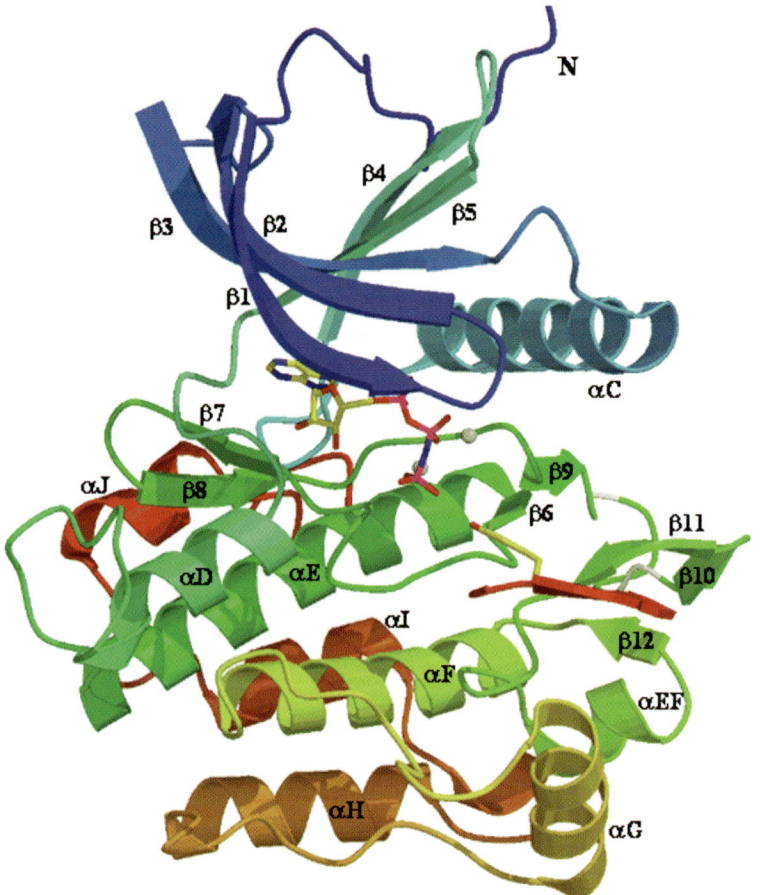

Fig. 3. The structure of insulin receptor kinase in the activated state in complex with AMP-PNP and a substrate peptide *(64).* The structure is colored from blue at the N-terminus to orange at the C-terminus, with the substrate peptide in red. Atoms of AMP-PNP in the cleft between the N- (top) and C-terminal (bottom) lobes and Tyr10 from the substrate peptide are drawn as ball-and-stick models. Structural elements are labeled according to the original nomenclature of the cyclic adenosine monophosphate-dependent protein kinase structure, and other elements mentioned in the text are as follows: P-loop (or Glycine-rich loop, β1–β2), hinge region (between β5 and αD), A-loop (including β9, β10, and β11), and catalytic loop (between β6 and β7). The figure was prepared using Molscript *(101)* from PDB entry 1IR3 *(64).*

of the PTK, facilitating transphosphorylation of the intracellular domain. The activation of nonreceptor PTKs is more complicated, involving interactions with other proteins, as well as oligomerization that allows transphosphorylation. Because of these differences, it is simpler to discuss the structures of the different families of PTKs separately.

3.1. Nonreceptor PTKs

3.1.1. THE SRC FAMILY KINASES

Many of the nonreceptor PTKs contain SH2 and/or SH3 domains (Fig. 2), although several, such as those from the focal adhesion kinase (FAK) and Janus kinase families, do not *(39)*. The domains are named after their homology with regions of the Src PTKs and they are responsible for interactions with activating and regulating proteins in the cell and for self-regulation of the kinase activity. C-Src exists in the cell in only one of two phosphorylated states: if phosphorylated at the C-terminus (Tyr527) it is inactive; if phosphorylated in the A-loop (Tyr416) it is active. The first structural insights into the role of the SH domains in negative regulation came with the structures of c-Src and Hck in 1997 *(34,40,41)*. The interactions of the SH2 domain with the phosphorylated C-terminal tail, and the SH3 domain with the linker between the SH2 domain and the kinase domain, are responsible for locking the protein into an inactive conformation (Fig. 4A). The ATP binding site is still capable of binding the cofactor, but the closure of the hinge between the N- and C-terminal lobes and the adoption of an A-loop conformation which helps to block the cleft *(42)*, means that helix C is pushed out of the cleft, which results in the breaking of the important Lys–Glu interaction. Also, the inactive conformation of the A-loop is not suitable for the binding of substrate. This form of downregulation has been described as a snap-lock mechanism *(43)*. The SH3 and SH2 domains have a flexible relative conformation in solution *(44)*, but when they bind to the back of the kinase domain they snap into a rigid conformation and lock the kinase domain so that it cannot breathe enough to perform its catalytic function. Evidence for this is that mutation of three residues in the linker between the SH3 and SH2 domains to glycine, which increases the flexibility between the two domains, gives rise to an active kinase despite phosphorylation at the C-terminus *(43)*.

The significance of the regulatory elements in the assembled inactive state of c-Src was obvious owing to the availability of the structure of the highly homologous Lck kinase domain phosphorylated in the A-loop, and therefore in an active conformation *(45)*. The Lck structure is typical of an active kinase domain as helix C is rotated in toward the catalytic site and the A-loop adopts an extended conformation that is stabilized by the presence of a phosphotyrosine. Recent structural studies of a totally unphosphorylated form of c-Src *(45b)* show that dephosphorylation of the C-terminal tail is enough to allow the kinase domain to adopt an active conformation (Fig. 4B). The SH3 domain remains attached to the SH2 kinase domain linker, but the whole unit swings away from the back of the kinase domain, presumably allowing the two lobes of the kinase domain to regain their relative flexibility. This release of the assembled inactive state could be induced by any event that leads to the

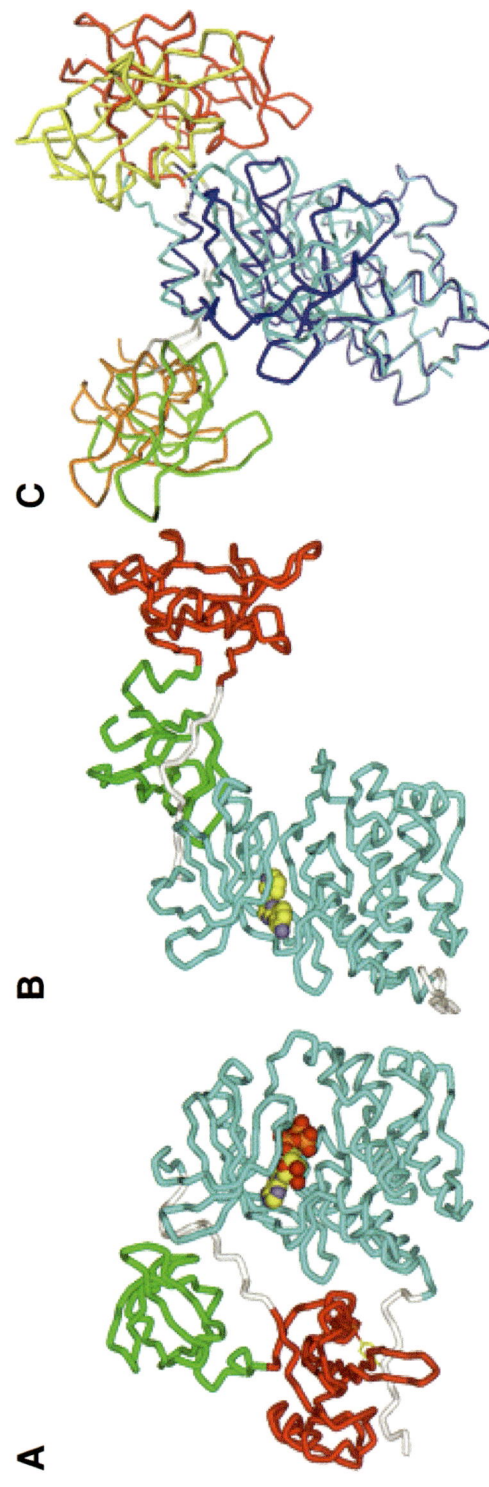

Fig. 4. Comparison of c-Src phosphorylated at Tyr527 in the C-terminal tail (**A**) and unphosphorylated (**B**) forms, plus two different states of Csk (active and inactive) superimposed (**C**). Src homology (SH)3, SH2, and kinase domains are colored green, red, and cyan respectively (as in Fig. 2) with the linker regions in white. The SH3, SH2, and kinase domains of the active conformation of Csk are shown in orange, yellow, and blue, respectively.

dissociation of the SH2-pTyr and/or SH3-linker interactions, such as the binding of activating proteins to the SH2 and SH3 domains (e.g., refs. *46* and *47*). Although the activation loop is not phosphorylated in this structure, it adopts an extended conformation similar to that of the active state, where Tyr416 is relatively exposed and accessible for phosphorylation. Interestingly, the unphosphorylated C-terminal tail folds back into the C-terminal lobe of the kinase domain where the final residue binds in a pocket analogous to that found to bind myristate in Abl (*see* Section 3.1.2) *(11)*. This suggests that c-Src could be further negatively regulated by the binding of its N-terminal myristate modification, although biochemical and structural evidence for this idea needs to be obtained. The myristoylation sequence in c-Src is responsible for plasma membrane location, where the kinase should be active. SH3 displacement and dephosphorylation of the C-terminal tail are just two possible mechanisms of activating the Src family kinases. Lerner and Smithgall have shown that through SH3-based activation, kinase activation can occur without tail release from the SH2 domain in vivo *(48)*. The unphosphorylated c-Src structure shows that activation by pTyr dephosphorylation leads to a conformation in which the SH2 binding site is free to interact with other proteins in the cell. These results suggest that different activating mechanisms may lead to different downstream signals *(48)*.

3.1.2. ABL KINASE

The N-terminal half of Abl is similar to the c-Src kinases, but Abl has a very long C-terminal extension containing DNA and actin-binding domains, without a phosphorylation site that would mimic the role of the c-Src tail (Fig. 2). Recent biochemical and structural studies have shown that Abl is regulated by the binding of its N-terminal myristate to the aforementioned pocket in the C-terminal lobe of the kinase domain *(11,23,49)* (Fig. 5A). This mechanism is probably reinforced by the interaction of residues in the N-terminal region of the sequence with the kinase domain, although these were not clearly seen in the structure. The binding of myristate is found to induce a conformational change in helix I, which allows the docking of the SH2 domain against the C-terminal lobe of the kinase. This partially occludes the pTyr binding site of the SH2 domain. Binding of SH2 ligands will prevent the interaction with the kinase domain, resulting in an opening of the assembled domain structure and a step toward activation of the enzyme. The assembled inactive state of the SH3, SH2, and kinase domains of c-Abl very clearly resembles that of the inactive c-Src structure (Fig. 5), but with some significant differences. The conformation of the Abl kinase domain is closer to an active conformation than in the c-Src structure. Helix C is close to the normal active position allowing formation of the Lys271-Glu286 salt bridge and the A-loop is in an extended conformation. The unusual conformation of the strictly conserved

A

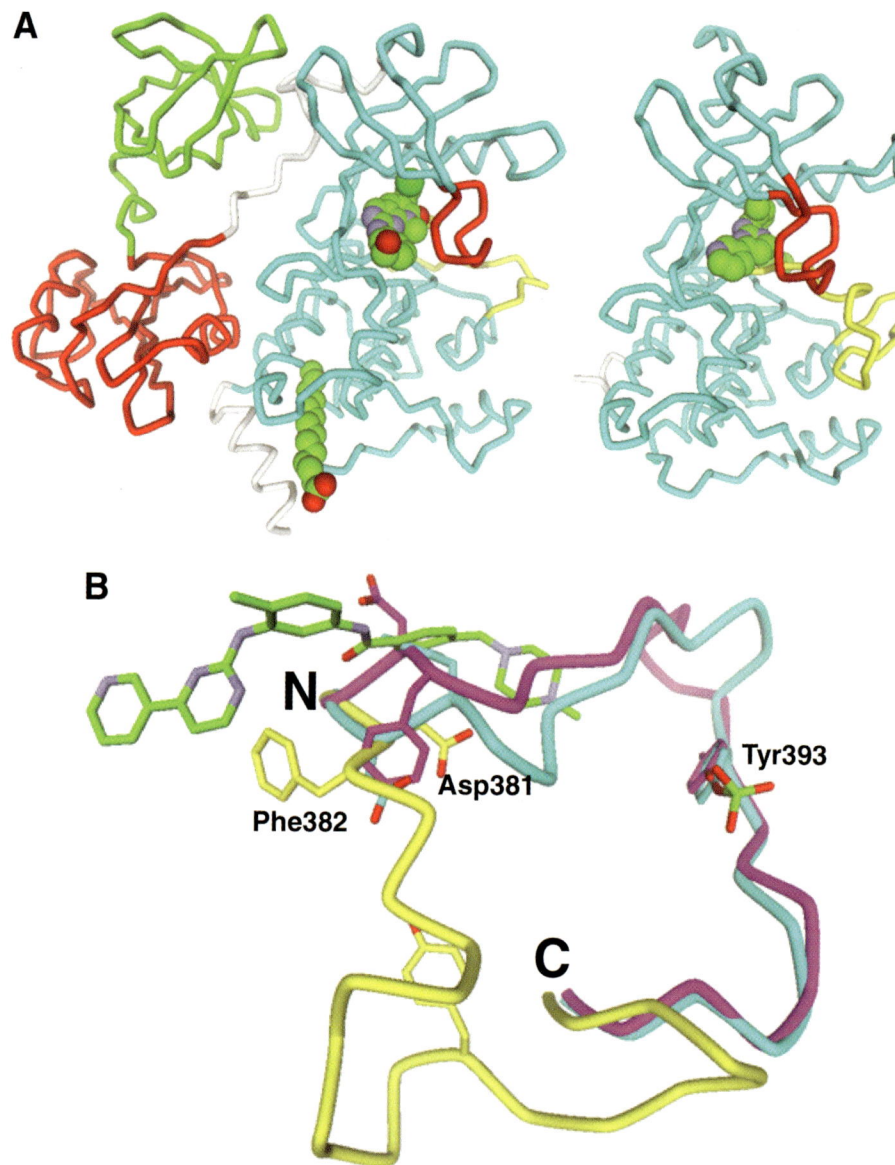

B

Fig. 5. Comparison of the different conformations of Abl kinase. **(A)** Overall with both the assembled inactive structure (left) and the isolated kinase domain inactive with Gleevec™ bound *(right),* and **(B)** A-loop showing the DFG motif in different structures (yellow: Gleevec complex, magenta: assembled inactive, cyan: active Lck for comparison).

DFG motif and the rigidity of the kinase lobes, which are locked into this conformation by the SH3–SH2 clamp (flexibility is necessary for the binding and release of cofactor and substrate; *see* ref. 50 for a review), are thought to be responsible for the lack of activity of Abl in this state. Activation of the structure is also caused by phosphorylation of Tyr245 *(51)*. This residue is located in the portion of the linker bound to the SH3 domain and is buried in a hydrophobic environment at the interface with the kinase domain. Phosphorylation would prevent it from binding in this site and disallow the assembled state.

An auto-inhibited conformation of Abl was also observed in structures of the Abl kinase domain in complex with inhibitors *(52,53)*. The structure of Abl in complex with Gleevec, which is successfully used in the treatment of chronic myelogenous leukemia (CML), showed that the drug binds to a unique inactive conformation of the kinase domain (Fig. 5). In most cases of CML, the disease is directly caused by what is known as the Philadelphia chromosome translocation where parts of the genes encoding the breakpoint cluster region (Bcr) and Abl are fused. This results in a fusion protein, Bcr-Abl, in which the N-terminal regulatory part of Abl is replaced by the N-terminal domains of Bcr, which include an oligomerization domain. Oligomerization plus the loss of the N-terminal regulatory region result in constitutive activation of the kinase domain. Gleevec appears to bind with an induced fit interaction, where the inhibitor binds in the adenine-binding site of the kinase in a similar fashion to many other ATP competitive kinase inhibitors, but extending beyond this region through the center of the molecule utilizing a hydrophobic pocket opened up by the displacement of the conserved DFG motif. Asp382 from this motif is flipped out of the position where it would be required to coordinate the Mg ions in the ATP phosphate binding site. The inhibitor blocks the path that the activated A-loop would adopt and the A-loop adopts an auto-inhibitory conformation in which the unphosphorylated tyrosine binds in and blocks the substrate binding site, similar to what is observed for Irk in the unphosphorylated state *(1)*. The P-loop also has a novel conformation in order to interact with the inhibitor. The necessity for the protein to be able to adopt these conformations in order to bind Gleevec, plus the requirement that the side chain of the residue in position 315 must be small (Thr in Abl) so that the inhibitor can reach into the hydrophobic pocket near DFG, are responsible for the excellent kinase inhibition profile and selectivity of this drug. Presumably, even highly homologous protein kinases such as c-Src and Hck cannot adopt the required conformation of the DFG motif. However, the fact that Gleevec makes contacts with residues that are not necessary for the function of the protein, and that it requires this special conformation for binding, a conformation that is not relevant for activity, means that it is susceptible to the formation of mutants that are insensitive to treatment with this drug (*see* ref. 55 for a review). These mutants are being observed in many relapsed patients that were first treated in the advanced stages of CML, although most patients treated early in the chronic phase are still responding very well.

3.1.3. BRUTON'S TYROSINE KINASE

Btk consists of an N-terminal pleckstrin homology domain, a proline-rich TEC homology domain, SH3 and SH2 domains, and the kinase domain (Fig. 2). Interactions of these domains with other proteins in the cell are responsible for regulation of the kinase activity as well as being important for the physiological functions of the protein. Phosphorylation of tyrosines 223 (SH3 domain ligand binding site) and 551 (kinase A-loop) activate the kinase. Low-resolution X-ray solution scattering studies (56) show that the full-length unphosphorylated Btk does not form an assembled, inactive structure like the Src and Abl kinases, although it contains a similar sequential arrangement of the SH3, SH2, and kinase domains. This suggests that these domains are not required for the negative regulation of the kinase domain. The crystal structure of the unphosphorylated kinase domain has been solved to 2.1 Å resolution and shows that helix C adopts an inactive conformation, even though the A-loop is in an extended conformation similar to that of active kinases (27). The P-loop and the DFG motif at the N-terminus of the A-loop also adopt inactive conformations. Based on the structure, it was suggested that phosphorylation of the A-loop could trigger a change in the hydrogen bonding pattern of Glu-445 from helix C such that the interaction with Arg544 would be replaced by Arg544–pTyr551 and Glu445–Lys-430 interactions, the latter being typical of active kinases. This would require correlated movements of the P-loop, the DFG motif, and helix C. It is not known at this stage whether in the complex environment of the cell Btk could adopt a tightly regulated assembled conformation such as observed for the Src and Abl kinases. There is some evidence that the SH3 domain has a negative regulatory role (57), and autophosphorylation of a conserved tyrosine in this domain is necessary to achieve full activity of Btk (58).

3.1.4. C-TERMINAL SRC KINASE

The phosphorylation of the regulatory tyrosine on the C-terminal tail of Src family kinases is controlled by C-terminal Src kinase (Csk). This kinase also has SH3, SH2, and kinase domains, but lacks an N-terminal acylation site, a tyrosine autophosphorylation site in the activation loop and a C-terminal regulatory tyrosine, suggesting that the method of regulation of Csk is different from the other cytoplasmic PTKs. Indeed, the structure of the full-length Csk has been determined (35), and it shows that the SH3 and SH2 domains adopt a novel arrangement with respect to the kinase domain (Fig. 4). The SH3 and SH2 domains lie on opposite sides of the top of the N-terminal lobe of the kinase domain, where they have no direct interactions with each other. The orientation of the SH3 domain resembles that of the SH3 domain in the inactive assembled state of c-Src or Hck, but this location of the SH2 domain has not yet been observed for other kinases. In this structure there are six independent molecules in the asymmetric unit of the crystal and two different orientations of the

SH2 domain are observed (four with the kinase domain in an active and two in an inactive conformation), suggesting that the linkers between the SH3 and SH2 domains and between the SH2 and kinase domains are flexible. Both of these linkers have many contacts with the kinase domain. Apparently, the position of the SH2 domain has an effect on the activation state of the kinase, as it interacts slightly differently with the N-terminal lobe in each case. Ligands such as Cbp/PAG that bind strongly to the SH2 domain are known to activate Csk (59), suggesting that the binding might stabilize the SH2 domain in the position that induces an active conformation of the kinase domain. The crystal structure of the isolated Csk domain with staurosporine bound has also been determined (13). This structure shows an inactive conformation of the kinase domain where helix C is bent and the critical ion pair (Glu236-Lys222) is not formed. The inactive conformation of the isolated kinase domain is different to that of the inactive full-length protein, so it may be possible that the presence of both the SH3 and SH2 domains is necessary for activity. Deletion of these domains significantly reduces Csk activity (60). A mutational analysis of the A-loop in Csk suggests that it has a role in the negative regulation of the kinase activity, but does not have a role in phosphorylation of substrates (61). The fold of the A-loop in Csk is rather similar to that of active Lck, even though the loop is four residues shorter.

3.1.5. Zap-70

The ζ-chain-associated protein of 70 kD, a kinase with two N-terminal SH2 domains, but no SH3 domain, plays a pivotal role in T-cell activation. One of the earliest events of T-cell activation is the phosphorylation of the T-cell receptor ζ-chain and the specific association and activation of the Syk family protein tyrosine kinase Zap-70 with the phosphorylated ζ-chain via its two SH2 domains (62). T cells are involved in transplant rejection, autoimmune diseases, and the initiation of inflammatory responses and are thus a prime target for pharmaceutical intervention. The structure of the kinase domain of Zap-70 in complex with staurosporine is very similar to that of activated Lck, even though the Zap-70 protein is unphosphorylated (Christian Ostermeier, PR, et al., unpublished data). The positions of helix C and the P-loop, the presence of the Lys369-Glu386 salt bridge, and the path of the A-loop all reflect an active conformation. The A-loop has a slightly different conformation in the C-terminal part, most likely because of the lack of phosphorylation of tyrosines 492 and 493. The conformation of the inactive state(s) of this protein have yet to be determined.

3.1.6. Focal Adhesion Kinase

Focal adhesion kinase has an N-terminal integrin-binding domain and a C-terminal focal adhesion binding domain. The structure of the catalytic domain of FAK in complex with ATP has been published recently (63). This is an example of one of the many new kinase structures that have appeared in the last couple of years from structural genomics based companies/consortiums. The availability of automation allows the testing of a large number of solutions for crystallization conditions with a small amount of protein (1–3 mg). Many different constructs can therefore be tried in parallel (*see* Section 1).

The 1.7 Å resolution structure shows an open conformation, where helix C is rotated away from the C-terminal lobe and the activation loop is disordered, probably consistent with the lack of phosphorylation of the kinase domain in this structure. Despite this open conformation, the salt bridge between Lys454 and Glu471 is formed. A novel feature of the structure is an intramolecular disulphide bond in the N-terminal lobe of the kinase, on the surface near helix C. This proximity suggests that the disulphide bridge might have an influence on the position of helix C and therefore a role in regulating the activity of the kinase. However, although the cysteines are completely conserved in vertebrate FAK sequences, disulphide bonds are extremely rare in cytoplasmic proteins and it is not known whether this bond is formed in vivo, especially since it lies near the surface. It is possible that the interaction with regulatory proteins in this area, as is observed for many other kinases (e.g., Cdk2), could protect the disulphide bond from the reducing environment of the cell *(63)*.

3.2. Receptor PTKs

3.2.1. IRK AND IGF-1R

The binding of insulin to the insulin receptor activates signaling pathways that regulate cellular metabolism and growth. Activation of the IGF-1R by binding of IGF1 (and to a lesser extent IGF2) is important for regulation of cell growth and differentiation, and protection from apoptosis. Elevated levels of IGF1R are observed in many human tumor types and inhibition of the kinase activity is actively pursued for the treatment of various types of cancer. IRK and IGF1R are type II membrane receptor tyrosine kinases consisting of two α (entirely extracellular) and two β chains (with extracellular, transmembrane, and cytosolic domains). The chains are linked together by disulphide bonds in the extracellular portion. Upon ligand binding, the receptors change their conformation such that trans-subunit autophosphorylation of the kinase domains occurs at residues Tyr1131, Tyr1135, and Tyr1136 of the A-loop (IGF-1R numbering). In vitro they are autophosphorylated in that order *(28)*. The ATP binding site of IGF-1R shares 100% sequence identity with that of IRK and they can form hybrid receptors. Because insulin has a role in controlling the blood glucose level, a side effect of inhibition of IRK in addition to IGF-1R with unselective inhibitors would be drug-induced diabetes. The structural studies of the two kinases therefore have the purpose of determining conformational differences between the two receptors that may be exploited in the design of selective inhibitors. The sequence identity of the kinase domain alone is 84%. The kinase domain is preceded by a juxtamembrane (JM) region and followed by a C-terminal tail, with sequence identities of 61% and 44%, respectively.

As already mentioned, the crystal structure of unphosphorylated IRK (Figs. 3 and 6) was the first structure of a PTK (1). It is similar to the previously determined Ser/Thr kinase structures; however, the conformation of the activation loop is significantly different. The N-terminal part of the A-loop showed an unusual conformation of the DFG motif in which Phe1151 is flipped out with a crankshaft-like motion of the main chain, partially occupying the adenine binding site. This results in an unsuitable conformation of Asp1150 for the correct positioning of the phosphate groups of ATP. In addition, Met1153 and Thr1154 block the binding site of ATP and Tyr1162, which is one of the three tyrosines in the A-loop that requires phosphorylation for full kinase activity, is found to occupy the substrate binding site. In summary, the A-loop adopts an autoinhibitory conformation that blocks the binding of cofactor and substrate and prevents appropriate positioning of catalytic residues. Cis-phosphorylation of Tyr1162 in this conformation is not possible because the binding of Mg-ATP is sterically hindered. However, transphosphorylation of Tyr1162 from a neighboring kinase is possible as this would not require the adoption of an A-loop conformation that would negatively affect the positioning of catalytic residues. Once phosphorylated, the equilibrium between the conformations of the A-loop would be shifted from the autoinhibited state to the extended state, in which the other tyrosines would be more accessible for phosphorylation. The structure of the fully activated IRK (Fig. 6) has also been determined (64). This structure reveals the conformations required for the binding of substrate and ATP and also shows the orientations of the three phosphorylated tyrosines in the activation loop. pTyr1163 is responsible for stabilizing the active conformation of the A-loop via interactions with N- and C-terminal regions of the loop, similar to what is observed for pTyr394 in Lck. In its unphosphorylated state, Tyr1162 stabilizes the inactive conformation, whereas it becomes exposed to solvent upon phosphorylation, as does pTyr1158. These two phosphotyrosines may be the target of downstream signaling proteins.

The crystal structure of IGF-1R kinase has been determined in three different states: trisphosphorylated in complex with an ATP analog and a specific peptide substrate (IGF-1R-3P; ref. 28), bisphosphorylated in complex with an ATP analog (IGF-1R-2P; ref. 31), and unphosphoryated in the apo form (IGF-1R-0P; ref. 6). The closure between the domains is greatest for the activated kinase domain (IGF-1R-3P), intermediate for the bisphosphorylated enzyme, and least for the unphosphorylated state. However, the greatest difference between these structures is seen in the conformation of the activation loop. In the IGF-1R-0P structure, the A-loop adopts a self-inhibitory conformation very similar to that seen for IRK; however, there are some slight differences in the proximal end containing the

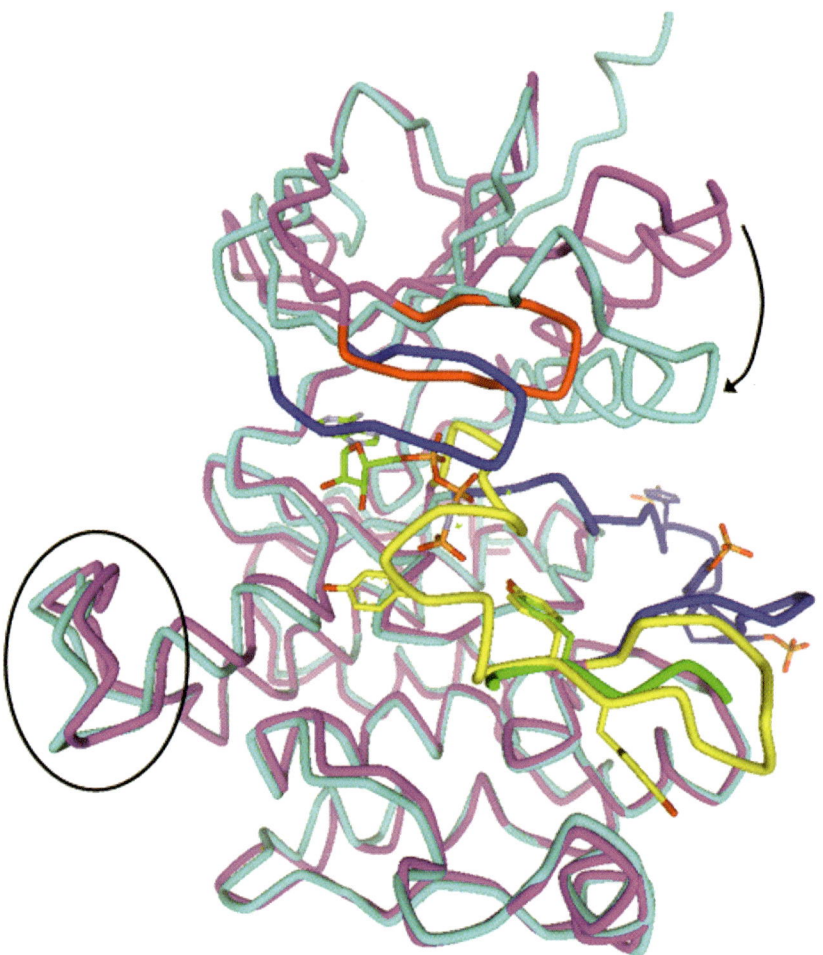

Fig. 6. Comparison of insulin receptor kinase in the active (cyan) and inactive (magenta) states. The glycine-rich loop (blue and red respectively), the activation loop (blue and yellow respectively), and helix C undergo large conformational changes. Peptide substrate as bound to the active (trisphosphorylated) form is shown in green, as is AMP-PNP bound in the cleft between the N- and C-terminal lobes. The shift of helix C between the inactive and active states is indicated by an arrow and the kinase insertion domain is circled.

totally conserved DFG motif. The A-loop in IGF-1R -2P is partially disordered in the middle, but resembles more the A-loop of the IGF-1R-3P structure than the IGF-1R-0P structure at the ends. The A-loop of the IGF-1R-3P structure is in an active conformation and the whole structure of this form is very similar to that of activated IRK *(64)*. A comparison of the unphosphorylated forms of IRK and IGF-1R shows slight differences in domain closure and in the conformation of the

P-loop as well as a small shift in the position of helix C. Although some of these differences can be justified by sequence changes, it cannot be ruled out that they could also be owing to the presence or absence of a ligand in the ATP binding site. Structures of the enzymes with inhibitors bound may bring extra information about how to improve selectivity in addition to the comparison of the structures in the apo or ATP-bound states. Like many other receptor PTKs, the IRK, and IGF-1R kinases have what is known as a kinase insertion domain (KID). The role of this region is not clear, but based on sequence motifs it seems that it contains protein interaction sites. The length and sequence of this region is not conserved between the different kinases. In the IRK structure the KID is ordered and contains a stabilizing salt bridge (Fig. 6) *(64)*, whereas in the IGF-1R structures this region is disordered. The flexibility of this region also argues for a role in protein interactions as its exposure to solvent means that it is accessible for binding.

3.2.2. MUSCLE-SPECIFIC KINASE

Muscle-specific kinase (MuSK, expressed exclusively in skeletal muscle) autophosphorylates in the JM region and in the activation loop after stimulation with agrin, a protein expressed in motor neurons. MuSK activation stimulates the clustering of acetylcholine receptors at the neuromuscular synapse. Supression of MuSK signaling in the absence of agrin is likely to be important to avoid spurious synapse formation. The crystal structure of MuSK *(12)* reveals that the A-loop adopts an autoinhibitory conformation like that observed in IRK, where it obstructs both the nucleotide and substrate binding sites. Like IRK, MuSK has three tyrosines in the A-loop and one of these (Tyr 754) binds in the position of a substrate tyrosine. Phosphorylation of tyrosines 754 and 755 presumably leads to a more open, extended conformation of the A-loop. Mutation of Tyr-553 in the JM region of MuSK abrogates autophosphorylation of the A-loop in vivo. The corresponding mutation in Irk (Tyr972 to Phe) has no effect on A-loop autophosphorylation, suggesting that the JM region of MuSK is an additional regulatory element. However, the Tyr-533 mutation to Phe has no effect on autophosphorylation of the A-loop in vitro. The portion of the JM region that is observed in the crystal structure (residues 560–566) forms a short α-helix that is bound in a groove between helix C and the N-terminal β-sheet. The rest of the JM region is disordered in the crystals, indicating significant flexibility. In analogy with the binding of FKBP12 to the JM region of the TGF-β receptor *(65)*, the binding of an inhibitory protein to the JM region of MuSK might stabilize a JM conformation that prevents kinase A-loop phosphorylation.

3.2.3. FLT3 AND cKIT

FLT3 and cKit both belong to a subfamily of class III receptor tyrosine kinases, which consist of an extracellular domain comprising five immunoglobulin domains, a transmembrane region, and a cytoplasmic domain with a JM region and a kinase domain with a large insertion (KID).

FLT3 has been found expressed in high levels in several leukemias. Many of these patients have been found to harbor either internal tandem duplication mutations in the JM region or point mutations (e.g., of Asp835, Ile836) in the A-loop, leading to constitutive activation of the receptor. The recently published structure *(66)* of FLT3 at 2.1 Å shows that the kinase adopts an inactive conformation very similar to that of VEGFR2 *(67)* and fibroblast growth factor receptor (FGFR)1 *(68)* (*see* Subheadings 3.2.7. and 3.2.8.). The unphosphorylated A-loop tyrosine (Tyr842) is found to bind in the substrate site of the kinase, thus inhibiting the binding of substrate and sequestering the tyrosine side chain that could otherwise be inadvertently phosphorylated leading to unwanted activation of the protein. The construct used for crystallization contains the JM region and the kinase domain, but the KID is deleted. Residues 572–600 are visible in the electron density, and show that the N-terminal part of the JM region has intimate contacts with elements of the kinase domain such as helix C, the P-loop, and the A-loop. It forms a wedge that seems to stabilize the inactive conformation of the kinase domain. The length of the C-terminal region of the JM domain is strictly conserved among members of the platelet-derived growth factor (PDGF) receptor PTK family. The FLT-3 structure shows that because of the rigidity of the linker between the two ends of the JM region, any change in the length of the C-terminal part will lead to less effective binding to the kinase domain and loss of autoregulation by this region. This explains the effect of mutations or internal tandem duplication observed in many leukemia patients. Mutations in the A-loop presumably shift the equilibrium between the auto-inhibited (regulated) and active states toward the active state by either destabilizing the inactive state or stabilizing the active conformation.

Mutations in the JM region and in the kinase domain of the stem cell factor receptor c-Kit also result in constitutive activation and are responsible for a number of cancers. The structure of the Kit kinase domain in complex with adenosine diphosphate and a magnesium ion reveals a fully active conformation of the kinase domain, despite the lack of phosphorylation in the A-loop *(69)*. The structure shows that the phosphate could easily be accommodated without major conformational changes, suggesting that phosphorylation is not really necessary for the activation of this kinase. Physical separation of the kinase domains may be more important than lack of phosphorylation as a criterion for activity. An interesting feature of this structure is that it is found as a dimer in the crystals, with the JM region of one monomer bound in the substrate binding site of a neighbor. This provides a picture of how oligomer formation allows trans phosphorylation.

Kit is also a target of Gleevec, and a recent structure of this kinase in complex with the drug at 1.65 Å (Cowan-Jacob, Fendrich, et al., unpublished data), which is used in the treatment of gastrointestinal stromal tumors, shows that it binds to Kit in a very similar manner to Abl kinase (Fig. 7). The Kit A-loop adopts an

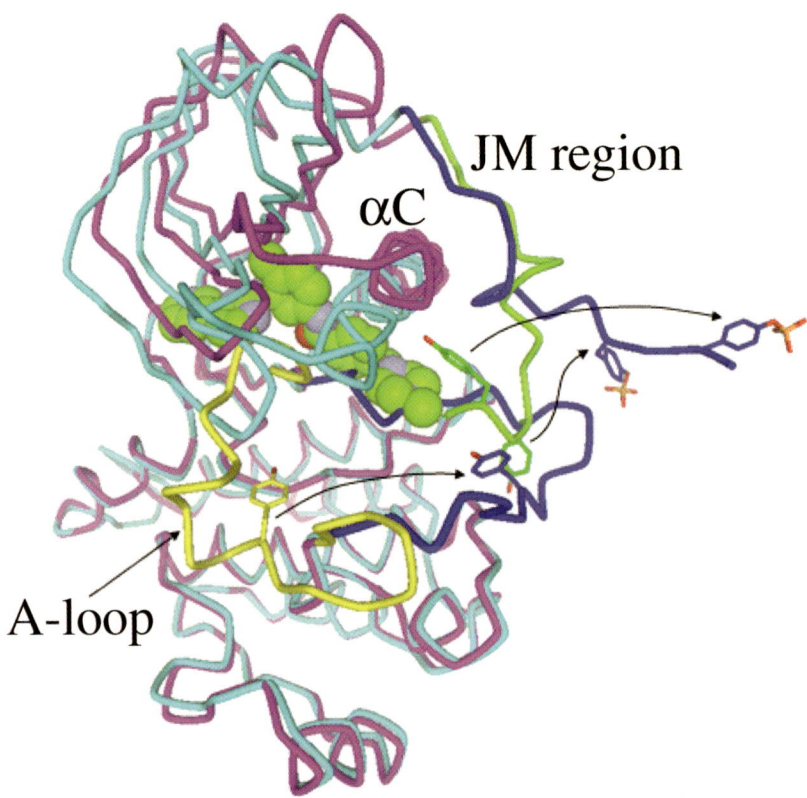

Fig. 7. Comparison of cKit in the active (cyan) and inactive (magenta) states. The arrows show the movements of the tyrosines in the juxtamembrane region (green to blue) and in the A-loop (yellow to blue) from the inactive to the active conformation. The shift in helix C can also easily be seen. Gleevec™ is shown as solid spheres in green; the A-loop of the active conformation (blue) passes through part of the inhibitor binding site.

autoinhibitory conformation where Tyr823 sits in the substrate binding site, as for Abl. The glycine-rich loop makes many interactions with the drug, although it adopts an active-like conformation. The DFG motif is displaced in a way different from the shift observed in Abl because of amino acid sequence differences in the residue just before DFG (Ala380 in Abl, Cys809 in Kit). This allows Phe811 to interact with Gleevec so that the glycine-rich loop does not have space to fold down over the ligand as in Abl. The C-terminal part of the JM region is visible and shows how the portion including Tyr568 and Tyr570 blocks the normal path of the A-loop in the active conformation. Phosphorylation of these residues leads to displacement of the JM region (Fig. 7), allowing the A-loop to adopt an active conformation. cKit mutations in patients not responding to treatment with Gleevec are located in the A-loop, where they destabilize the inactive conformation, or in the hinge region where they sterically hinder binding of the drug. For example, mutants of Asp816 in the A-loop are not able to form an H-bond with Asn819 that stabilizes the inactive conformation. Mutants located earlier in the sequence of the JM region presumably have a similar structural effect to that described for FLT3 (*see* Section 3.2.3), where a displacement of the JM regulatory element occurs.

3.2.4. cMET

The hepatocyte growth factor/scatter factor receptor (cMet) is a target in oncology particularly due to its role in the development of invasive and metastatic tumor phenotypes. It is an α-β heterodimer with an extracellular α-chain linked by a disulphide bond to a membrane-spanning β-chain. Phosphorylation of tyrosines 1234 and 1235 in the A-loop activate the kinase, which then leads to autophosphorylation of tyrosines 1349 and 1356 in the C-terminal tail. The latter become binding sites for SH2, Gab1, and PTB binding domains of a variety of transducer and adaptor proteins. The structure of the unphosphorylated kinase domain and C-terminal tail has been determined in the apo form and also in complex with a staurosporine analog (*10*). These structures both show an inactive conformation of the kinase domain where helix C is displaced, the Lys1110-Glu1127 salt bridge is not formed, and the A-loop obstructs the substrate binding site. Part of the A-loop also passes through the phosphate binding site of ATP and disallows proper positioning of helix C. It can, however, interact with the inhibitor, which is interesting for designing selective inhibitors. There is also an extra α-helix (helix A) at the N-terminus of the kinase domain that interacts with helix C. It has been suggested that this helix might have a role in the activation of the kinase domain when an extracellular ligand binds, because of the fact that a concerted movement of helices A and C will be required (*10*). Activating mutations in the A-loop and the JM region, like in the cases of Abl and cKit, are found in patients with various forms of cancer. The C-terminal tail structure shows that residues 1349–1352, 1353–1356 and 1356–1359 have similar conformations to peptides bound to SH2, PTB, and Grb2SH2 domains, respectively. The binding of these proteins to the first two motifs would be prevented by clashes with the C-terminal lobe of the kinase, but the third motif is accessible. In the structure the tyrosine residues in these motifs are unphosphorylated. Phosphorylation would presumably make them all accessible for binding and thus downstream signaling.

3.2.5. EphA2 and EphB2

The Eph receptors consist of an extracellular portion containing a ligand-binding domain, a cysteine-rich domain, and two fibronectin III repeats, and an intracellular portion containing a JM region, a tyrosine kinase domain, a sterile-α-motif (involved in oligomerization), and a PDZ binding motif. Like most other receptor tyrosine kinases, the catalytic domain of the Eph receptors is regulated by the JM region. In the structure of EphB2 (*14*), the unphosphorylated JM region contains a helix that interacts extensively with the N-terminal lobe of the kinase domain. The interactions cause distortion of helix C in the N-terminal lobe, which means that the conserved glutamate side chain from this structural element is slightly displaced from the active site and the kinase is inactive. The JM region probably also blocks the A-loop from adopting an active conformation. The overall conformation of the kinase domain is closed and resembles that of an active conformation, despite the fact that helix C is bent and the A-loop has an inactive fold. Phosphorylation of the tyrosine residues in the JM region would keep it away from the kinase domain owing to steric and electrostatic forces as observed for cKit. Helix C and the A-loop could then adopt an active conformation and the JM region would be available to interact with signaling proteins that have pTyr binding sites. However, the structure of the isolated kinase domain of the homologous EphA2 shows that in the absence of the JM region, helix C can still be bent and the kinase remains in an inactive conformation *(63)*. This suggests that there is an inherent structural flexibility in helix C of the Eph receptor kinases that facilitates autoinhibition. The A-loop in both of these structures is disordered in the crystals, indicating that it is exposed to solvent and flexible. It is possible that the JM region can stabilize the autoinhibited conformation of the Eph receptor kinase, whereas phosphorylation of the A-loop in the absence of the JM region destabilizes the bent conformation of helix C and shifts the equilibrium toward the active conformation.

3.2.6. EPIDERMAL GROWTH FACTOR RECEPTOR

In contrast to other members of the receptor tyrosine kinase (RTK) family, there is evidence that epidermal growth factor receptors (EGFRs) exist as preformed dimers and form higher order oligomers, and heterooligomers with other members of the subfamily (ErbB2, ErbB3, ErbB4) for signaling *(70,71)*. Autophosphorylation of tyrosine residues in the cytoplasmic part of the receptor provides sites for the binding of SH2 proteins and signaling. EGFR (and ErB2 and ErbB4) does not require phosphorylation of residues in the kinase domain for activity. In addition, the EGFR subfamily members have a dimerization domain between the kinase domain and the C-terminal phosphorylation sites.

The structure shows an active conformation for the kinase domain despite the lack of phosphorylation *(72)*. The A-loop superimposes very well with those of activated kinase structures, such as trisphosphorylated IRK and monophosphory-lated Lck. Although Tyr845 is in the same position as tyrosines in other kinases that need to be phosphorylated for activity, mutation of this residue in EGFR has no effect on the activity *(73)*. This may be because of the position of a neighbour-ing glutamate side chain (Glu848), which could structurally mimic the phosphate group of a phosphorylated A-loop tyrosine, stabilizing the active conformation of the A-loop. In the crystals there is a large packing interface between two kinase domains with part of the C-terminal tail sandwiched between them. It is possible that this type of interaction has a role in the regulation of the activity of the kinase. A C-terminal motif (Leu-Val-Ile) has previously been shown to have a role in the regulation of the phosphorylation of substrate tyrosines, but it interacts tightly with the C-terminal lobe and is not available for binding in the conformation observed in the crystal structure. These residues may become more accessible on disruption of the dimer, an event that has been observed in vitro on the addition of ATP. The binding of ATP would require a change in the relative orientation of the N- and C-terminal lobes, which would disturb the dimer interaction seen in the crystals.

3.2.7. Fibroblast Growth Factor Receptor

Dimerization of the fibroblast growth factor receptor (FGFR) requires the binding of heparin sulphate proteoglycans in addition to the growth factor. The subsequent activation allows autophosphorylation of up to seven tyrosines in the cytosolic domain, including two in the kinase domain that are critical for activation, and one in the C-terminal tail that is a site for binding of phospholipase. The crystal structure of the unphosphorylated kinase domain of FGFR1 shows an autoinhibited state, where interactions between the conserved DFG motif at the N-terminus of the A-loop and residues from helix C hold the kinase in a rather open conformation, although less open than the inactive IRK structure *(9)*. The path of the A-loop in FGFR1 lies halfway between that of an active conformation like in trisphosphorylated IRK and an inactive conformation like in unphosphorylated IRK or Abl. The DFG motif is in an active conformation, but the binding of ATP is blocked because the P-loop has folded down into the ATP binding site. The C-terminal part of the A-loop blocks the substrate binding site, but the unphosphorylated tyrosines of the A-loop do not bind in the substrate site in the way that was observed for IRK. Structures of FGFR1 have also been determined in complex with inhibitors *(74)*. It is interesting to note that the most selective of these induced conformational changes in the P-loop. As for Abl kinase, the ability of some kinase domains and not others to adapt to the binding of certain inhibitors contributes to the selectivity of these compounds.

3.2.8. Vascular Endothelial Growth Factor Receptor and Tie2

The Tie receptor PTKs (Tie1 and Tie2, or Tek) as well as the vascular endothelial growth factor receptor (VEGFR) PTKs have a role in normal vascular development. They also have a critical role in a number of diseases, such as ischemic coronary artery disease, cancer, diabetic retinopathy, and rheumatoid arthritis. The VEGFRs are believed to act in the early stages of vascular development, whereas the Tie receptors are involved in vascular remodeling and maturation.

Tie2 is tightly regulated by agonistic and antagonistic extracellular ligands. Activation requires phosphorylation in response to the binding of agonists. The structure of the unphosphorylated Tie2 kinase domain has been determined using a construct that also contains the KID and the C-terminal tail, but does not contain the JM region *(29)*. The structure was solved in four different crystal forms and all of them show the same conformation, which rules out the possibility that the structures of features like the P-loop, the C-terminal tail, and the A-loop would be owing to crystal packing artefacts, because the different crystal forms have different crystal packing arrangements. This structure shows some novel means of self-regulation including the P-loop, which binds in the ATP binding cleft and blocks the binding of this cofactor, and the C-terminal tail, which partially blocks the substrate binding site. The A-loop adopts an active-like conformation, despite the fact that it is not phosphorylated, but the DFG motif has an inactive conformation similar to that seen for IRK, which also contributes to block the binding of ATP. The KID is ordered in Tie2 and forms 2 short α-helical segments that pack against the C-terminal lobe (Fig. 8). The C-terminal tail packs under the KID, runs along αI, αF, and αE and ends near the substrate binding site. Two tyrosines in this tail (Tyr1101, Tyr1112) are buried in this structure, but are known to interact with a number of proteins containing SH2 and PTB domains when phosphorylated. Deletion of the C-terminus has been shown to significantly enhance activity of Tie2 (32). Helix C from the N-terminal lobe is shifted compared to an active PTK conformation, not dramatically, but enough to prevent the formation of the Glu872 to Lys855 salt bridge required for the correct positioning of the α and β phosphates of ATP for catalysis.

Phosphorylation of Tyr1212 in the VEGFR2 (KDR) C-terminal tail is also crucial for the activation of this receptor and subsequent VEGFR2-mediated angiogenesis *(75)*. Other tyrosine residues that become phosphorylated are Tyr1054 and Tyr1059 in the activation loop, and Tyr799 and Tyr1173 that are binding sites for PI3 kinase. The structure of KDR phosphorylated on Tyr1059 *(67)* has an open, inactive conformation very similar to that of FGFR1 *(9)*. The A-loop is largely disordered, but the positions of the P-loop and helix C superimpose well with those of FGFR1. The start and end portions of the KID are visible in the electron density and form a two-stranded β-sheet between them. Phe935, which is conserved in many KIDs, is important for the structural integrity of this motif. Therefore, this structural feature may be well conserved among other receptor PTKs. The structure of KDR phosphorylated on Tyr1054 and Tyr1059 has also been determined in complex with a small-molecule inhibitor *(76)* and revealed that the DFG motif is displaced to allow binding of the compound in a very similar fashion to what is observed for the binding of Gleevec to Abl kinase. The relative orientation of the N- and C-terminal lobes is very similar to that observed for Abl, but slightly more closed than in the apo-structure. The position of helix C is also similar in Abl. However, the P-loop in

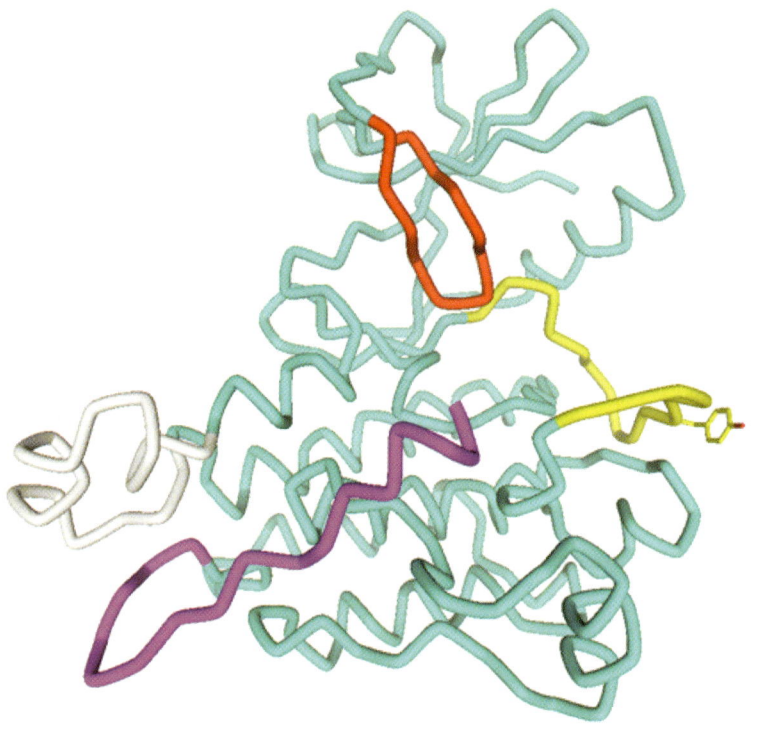

Fig. 8. A view of the Tie2 structure (cyan) showing the position of the A-loop (yellow), the P-loop (red), the kinase insertion domain (white), and the C-terminal tail (magenta).

KDR adopts an active-like conformation, suggesting that the unusual conformation of this loop in Abl is very specific to Abl in complex with Gleevec.

3.3. Common Structural Features, Their Role, and the Utility for Drug Design

The structural descriptions of the various domains of the nonreceptor and receptor tyrosine kinases highlight their roles in the regulation of the kinase activity. There are general themes that span the two groups, such as kinases in which autoinhibition involves the positioning of the A-loop such that the substrate binding site is blocked (e.g., Abl, IRK, IGF-1R, MuSK, FLT3, KIT, Met, FGFR), in which case the DFG motif is usually displaced rendering the kinase inactive, and yet others for which helix C is displaced by intereactions (or the lack of interactions) with other domains or with cellular proteins (cSrc, FAK, Btk, Met, EphB2, EphA2, Tie2). There are also themes restricted to each family such as the assembled inactive state involving SH3 and SH2 domains of nonreceptor PTKs (cSrc, Hck, Abl) and variations of these themes involving the same domains to help activate the kinase (e.g.,

Csk). A binding site for the myristate regulatory element in Abl is also found in cSrc where it is used by the unphosphorylated C-terminal tail. The receptor PTKs have regions such as JM domains (e.g., cKit, FLT-3, IRK, EphB2) and/or C-terminal tails (e.g., Tie, VEGFR) to regulate their activity. Many of these regulatory regions, in both families of kinases, are also involved in downstream signaling.

Each of the structures provides detailed information about intereactions of residues that stabilize or destabilize certain conformations. Many small-molecule PTK inhibitors have up till now been targeted against the well-conserved ATP binding site in the active conformation (77), exploiting subtle differences in this region to attain specificity. The most obvious example of this is the gatekeeping residue for the so-called selectivity pocket (Thr315 in Abl kinase, Thr670 in cKit, Phe654 in MuSK, Leu1157 in Met). When this residue is small, inhibitors can be designed that reach past this side chain into a hydrophobic pocket. These inhibitors will not bind to kinases with a large side chain in this position. Others have tried to target other regions of the kinase such as the substrate binding site (e.g., ref. 78). The development of a drug targeted against Abl kinase showed that an extremely high degree of selectivity can be obtained by inhibiting the unique inactive conformation of the kinase domain (79). Differences in the regulatory mechanisms, as summarized above for many of the known kinase structures, mean that they can have very different inactive conformations, resulting in binding sites that are unique for the kinase in question (Fig. 9). Such conformational differences can also be caused by local sequence differences, and even by residues outside the binding sites. Perhaps the best example of this is the effect of mutants of Abl kinase that do not contact the ligand, that do not lie in regulatory elements, but still cause insensitivity to Gleevec (55). In the Gleevec case, targeting of the inactive conformation led to the selection of resistant mutants in patients that were treated in the late stages of CML. This is because of the fact that the inhibitor contacts residues that are not important for the binding of the cofactor or for the catalytic function of the enzyme as a whole. This does not invalidate the approach of targeting an inactive conformation because Gleevec is a very successful drug in the majority of CML patients.

In summary, structural studies of inhibitors in complex with the kinase domain are very useful for discovering possible determinants of selectivity in addition to optimizing potency.

4. NMR AND PROTEIN KINASES

Complementing X-ray crystallography, NMR spectroscopy is another powerful technique to determine protein structures. NMR investigates proteins in solution, without the necessity to crystallize them. In addition to structure determination, NMR can determine protein dynamics. Flexibility of atom groups, or even larger movements of entire protein segments can be detected for both their frequencies and amplitudes of motion, on a timescale of picoseconds to

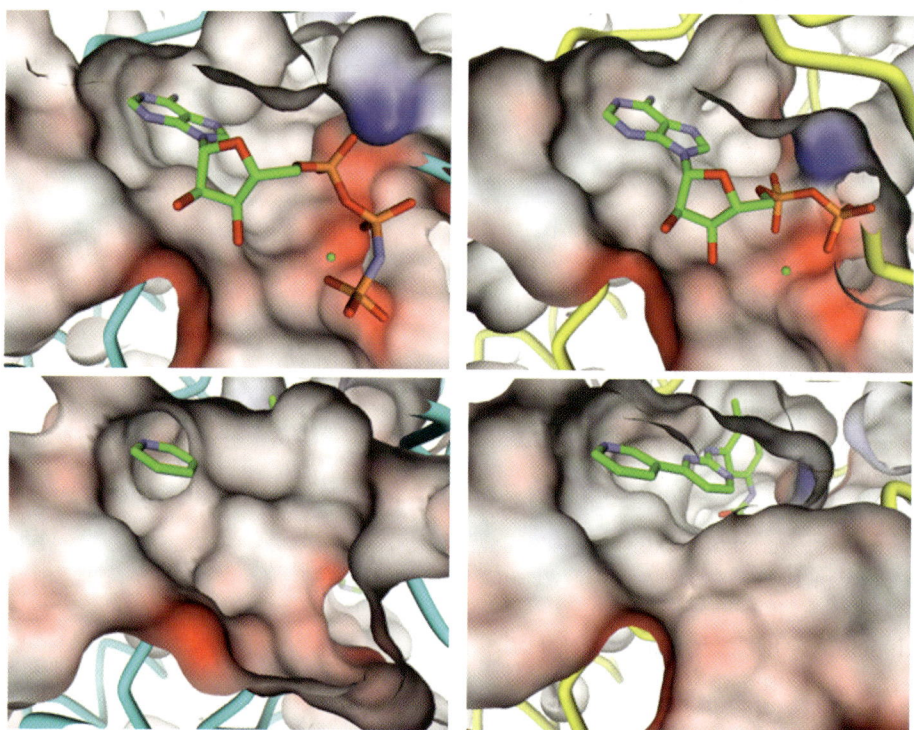

Fig. 9. Comparison of the solvent accessible surfaces of the adenosine triphosphate sites for active (top left) and inactive insulin receptor kinase (IRK) (bottom left) and for active (top right) and inactive cKit (bottom right). All structures were obtained in complex with an inhibitor (shown) except for the inactive IRK (bottom left). The ligand from the cKit complex is included in the picture of the inactive IRK structure to show that it does not fit, because the shapes of the two binding sites in the inactive proteins are so different.

milliseconds *(80–82)*. Another feature of NMR, however, which is particularly important for drug discovery, is its ability to detect and characterize weak interactions between proteins and ligands. In fact, NMR can be used to screen compound libraries for their interaction with a target protein. Because NMR can detect even weak interactions with affinities in the millimolar range, this technique is well-suited for fragment-based strategies, in which a lead candidate is built up piece by piece in a modular way *(83–85)*. This section will highlight both of these aspects of NMR and their application to protein kinases.

4.1. Structure and Dynamics of Kinases by NMR

Apart from solving solution structures of proteins and protein–ligand complexes, NMR can determine and characterize flexibility and motion ("dynamics") in proteins. These dynamical features can be resolved to the level of individual amino acid residues. They include very fast motions on the picosecond to nanosecond timescales, but also slower motions on the microsecond to millisecond timescale. Whereas fast motions are sometimes correlated to increased temperature factors in crystallographic measurements, slower motions are hardly detectable by crystallography, but are often biologically most relevant *(81,86,87)*. In fact, conformational flexibility on the microsecond to millisecond timescale is essential for the catalytic action of many enzymes, and during enzymatic activity, these very motions have recently been detected *(88,89)*.

Kinases are enzymes, and their ability to perform their catalytic action depends on conformational flexibility. There may even be more importance in kinase dynamics: given the substrate specificity of protein kinases in spite of their abundance, dynamics could be the key to achieving selectivity. It is conceivable that kinases could not attain such a high degree of selectivity if their structure was static, and that only conformational freedom and motions accommodate or reject substrates according to the biological function of the kinase. Analogous considerations may be true for the specificity of kinase inhibitors.

Given the high relevance of protein dynamics for the control of biological processes, it is surprising that protein kinases have not yet been subjected to dynamical investigations by NMR. To the best of our knowledge, there are no literature reports on solution structure or dynamics of protein kinases (status January 2004). Therefore, we have recently initiated efforts to investigate by NMR the solution structure and dynamics of c-Abl kinase, currently one of the most important kinases for pharmaceutical research, and the molecular target for the most innovative therapeutical treatment, Gleevec. The development of resistance in late-stage CML patients, largely because of mutations in Abl kinase, has led to a desire to develop follow-up compounds for the treament of these cases. Knowledge about the solution structure and dynamics of Abl in the presence and absence of inhibitors can be a valuable complement to the X-ray structures (*see* Subheading 3.1.2.)

Knowledge about protein structure and dynamics can come from NMR, but needs isotopically labeled proteins. For a detailed analysis, double or triple isotopic labeling is necessary, because assignment of the protein resonance is a prerequisite. Resonance assignment is the process of identifying which resonances in the NMR spectrum originate from which amino acid residue in the protein. Typically, protein resonance assignment is achieved from uniformly $^{15}N,^{13}C$- (for proteins up to 25 kDa) or $^{15}N,^{13}C,^{2}H$-labeled- (for larger proteins) proteins, using the well-known sequential resonance assignment method on the basis of triple resonance NMR experiments *(90,91)*. This poses severe problems for NMR work with kinases: most kinase catalytic domains cannot be readily expressed in *E. coli,* so that isotopic labeling could be achieved by growing the bacteria in isotopically enriched minimal medium *(92)*. Instead, eukaryotic expression systems have to be chosen, such as Baculovirus-infected insect cells.

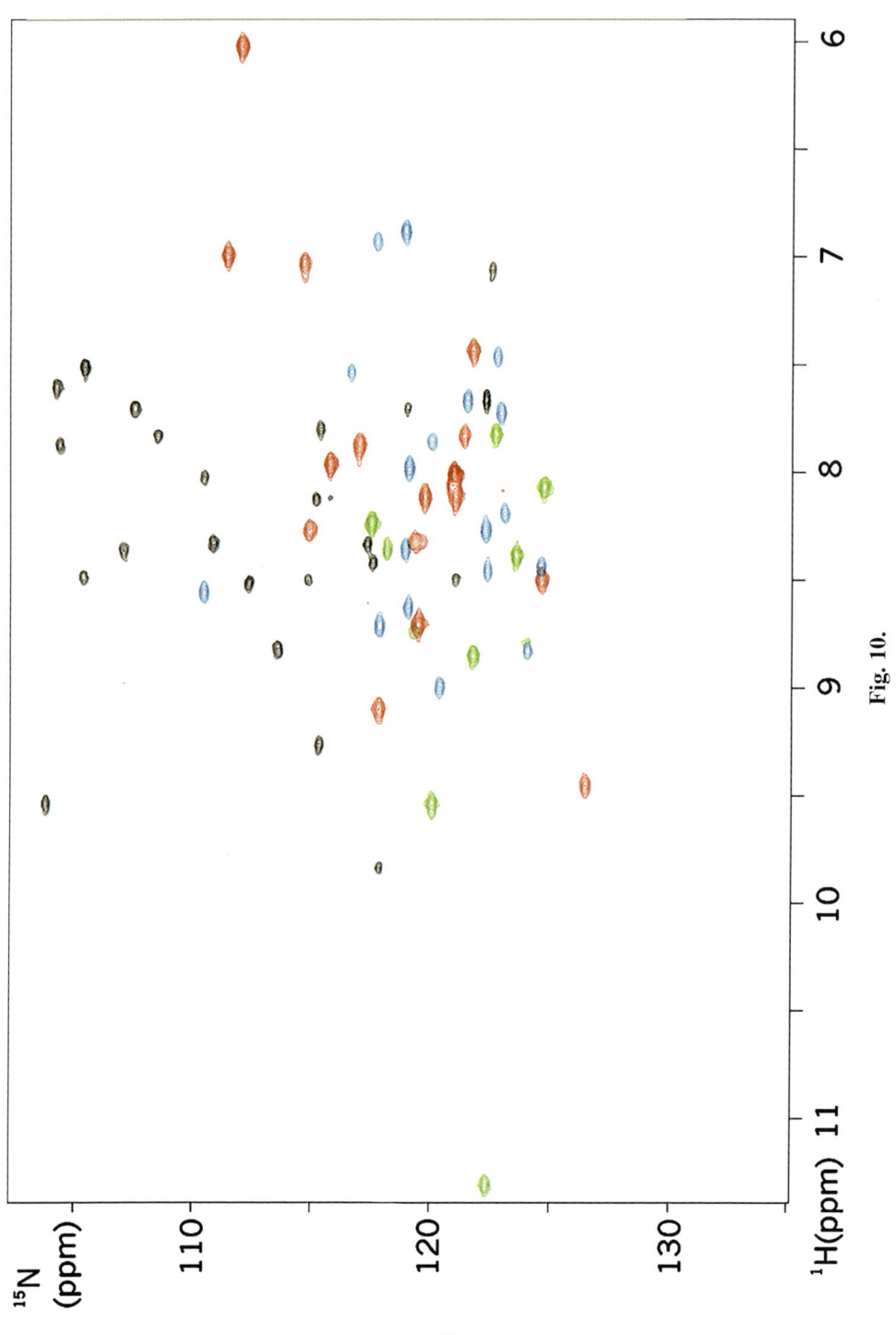

Fig. 10.

The drawback of this expression system is that media for protein isotope labeling are not yet commercially available and are expected to be quite expensive.

In order to facilitate NMR studies of kinase catalytic domains anyway, we have developed a novel protocol for amino-acid type selective isotope labeling in Baculovirus-infected insect cells *(24)*. By this method, all residues of a particular amino acid type (e.g., Phe, Tyr, or Val) are isotopically labeled, whereas all other residues in the expressed protein are unlabeled. This is achieved by establishing culture conditions in which one labeled amino acid is added as ^{15}N- (or ^{13}C-), and all others are unlabeled. The insect cells then incorporate the labeled amino acid into the protein. This method was successfully applied to c-Abl, and four samples of the c-Abl/Gleevec complex were obtained in which Phe, Tyr, Val, or Gly were selectively ^{15}N-labeled (*see* Fig. 10). Isotope costs for several milligrams of these selectively labeled samples are often only a few hundred dollars. All samples were of high purity with a high label incorporation rate.

Full backbone dynamics data can be extracted from these samples. However, resonance assignment is not straightforward because the sequential assignment procedure does not work in selectively labeled proteins. A possible solution is the preparation of dual-labeled samples in which amino acid Yyy is ^{15}N-labeled, and amino acid Xxx is ^{13}CO-labeled. If Xxx and Yyy are chosen so that an Xxx-Yyy pair occurs at an interesting position, the adjacent ^{13}CO-^{15}N pair can be detected in an HNCO experiment, so that the ^{15}N/^{1}H resonances of Yyy are assigned *(93)*. Alternatively, we developed a strategy for assignment of selectively labeled protein complexes. This strategy is based on the strong relaxation enhancement of protons that are in vicinity to a paramagnetic center ("spin label"). In practice, the ligand (Gleevec in our case) is chemically modified to incorporate a paramagnetic moiety such as TEMPO at a defined position. Measurement of paramagnetic relaxation enhancements are then used to measure distances between protein protons and the paramagnetic center. Because the structure of the complex is known, resonance assignments can be deduced *(94)*.

These novel methods are currently applied in our laboratories to characterize solution structure and dynamics of c-Abl complexes. Detailed results will be presented elsewhere.

4.2. NMR-Based Screening

Nowadays, high-throughput screening (HTS) of large compound libraries is generally considered the approach to identify lead compounds for novel targets. Depending on the nature of the compound libaries and the target protein, this

---

Fig. 10. ^{15}N/^{1}H HSQC spectra of c-Abl/Gleevec™ complexes in which one amino acid at a time is ^{15}N-labeled. Spectra shown are for ^{15}N-Phe (green), ^{15}N-Tyr (red), ^{15}N-Val (blue), and ^{15}N-Gly (black). For details, *see* ref. *24*.

approach is often, but by far not always successful. Recently, fragment-based screening has developed into a novel and alternative lead-finding approach in drug discovery research. In fragment-based screening, a lead compound is built up piece by piece, starting from small initial hits. These fragments are typically less than 300 Da in mass and show only weak binding affinity to the target protein. A well-chosen fragment, however, has the potential to be linked, merged, or decorated to increase affinity into the submicromolar range *(85,95–97)*. Fragment-based screening libraries are often relatively small (10^2–10^4 compounds), but much effort is devoted to make them drug-like, diverse, and chemically accessible *(98)*. The low affinity of fragments to target protein requires special techniques to detect them *(85)*. One such technique is NMR spectroscopy, which can detect in a robust manner weak affinities with K_D in the millimolar range. Additionally, NMR can characterize structures of protein–ligand complexes and investigate competition with known ligands. Fragment-based screening is most powerful when not just a single method is applied, but in combination with NMR, X-ray crystallography, and molecular modelling. Fragment-based screening by X-ray involves soaking crystals in mixtures of compounds and identifying the binder(s) based on the form of the electron density. The X-ray structures are an ideal starting point for optimization of weak hits to design novel high-affinity ligands, but not all proteins can be crystallized in the apo form.

NMR-based screening employing the SHAPES strategy *(99)* has been applied to identify lead candidates that target protein kinases. Most known kinase inhibitors compete with ATP for the ATP binding site in the catalytic domain. Chemically, many ATP competitive inhibitors are based on a phenylamino-pyrimidine scaffold, and novel classes of kinase inhibitors are highly desirable *(100)*. In a recent paper, the discovery of submicromolar ATP-competitive inhibitors of Jnk3 kinase by NMR-based approaches was described *(97)*. These inhibitors contain novel scaffolds and were discovered after HTS had failed to generate viable hits. The underlying compound collection is the SHAPES library, which consists of a few hundred small compounds that comprise the most common drug scaffolds and are synthetically accessible *(98)*. This library was screened against Jnk3 kinase using ligand-observation techniques *(85,97,99)*. 17 weakly binding ligands were identified and there was evidence that 13 of them bind specifically to the ATP binding site. Some of these primary NMR hits are shown in Fig. 11. Because no crystal structures with these weakly binding ligands could be solved, broad substructure and similarity searches were carried out. The analogs were prioritized by virtual screening, and about 100 high-scoring compounds were selected around one of the various scaffolds of primary NMR hits. In addition, about 200 compounds were selected that comprised two or more of these scaffolds. Substructure search and virtual screening with one of the weakly binding primary NMR hits (a pyrazole), resulted in

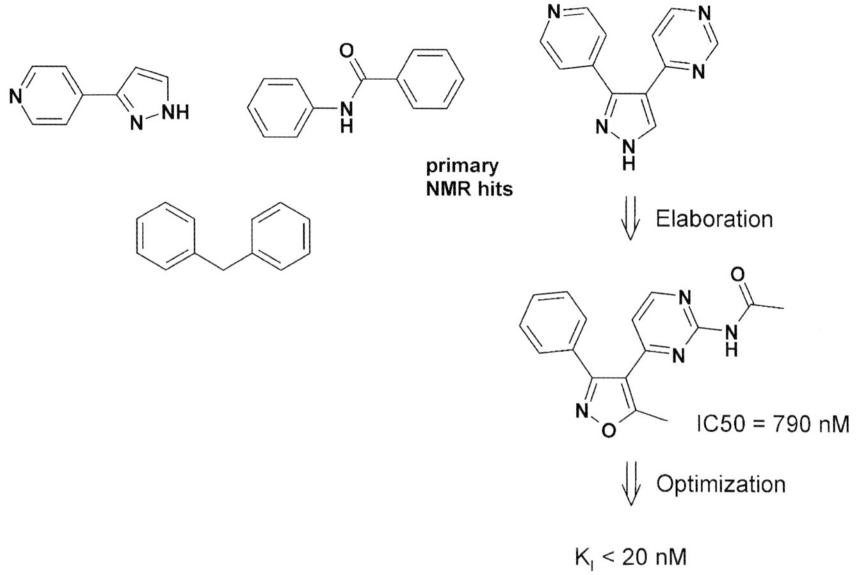

primary
NMR hits

⇓ Elaboration

IC50 = 790 nM

⇓ Optimization

$K_I < 20$ nM

Fig. 11. Elaboration and optimization of nuclear magnatic resonance SHAPES hits that resulted in high-affinity ligands for Jun N-terminal kinase 3 *(97)*.

similar compounds (isoxazoles) with more than 1000-fold higher potency (K_I=790 nM; Fig. 11). Chemical optimization of this lead compound could take advantage of the X-ray structures of Jnk3 complexed with several lead compounds identified in this manner. In particular, a subsite was detected that was occupied by some lead compounds and by the noncleavable ATP analog AMP-PNP, but not by compounds of the isoxazole class. Exploiting this sub-site in a medicinal chemistry program led to a drastic increase in potency, so that the compounds could eventually be optimized to a K_I of less than 20 nM *(97)*. This example shows that NMR-based lead-finding approaches can be successfully applied to identify potent kinase inhibitors containing novel scaffolds.

ACKNOWLEDGMENTS

We would like to thank Eva Povondra for help in the research for this document and John Priestle and Nikolaus Schiering for critical reading of the manuscript.

REFERENCES

1. Hubbard SR, Wei L, Ellis L, Hendrickson WA. Crystal structure of the tyrosine kinase domain of the human insulin receptor. *Nature* 1994; 372:746–753.
2. Pritchard ML, Rieman D, Field J, et al. A truncated v-abl-derived tyrosine-specific tyrosine kinase expressed in *Escherichia coli*. *Biochem J* 1989; 257:321–329.
3. Wang D, Huang X-Y, Cole PA. Molecular determinants for Csk-catalysed tyrosine phosphorylation of the Src tail. *Biochemistry* 2001; 40:2004–2010.

4. Chang C-I, Xu B-E, Akella R, Cobb MH, Goldsmith EJ. Crystal structure of MAP kinase p38 complexed to the docking sites on its nuclear substrate MEF2A and activator MKK3b. *Mole Cell* 2002; 9:1241–1249.

5. Steussey CN, Popov KM, Bowker-Kinley MM, Sloan RB, Harris RA, Hamilton JA. Structure of pyruvate dehydrogenase kinase. *J Biol Chem* 2001; 276:37443–37450.

6. Munshi S, Kornienko M, Hall DL, et al. Crystal structure of the apo, unphosphorylated insulin-like growth factor-1 receptor kinase. *J Biol Chem* 2002; 277:38797–38802.

7. Wei L, Hubbard SR, Hendrickson WA, Ellis L. Expression, characterisation, and crystallisation of the catalytic core of the human insulin receptor protein-tyrosine kinase domain. *J Biol Chem* 1995; 270:8122–8130.

8. Ellis L, Levitan A, Cobb MH, Ramos P. Efficient expression in insect cells of a soluble, active human insulin receptor protein-tyrosine kinase domain by use of a baculovirus vector. *J Virology* 1988; 62:1634–1639.

9. Mohammadi M, Schlessinger J, Hubbard SR. Structure of the FGF receptor tyrosine kinase domain reveals a novel autoinhibitory mechanism. *Cell* 1996; 86:577–587.

10. Schiering N, Knapp S, Marconi M, et al. Crystal structure of the tyrosine kinase domain of the hepatocyte growth factor receptor c-Met and its complex with the microbial alkaloid K-252a. *Proc Nat Acad Sci USA* 2003; 100:12654–12659.

11. Nagar B, Hantschel O, Young MA, et al. Structural basis for the autoinhibition of c-Abl tyrosine kinase. *Cell* 2003; 112:859–871.

12. Till JH, Becerra M, Watty A, et al. Crystal structure of the MuSK tyrosine kinase: insights into receptor autoregulation. *Structure* 2002; 10:1187–1196.

13. Lamers MB, Antson AA, Hubbard RE, Scott RK, Williams DH. Structure of the protein tyrosine kinase domain of C-terminal Src kinase (CSK) in complex with Staurosporine. *J Mol Biol* 1999; 285:713–725.

14. Wybenga-Groot LE, Baskin B, Ong SH, Tong J, Pawson T, Sicheri F. Structural basis for autoinhibition of the EphB2 receptor tyrosine kinase by the unphosphorylated juxtamembrane region. *Cell* 2001; 106:745–757.

15. Yaqub S, Abrahamsen H, Zimmerman B, et al. Activation of C-terminal Src kinase (Csk) by phosphorylation at serine-364 depends on the Csk-Src homology 3 domain. *Biochem J* 2003; 372:271–278.

16. Shaffer J, Sun G, Adams JA. Nucleotide rlease and associated conformational changes regulate function in the COOH-terminal Src kinase, Csk. *Biochemistry* 2001; 40:11149–11155.

17. Sun G, Budde RJA. Affinity purification of Csk protein tyrosine kinase based on its catalytic requirement for divalent metal cations. *Prot Expr Pur* 2001; 21:8–12.

18. Kashiwakura J-I, Suzuki N, et al. Takeno M, Evidence of autophosphorylation in Txk: Y91 is an autophosphorylation site. *Biol Pharm Bull* 2002; 25:718–721.

19. Till JH, Chan PM, Miller WT. Engineering the substrate specificity of the Abl tyrosine kinase. *J Biol Chem* 1999; 274:4995–5003.

20. Caspers P, Stieger M, Burn P. Overproduction of bacterial chaperones improves the solubility of recombinant protein tyrosine kinases in Escherichia coli. *Cell Mol Biol* 1994; 40:635–44.

21. Amrein KE, Takacs B, Stieger M, Molnos J, Flint NA, Burn P. Purification and characterisation of recombinant human p50[csk] protein-tyrosine kinase from an *Escherishia coli* expression system overproducing the bacterial chaperones GroES and GroEL. *Proc Natl Acad Sci USA* 1995; 92:1048–1052.

22. Radziejewski C, Miller WT, Mobashery S, Goldberg AR, Kaiser ET. Purification of recombinant pp60[v-src] protein tyrosine kinase and phosphorylation of peptides with different secondary structure preference. *Biochemistry* 1989; 28:9047–9052.

23. Pluk H, Dorey K, Superti-Furga G. Autoinhibition of c-Abl. *Cell* 2002; 108:247–259.

24. Strauss A, Bitsch F, Cutting B, et al. Amino-acid-type selective isotope labeling of proteins expressed in Baculovirus-infected cells useful for NMR studies. *J Biomol NMR* 2003; 26:367–372.

25. Chan PM, Ilangumaran S, La Rose J, Chakrabartty A, Rottapel R. Autoinhibition of the Kit receptor tyrosine kinase by the cytosolic juxtamembrane region. *Mol Cell Biol* 2003; 23:3067–3078.

26. Tanis Q, Veach D, Duewel HS, Bornmann WG, Koleske AJ. Two distinct phosphorylation pathways have additive effects on Abl family kinase activation. *Mol Cell Biol* 2003; 23:3884–3896.

27. Mao C, Zhou M, Uckun FM. Crystal structure of Bruton's tyrosine kinase domain suggests a novel pathway for activation and provides insights into the molecular basis of X-linked Agammaglobulinemia. *J Biol Chem* 2001; 276:41435–41443.

28. Favelyukis S, Till JH, Hubbard SR, Miller WT. Structure and autoregulation of the insulin-like growth factor 1 receptor kinase. *Nature Struct Biol* 2001; 8:1058–1063.

29. Shewchuk LM, Hassell AM, Ellis B, et al. Structure of the Tie2 RTK domain: self-inhibition by the nucleotide binding loop, activation loop, and C-terminal tail. *Structure* 2000; 8,1105–1113.

30. Schindler T, Sicheri F, Pico A, Gazit A, Levitzki A, Kuriyan J. Crystal structure of Hck in complex with a Src family-selective tyrosine kinase inhibitor. *Mol Cell* 1999; 3:639–648.

31. Pautsch A, Zoephel A, Ahorn H, Spevac W, Hauptmann R, Nar H. Crystal structure of bis-phosphorylated IGF-1 receptor kinase: insight into domain movements upon kinase activation. *Structure* 2001; 9:955–965.

32. Niu X-L, Peters KG, Kontos CD. Deletion of the carboxyl terminus of Tie2 enhances kinase activity, signaling, and function. *J Biol Chem* 2002; 277:31768–31773.

33. Haystead CMM, Gregory P, Sturgill TW, Haystead TA. γ-Phosphate-linked ATP-Sepharose for the affinity purification of kinases. *Eur J Biochem* 1993; 214: 459–467.

34. Xu W, Harrison SC, Eck MJ. Three-dimensional structure of the tyrosine kinase c-Src. *Nature* 1997; 385:595–602.

35. Ogawa A, Takayama Y, Sakai H, et al. Structure of the carboxyl-terminal Src kinase, Csk. *J Biol Chem* 2002; 277:14351–14354.

36. Knighton DR, Zheng J, Ten Eyck LF, et al. Crystal structure of the catalytic subunit of cyclic adenosine monophosphate-dependent protein kinase. *Science* 1991; 253:407–414.

37. Johnson LN, Noble MEM, Owen DJ. Active and inactive protein kinases: structural basis for regulation. *Cell* 1996; 85:149–158.

38. Huse M, Kuriyan J. The conformational plasticity of protein kinases. *Cell* 2002; 109:275–282.

39. Robinson DR, Wu Y-M, Lin S-F. The protein kinase family of the human genome. *Oncogene* 2000; 19:5548–5557.

40. Sicheri F, Moarefi I, Juriyan J. Crystal structure of the Src family tyrosine kinase Hck. *Nature* 1997; 385:602–609.

41. Williams JC, Weijland A, Gonfloni S, et al. The 2.35 Å crystal structure of the inactivated form of chicken Src: a dynamic molecule with multiple regulatory interactions. *J Mol Biol* 1997; 274:757–775.

42. Xu W, Doshi A, Lei M, Eck MJ, Harrison SC. Crystal structures of c-Src reveal features of its autoinhibitory mechanism. *Mol Cell* 1999; 3:629–638.

43. Young MA, Gonfloni S, Superti-Furga G, Roux B, Kuriyan J. Dynamic coupling between the SH2 and SH3 domains of c-Src and Hck underlies their inactivation by C-terminal tyrosine phosphorylation. *Cell* 2001; 105:115–126.

44. Fushman D, Xu R, Cowburn D. Direct determination of changes of interdomain orientation on ligation: use of orientational dependence of ^{15}N NMR relaxation in Abl SH(32). *Biochemistry* 1999; 38:10225–10230.

45. Yamaguchi H, Hendrickson W. Structural basis for activation of human lymphocyte kinase Lck upon tyrosine phosphorylation. *Nature* 1996; 384:484–489.

45b. Cowan-Jacob SW, Fendrich G, Liebetanz J, Fabbro D, Manley PW. Crystal structure of unphosphorylated c-Src in complex with an analogue of Gleevec™ reveals relative orientations of the SH3, SH2 and kinase domains in the active conformation. In: Abstracts of papers, 226ᵗʰ ACS National Meeting, New York; Washington, DC: American Chemical Society, 2003:COMP–078.

46. Erpel T, Courtneidge SA. Src family protein tyrosine kinases and cellular signal transduction pathways. *Current Opin Struct Biol* 1995; 7:176–182.

47. Thomas JW, Ellis B, Boerner RJ, Knight WB, White GC, Schaller MD. *J Biol Chem* 1998; 273:577–583.

48. Lerner EC, Smithgall TE. SH3-dependent stimulation of Src-family kinase autophosphorylation without tail release from the SH2 domain in vivo. *Nature Struct Biol* 2002; 9:365–369.

49. Hantschel O, Nagar B, Guettler S, et al. A myristoyl/phosphotyrosine switch regulates c-Abl. *Cell* 2002; 112:845–857.

50. Harrison S. Variation on a Src-like theme. *Cell* 2003; 112:737–740.

51. Brasher BB, van Etten RA. C-Abl has high intrinsic tyrosine kinase activity that is stimulated by mutation of the Src homology 3 domain and by autophosphorylation at two distinct regulatory tyrosines. *J Biol Chem* 2000; 275:35631–35637.

52. Schindler T, Bornmann W, Pellicena P, Miller WT, Clarkson B, Kuriyan J. Structural mechanism for STI-571 inhibition of abelson tyrosine kinase. *Science* 2000; 289:1938–1942.

53. Nagar B, Bornmann WG, Pellicena P, et al. Crystal structures of the kinase domain of c-Abl in complex with the small molecule inhibitors PD173955 and Imatinib (STI-571). *Cancer Res* 2002; 62:4236–4243.

54. Manley PW, Cowan-Jacob SW, Buchdunger E, et al. Imatinib: a selective tyrosine kinase inhibitor. *Eur J Cancer* 2002; 38:S19–S27.

55. Cowan-Jacob SW, Guez V, Fendrich G, et al. Imatinib (STI-571) resistance in chronic myelogeous leukemia: molecular basis of underlying mechanisms and potential strategies for treatment *Mini Rev Med Chem* 2004; 4:285–289.

56. Marquez JA, Smith CI, Petoukhov MV, et al. Conformation of full-length Bruton tyrosine kinase (Btk) from synchrotron x-ray solution scattering. *EMBO J* 2003; 22:4646–4624.

57. Park H, Wahl MI, Afar DE, et al. Regulation of Btk function by a major autophosphorylation site within the SH3 domain. *Immunity* 1996; 4:515–525.

58. Nore BF, Mat/son PT , Antonsson P, et al. Identification of phosphorylation sites within the SH3 domains of Tec fmaily tyrosine kinases. *Biochem Biophys Acta* 2003; 1645:123–132.

59. Takeuchi S, Takayama Y, Ogawa A, Tamura K, Okada M. Transmembrane phosphoprotein Cbp positively regulates the activity of the carboxyl-terminal Src kinase, Csk. *J Biol Chem* 2000; 275:29183–29186.

60. Shekhtmann A, Ghose R, Wang D, Cole PA, Cowburn D. Novel mechanism of the non-receptor protein tyrosine kinase Csk: insights from NMR mapping studies and site-directed mutagenesis. *J Mol Biol* 2001; 314:129–138.

61. Lin X, Lee S, Sun G. Functions of the activation loop in Csk protein-tyrosine kinase. *J Biol Chem* 2003; 278:24072–24077.

62. Chu DH, Morita CT, Weiss A. The Syk family of protein tyrosine kinases in T-cell activation and development. *Immunol Rev* 1998; 165:167–180.

63. Nowakowski J, Cronin CN, McRee DE, et al. Structures of the cancer-related Aurora-A, FAK, and EphA2 protein kinases from nanovolume crystallography. *Structure* 2002; 10:1659–1667.

64. Hubbard SR. Crystal structure of the activated insulin receptor tyrosine kinase in complex with peptide substrate and ATP analog. *EMBO J* 1997; 16:5573–5581.

65. Huse M, Chen Y-G, Massague J, Kuriyan J. Crystal structre of the cytoplasmic domain of the type I TGFβ receptor in complex with FKBP12. *Cell* 1999; 96:425–436.

66. Griffith J, Black J, Faerman C, et al. The stroctural basis for autoinhibition of FLT3 by the juxtamembrane domain. *Mol Cell* 2004; 13:169–178.

67. McTigue MA, Wickersham JA, Pinko C, et al. Crystal structure of the kinase domain of human vascular endothelial growth factor receptor 2; a key enzyme in angiogenesis. *Structure* 1999; 7:319–330.

68. Mohammadi M, Froum S, Hamby JM, et al. Crystal structure of an angiogenisis inhibitor bound to the FGF receptor tyrosine kinase domain. *EMBO J* 1998; 17:5896–5904.

69. Mol CD, Lim KB, Sridhar V, et al. Structure of a c-Kit product complex reveals the basis for kinase transactivation. *J Biol Chem* 2003; 278:31461–31464.

70. Schlessinger J. Cell signalling by receptor tyrosine kinases. *Cell* 2000; 103:211–225.

71. Moriki T, Maruyama H, Maruyama IN. Activation of preformed EGF receptor dimers by ligand-induced rotation of the transmembrane domain. *J Mol Biol* 2001; 311:1011–1026.

72. Stamos J, Sliwkowski MX, Eigenbrot C. Structure of the epidermal growth factor receptor kinase domain alone and in complex with a 4-anilinoquinazoline inhibitor. *J Biol Chem* 2002; 277:46265–46272.

73. Gotoh N, Tojo A, Hino M, Yazaki Y, Shibuya M. A highly conserved tyrosine residue at codon 845 within the kinase domain is not required for the transforming activity of human epidermal growth factor receptor. *Biochem Biophys Res Commun* 1992; 186:768–774.

74. Mohammadi M, McMahon G, Sun L, et al. Structure of the tyrosine kinase domain of fibroblast growth factor receptor in complex with inhibitors. *Science* 1997; 276:955–960.

75. Meyer RD, Dayanir V, Majnoun F, Rahimi N. The presence of a single tyrosine residue at the carboxyl domain of vascular endothelial growth factor receptor-2/FLK-1 regulates its autophosphorylation and activation of signalling molecules. *J Biol Chem* 2002; 277:27081–27087.

76. Manley PW, Bold G, Fendrich G, et al. Advances with VEGF-R kinase inhibitors for the treatment of angiogenesis. *Cell Mol Biol Lett* 2003; 8(2A), 532–533.

77. Traxler P, Furet P. Strategies toward the design of novel and selective protein tyrosine kinase inhibitors. *Pharmacol Ther* 1999; 82:195–206.

78. Parang K, Till JH, Ablooglu AJ, Kohanski RA, Hubbard SR, cole PA. Mechanism-based design of a protein kinase inhibitor. *Nature Struct Biol* 2001; 8:37–41.

79. Fabbro D, Ruetz S, Buchdunger E, et al. Protein kinases as targets for anitcancer agents: from inhibitors to useful drugs. *Pharmacol Ther* 2002; 93:79–98.

80. Peng JW, Wagner G. Investigation of protein motions via relaxation measurements. *Methods Enzymol* 1994; 239:563–596.

81. Palmer AG, III, Kroenke CD, Loria JP. Nuclear magnetic resonance methods for quantifying microsecond-to-millisecond motions in biological macromolecules. *Methods Enzymol* 2001; 339:204–238.

82. Stone MJ. NMR relaxation studies of the role of conformational entropy in protein stability and ligand binding. *Acc Chem Res* 2001; 34:379–388.

83. Hilpert K, Ackermann J, Banner DW, et al. Design and synthesis of potent and highly selective thrombin inhibitors. *J Med Chem* 1994; 37:3889–901.

84. Murray CW, Verdonk ML. The consequences of translational and rotational entropy lost by small molecules on binding to proteins. *J Comput Aided Mol Des* 2002; 16:741–753.

85. Jahnke W, Widmer H. Protein NMR in biomedical research. *Cell Mol Life Sci* 2004; 61:580–599.

86. Feher VA, Cavanagh J. Millisecond-timescale motions contribute to the function of the bacterial response regulator protein SpoOF. *Nature (London)* 1999; 400:289–293.

87. Kern D, Zuiderweg ERP. The role of dynamics in allosteric regulation. *Curr Opin Struct Biol* 2003; 13:748–757.

88. Volkman BF, Lipson D, Wemmer DE, Kern D. Two-state allosteric behavior in a single-domain signaling protein. *Science* 2001; 291:2429–2433.
89. Eisenmesser EZ, Bosco DA, Akke M, Kern D. Enzyme dynamics during catalysis. *Science* 2002; 295:1520–1523.
90. Wüthrich K. NMR—this other method for protein and nucleic acid structure determination. *Acta Crystallogr D Biol Crystallogr* 1995; D51:249–270.
91. Clore GM, Gronenborn AM. NMR structure determination of proteins and protein complexes larger than 20 kDa. *Curr Opin Chem Biol* 1998; 2:564–570.
92. Goto NK, Kay LE. New developments in isotope labeling strategies for protein solution NMR spectroscopy. *Curr Opin Struct Biol* 2000; 10:585–592.
93. Weigelt J, Van Dongen M, Uppenberg J, Schultz J, Wikstroem M. Site-selective screening by NMR spectroscopy with labeled amino acid pairs. *J Am Chem Soc* 2002; 124:2446–2447.
94. Cutting B, Strauss A, Fendrich G, Manley PW, Jahnke W. NMR resonance assignment of selectively labeled proteins by the use of paramagnetic ligands. *J Biomol NMR* 2004; 30:205–210.
95. Shuker SB, Hajduk PJ, Meadows RP, Fesik SW. Discovering high-affinity ligands for proteins: SAR by NMR. *Science* 1996; 274:1531–1534.
96. Stockman BJ, Dalvit C. NMR screening techniques in drug discovery and drug design. *Prog NMR Spectrosc* 2002; 41:187–231.
97. Fejzo J, Lepre C, Xie X. Application of NMR screening in drug discovery. *Curr Top Med Chem* 2003; 3:81–97.
98. Lepre C. Strategies for NMR screening and library design. In: Zerbe O, ed. *BioNMR in Drug Research*. Vol. 16. Weinheim: Wiley-VCH, 2003:391–415.
99. Fejzo J, Lepre CA, Peng JW, et al. The SHAPES strategy: an NMR-based approach for lead generation in drug discovery. *Chem Biol* 1999; 6:755–769.
100. Meyer B, Peters T. NMR spectroscopy techniques for screening and identifying ligand binding to protein receptors. *Angew Chem Int Ed Engl* 2003; 42:864–890.
101. Kraulis PJ. MOLSCRIPT: a program to produce both detailed and schematic plots of protein structures. *J Appl Cryst* 1991; 24:946–950.
102. Zhu X, Kim JL, Newcomb JR, et al. Structural analysis of the lymphocyte specific kinase Lck in complex with non-selective and Src family selective kinase inhibitors. *Structure* 1999; 7:651–661.
103. Witucki LA, Huang X, Shah K, et al. Mutant tyrosine kinases with unnatural nucleotide specificity retain the structure and phospho-acceptor specificity of the wild-type enzyme. *Chem Biol* 2002; 9:25–33.
104. Till JH, Ablooglu AJ, Frankel M, Bishop SM, Kohanski RA, Hubbard SR. Crystallographic and solution studies of an activation loop mutant of the insulin receptor tyrosine kinase. Insights into kinase mechanism. *J Biol Chem* 2001; 276:10049–10055.
105. Li S, Covino ND, Stein EG, Till JH, Hubbard SR. Structural and biochemical evidence for an autoinhibitory role for tyrosine 984 in the juxtamembrane region of the insulin receptor. *J Biol Chem* 2003; 278:26007–26014.
106. Hubbard SR, Till JH. Protein tyrosine kinase structure and function. *Ann Rev Biochem* 2000; 69:373–398.

10 Testing of Signal Transduction Inhibitors in Animal Models of Cancer

Terence O'Reilly, PhD
and Robert Cozens, PhD

CONTENTS

INTRODUCTION
PHARMACOKINETICS AND ADMINISTRATION OF COMPOUNDS
EXPRESSION OF TARGET IN VIVO
EVALUATION OF EGFR INHIBITORS IN PRECLINICAL IN VIVO
    MODELS
IDENTIFICATION AND PROFILING OF BIOMARKERS
    AND SURROGATE MARKERS
PRECLINICAL TESTING OF EGFR INHIBITORS AND PREDICTION
    OF CLINICAL OUTCOME
CONCLUSIONS AND RECOMMENDATIONS
REFERENCES

1. INTRODUCTION

Animal models of human disease are used to investigate, describe, explain, and predict biological phenomena and drug effects. In vitro studies provide important information on the potency and spectrum of activity of new compounds by focusing on their interaction with molecular targets in cell-free, cellular, or tissue-based systems. However, in vivo testing exposes compounds to a variety of host factors and allows for potential interaction with other targets than those selectively tested in vitro. Thus it may not be surprising that many agents active in vitro are inactive or deviant from expectations in vivo. Animal

From: *Cancer Drug Discovery and Development:*
Protein Tyrosine Kinases: From Inhibitors to Useful Drugs
Edited by: D. Fabbro and F. McCormick © Humana Press Inc., Totowa, NJ

models have been extensively used in the preclinical testing of novel anticancer compounds *(1–8)*, and they remain crucial for determining the safety of novel agents *(9,10)*. In the space available it is impossible, and probably not desirable, to try to describe all the tumor models that have been used for testing kinase inhibitors. We will concentrate on some general principles that we have applied and illustrate some of the points by specific examples. Many of these principles not only will apply to the testing of kinase inhibitors; they can equally be applied, after suitable adaptation, to testing of many anti-tumor agents. Indeed, the generalities rather than the specifics probably apply to the testing of agents for many diseases. The focus will be on the preclinical profiling of epidermal growth factor receptor (EGFR) inhibitors in animal models, many of which are in advanced clinical testing (for general reviews, *see* refs. *11–15*). Specific reviews are available on the development of erlotinib *(16,17)* and gefitinib *(18)*.

1.1. General Aspects of Animal Tumor Models for Pharmacology

Animal models as used in cancer can be divided conveniently into two groups. The first group consists of those models designed to understand the natural history of cancer. These models are used to answer questions such as the causes of cancer and the way cancers develop from precancerous lesions to invasive and metastatic disease. The second group consists of those models that are useful for the testing, selection, and profiling of new anticancer treatment modalities. Whereas in many cases models could be used for either purpose, it is important to remember that models for studying drugs require some special characteristics. The need for these characteristics are dictated by the need to be able to compare drugs one with another and to compare different treatment regimens. This means that the models should be reproducible from animal to animal both within an experiment and between experiments. We will see later that variability may be useful in certain circumstances, but even then the degree of variability should be well documented and wherever possible the reasons for variability understood even if they cannot be well controlled.

Animal models of disease should be designed to reproduce as closely as possible the disease as seen in patients. Although this statement may be indisputable, when applied to cancer and particularly to the testing of new treatment modalities, there are some obvious difficulties associated with trying to meet this goal. Any survey of the literature will immediately make it obvious that the vast majority of experiments evaluating the efficacy of anticancer agents in animals use human tumors growing as subcutaneous xenografts in immunodeficient athymic nude (nu/nu) mice. However, although these remain the mainstay for pharmacological studies, it is apparent that they may not be the best type of model. Several other possibilities exist and many of these models are presented in some detail in a recent book *(19)*. Although many strains of immunodeficient mice have been used for xenografts (bg, xid, scid, and combinations), the nu/nu remains the most commonly used mouse host for xenografts (reviewed in refs *19* and *20*). Tumor xenografts are usually placed subcutaneously into the flank, either as an injected cell suspension or as a tumor fragment from a donor animal; location of the subcutaneous xenograft can modify tumor take rate and growth rate, with an anterior site preferable to a posterior one *(21)*. The mutations that cause immunodeficiency have been introduced into several genetic backgrounds and this can result in differences in the take rate of xenografts *(19–23)*, tumor growth rate *(24)*, and tumor sensitivity and mouse tolerability to anticancer therapy *(25)*. Tumor growth rate increases after serial transplantation, and in some cases the histological and biochemical profiles also change *(22,26)*. Correlation analysis suggested xenograft tumor growth rate was an important determinant of drug response for several conventional anticancer agents *(27)*. Background strain also appears to affect the formation of stromal component of tumors, and hence the invasive nature of the experimental tumor *(19,28,29)*. Different strains of inbred mice differ in response to both growth factor-stimulated angiogenesis in vivo and the in vitro migratory activity of endothelial cells, and also show a differential response to angiogenesis inhibitors *(30)*. The influence of mouse strain on metabolism *(31)* and tissue distribution *(32)* of the anticancer agent can certainly be a significant factor in determining the outcome of therapy.

Orthotopic models are designed to implant the tumor within the tissues where the primary tumor is located in humans, and unlike most subcutaneous tumor implantation models, metastases frequently arise *(28,33–36)*. Orthotopic tumor implantation places tumor cells in an environment better mimicking the location from which they arose. For example, placing human breast tumor cell lines into the mammary fat pad of nu/nu mice consistently improves tumor growth as compared with subcutaneous injection, which did not occur with human colon, renal, and melanoma cells injected in the same location *(37)*. Orthotopic growth of the tumors often results in differential sensitivity of the tumors to chemotherapy, at least with conventional agents. Human small-cell lung cancer growing in the lungs of nude mice is susceptible to cisplatin, but not to mitomycin C (the reverse of the situation with subcutaneously growing tumor), better reflecting the clinical situation *(38)*. Although primary orthotopic tumors respond to high-dose chemotherapy, metastases of pancreatic *(39)* and gastric tumors *(40)* are apparently more resistant to therapy. Human colon carcinoma is susceptible to doxorubicin when growing subcutaneously, but lung and liver metastases arising from orthotopic tumors were resistant *(41,42)*; however, little difference in sensitivity to 5-fluorouracil was observed. This may be owing in part to the fact that expression of the multidrug-resistant phenotype (mdr1/P-glycoprotein mediated) by the metastases is facilitated by the organ-specific environment *(43)*. Taxol® displays site-dependent pharmacodynamics (PD) in an orthotopic rat prostate tumor model (MAT-LyLu primary vs lymph node metastateses) *(44)*.

Genetically engineered models in which the genetic changes predispose an animal to cancer have shown themselves to be useful for studying the biology of cancer (e.g., refs. *1* and *45–47*). Although some authorities suggest that use of genetically engineered models for pharmacology should increase *(1)*, their use for this purpose has been limited. This may be caused by the extra costs involved as a result of the often long latency periods and stochastic and sometimes low rates of tumor formation. This latter factor also means that the degree of reproducibility needed for pharmacological studies is low. Brandt et al. *(48)* have recently described a genetically engineered model in which HC-11 mouse mammary epithelial cells are rendered tumorigenic in syngenic BALB/c mice by introduction of the neuT oncogene. When introduced into the mammary fat pad, tumors arise in a predictable manner with a short latency. However, response to chemotherapy is more variable than conventional sc xenografts, and therefore may be a better predictor of clinical efficacy than subcutaneous xenografts.

However, no model or even type of model seems to be fully predictive of clinical utility. Cancer is not one disease but rather a grouping of many hundreds of diseases, many of which are superficially similar but often genetically and molecularly quite diverse *(49)*. Thus the use of any one or even a small selection of tumor models may at best reflect only a very small proportion of the cancers seen in patients. Therefore, good activity in one model may not translate into good activity in the clinical setting, where a large number of genetically distinct cancers may be targeted. This may be one factor, among, many others, that leads to the common observation that preclinical efficacy in animal models is often a poor predictor of clinical benefit. There are other factors that could also be important and some of these will be discussed with particular reference to kinase inhibitors.

1.2. In Vitro Compound Preselection

From both an ethical and an economic point of view, the selection of compounds for testing of compounds in vivo should be carefully made. The entire process of a medicinal chemistry program should carefully integrate the later stages of in vivo testing. Prior to the initiation of animal experiments, and in particular, long-term efficacy studies, a set of information is needed. In vitro experiments should utilize thoroughly characterized tumor cell lines that have the capacity to form tumors in vivo. Proof of mechanism in vitro should be convincing with demonstration of target inhibition/activation consistent with inhibition of proliferation or another marker of tumor inhibition; however, this is may be done using techniques not involving small molecules (antisense, small interfering RNA, transient transfection using dominant-negative genes, etc.). Compounds should be selected based on adequate in vitro evidence of activity and selectivity for the proposed target, and suitable physicochemical properties. When possible, knowledge of the duration of effect after removal of the compound should be available and may be useful in planning administration regimens. Information on stability in plasma, plasma protein binding, and metabolic stability of the compound in microsomes is needed if compounds are to be compared and prediction of likely clinical utility made. In vitro pharmacokinetic (PK) data could also be included in the preselection criteria; for example, using CaCo-2 cells to select compounds for increased oral bioavailability *(50)*. As part of the in vitro screening process, some indication of the toxicity of the compound inhibition of proliferation against tumor/normal cell lines not expressing/overexpressing target or bearing wild-type form. Observations giving early indications of toxicity should be incorporated into in vivo preclinical efficacy testing *(51)*.

2. PHARMACOKINETICS AND ADMINISTRATION OF COMPOUNDS

One aspect of in vivo testing of anticancer agents that receives less attention than it deserves is that of PK. The first trials of a new drug in patients are primarily designed to determine the PK of the compound and its toxicity. It seems that in many cases the first experiments in animals are conducted using doses at or close to the maximum tolerated dose with little regard to the PK properties of the compound in the mice. This may mean that compounds are dosed at levels that would be difficult or even impossible to achieve in the clinic. Kerbel *(52)* has pointed out that most preclinical experiments are conducted at or close to the maximum tolerable dose (MTD) in mice rather than the MTD in humans. This may result in too good activity in the animals that can never be realized in humans because of the need to dose at lower levels in patients. Similarly, formulations (which can profoundly affect the PK of a compound) should be selected carefully. It is not useful to find a formulation for a compound that results in good PK of the compound if the formulation, or something similar, cannot be used in patients.

In Figure 1 we show how formulation can affect the PK of PKI166, an inhibitor of the EGFR kinase, after oral administration to mice. Two formulations were used: dimethyl sulfoxide/Tween-80/NaCl, in which the compound is first dissolved in dimethyl sulfoxide then diluted 1:20 with physiological saline containing 0.5% Tween-80 to produce a finely divided homogeneous suspension, and *N*-methylpyrolidine (NMP)/PEG300, in which the compound is dissolved in pyrolidinone and then diluted 1:3.33 with PEG300 to produce a clear solution. Clearly the latter formulation produces greater exposure; however, it is perhaps pertinent to note that the ratio of tumor concentration to plasma concentration is similar with both formulations. Not surprisingly, it seems that bioavailability is altered rather than uptake and/or retention by the tumor.

If animals are to receive a compound already available for human use, it may be possible, (and probably desirable) to use the compound in its clinical formulation. However, care must be taken to ensure that excipients are suitable for administration to animals and it must be remembered that it is seldom possible

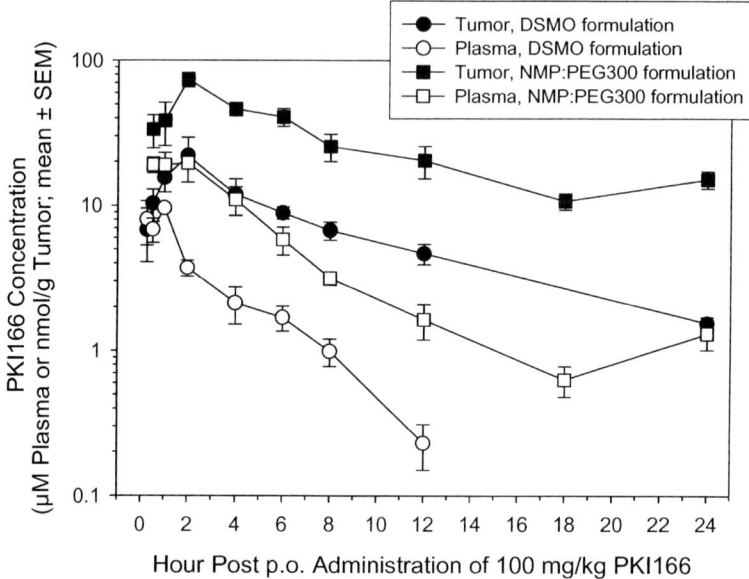

Fig. 1. Effect of formulation on plasma and tumor levels of PKI166. PKI166 was formulated in 5% DMSO/95% Tween-80 (0.5%)/NaCl (0.9%) or 30% NMP/70% PEG300. 100 mg/kg was administered to mice bearing subcutaneous A431 tumors. At selected times mice were sacrificed ($n = 4$ per data point) and plasma and tumor concentrations of PKI166 determined by reversed-phase high-performance liquid chromatography (T. O'Reilly and J. Brueggen, unpublished data). DMSO, dimethyl sulfoxide.

to give the dosage form directly to animals without some further preparation. For example, material presented in a capsule for human use must, in most cases, be removed from the capsule and dissolved or suspended in a vehicle for administration. Experimental compounds must usually be formulated by the experimenter. The simplest and often the best way is to dissolve or suspend the compound in an aqueous solvent. However, few anticancer agents are adequately soluble in pure aqueous solvents and other solvents or methods may need to be employed. Space does not permit an extensive review of the possibilities. The formulation of anti-infective compounds for use in animal experiments has been reviewed (53) and much of what is described there can be applied for anticancer compounds.

PK parameters can be used to predict doses to be used in animal models or at least to better interpret the data from animal models particularly where compounds or dosing regimens are to be compared. Considering the large and difficult to predict differences between PK in animals and humans, it is not justified to compare the anti-tumor efficacy of compounds without knowing anything about their PK properties. The best comparisons are made on the basis of equal exposure, although to achieve this can be difficult and time consuming. The PK parameters that are most useful for the comparison of compounds are: elimination half-life, area-under-the-concentration/time curve, apparent volume of distribution, total plasma clearance, absolute bioavailability (relative bioavailability, e.g., between formulations, can also be useful), and free fraction of drug. All of these, with the exception of free fraction of drug, can be determined from plasma concentrations obtained at various time points after administration of the drug. With these parameters and a reasonable knowledge of PK, it is possible to perform simple modeling to predict doses and regimens likely to be effective. More extensive modeling to include PK and PD relationships, and even interspecies scaling *(54)* require more extensive knowledge. As an example of the benefits of simple modeling to predict efficacious doses, we have shown that STI571, an inhibitor of platelet-derived growth factor receptor (PDGFR) kinase, requires dosing at a regimen that results in rapid attainment and maintenance of trough levels in tumors that are at least $10 \times IC_{50}$ (determined in vitro) to produce optimal antitumor activity *(55)*. Figure 2 shows the PK profile of STI571, simulated tumor concentrations (based on data after a single dose of compound), and the efficacy attained with various regimens. Regimens that did not result in rapid or sustained trough levels in tumors produced less activity (data not shown) than regimens that resulted in levels that rapidly reached $10 \times IC_{50}$ and consistently met or exceeded this concentration. The determination of PK parameters and some basic aspects concerning the bioanalysis of drugs in preclinical efficacy models and experimental design have been described *(53)*.

The determination of drug concentrations in tumors is, of course, important for a full understanding and in this regard it is perhaps important to note that drug concentrations can vary depending on the location of the tumor. That is that a subcutaneous tumor may exhibit quite different drug concentrations from the same tumor growing orthotopically. Solorazano et al. *(56)* used immunohistochemical detection of both EGFR and its phosphorylated form and demonstrated that inhibition of EGFR phosphorylation could be maintained for up to 2–3 d after cessation of daily therapy. This was in agreement with data showing that daily or three times weekly dosing of PKI166 produced similar and significant inhibition of tumor growth. Moreover, the less frequent dosing schedule was better tolerated. This is in contrast to other data in subcutaneous xenografts *(57)*, in which daily treatment was required to produce optimal effects. These observations suggest that PK in pancreatic tumors may be different from that in subcutaneous tumors.

Food can have an influence on the PK of orally administered compounds. Therefore, it must be decided whether the animals should be fasted or not. Clearly, if the intention is to use PK to aid planning and interpretation of efficacy experiments, then the PK experiments will be done in animals held under the same conditions as those employed in the efficacy experiments. With rodents it is often difficult to ensure that all animals employed have the same

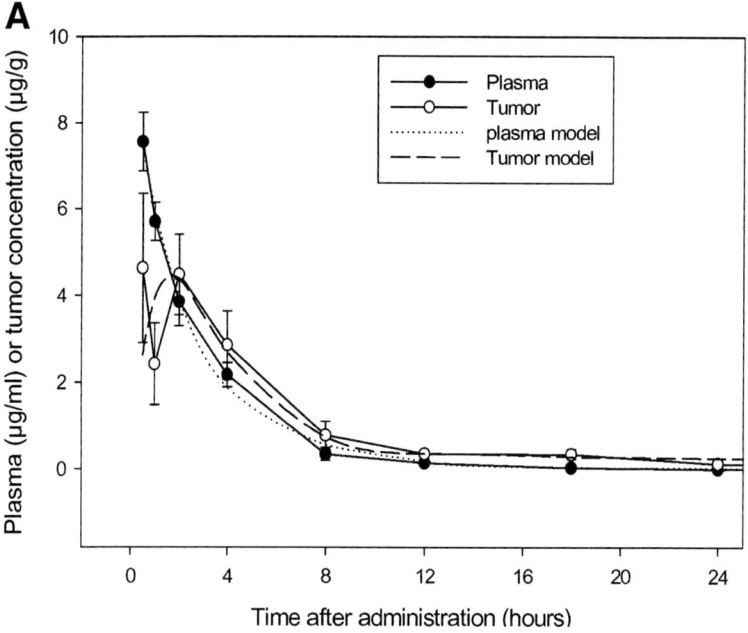

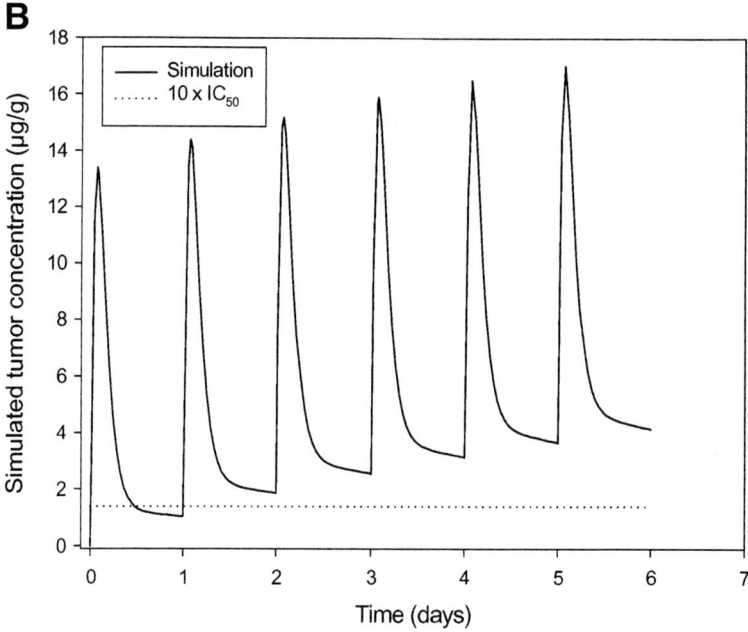

Fig. 2. Anti-tumor effect of ST1571 against subcutaneously transplanted v-*sis* fibroblast tumors in female BALB/c nude mice. (**A**) Female nude mice bearing subcutaneous implanted BALB/c 3T3 v-*sis* tumors received a 50 mg/kg oral dose of STI 571. At various times after

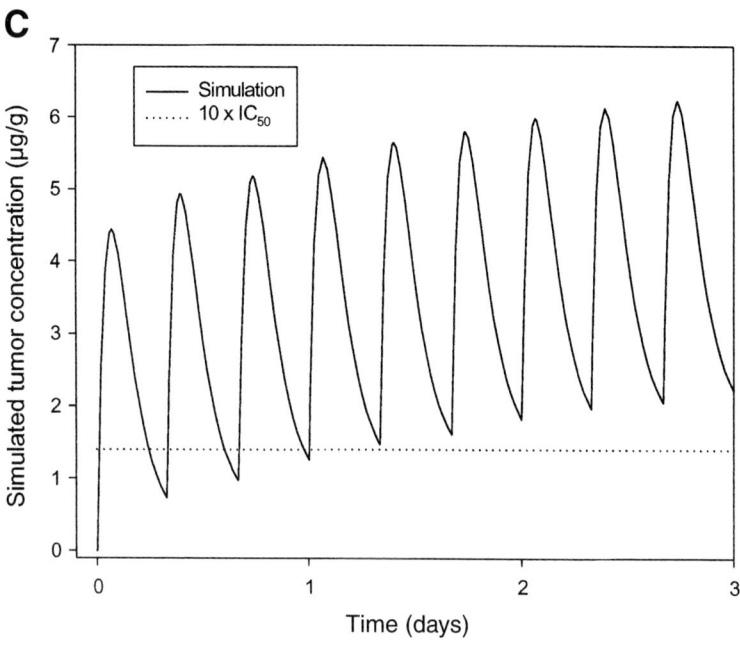

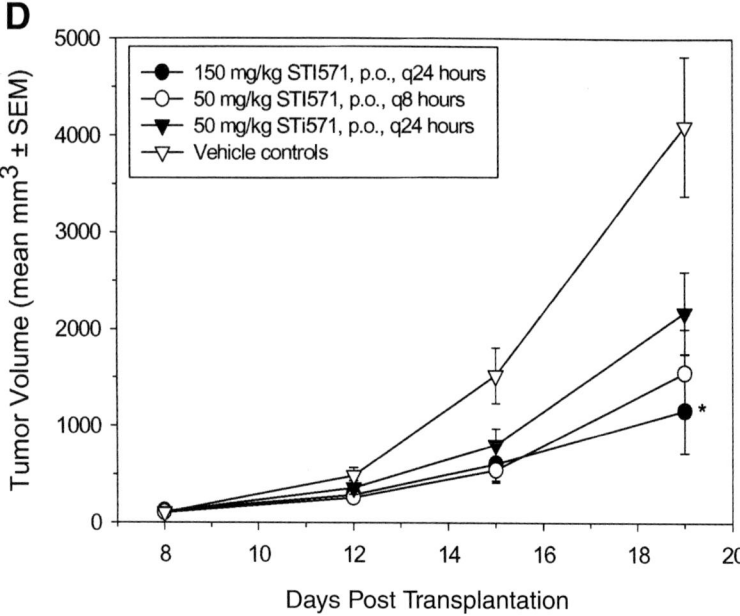

Fig. 2. *(Continued)* administration mice were sacrificed plasma and tumors prepared for determination of drug concentrations by reverse-phase high-performance liquid chromatography. The data were fitted to a three-compartment model. Bars are sem, *n* = 3. For clarity, only the

nutritional status. This implies that in nonfasted rodents a larger variation of PK results is found than in fasted animals if the oral bioavailability of the test compound is markedly affected by food intake.

A full treatise on PK is well beyond the scope and space limitations of this chapter. PK is a science in itself and several monographs serve to introduce the reader to the basic and more advanced concepts. The book by Rowland and Tozer *(58)* has become a standard reference work for those involved in PK, although other volumes are also useful for a less comprehensive coverage (e.g., refs *59* and *60*).

3. EXPRESSION OF TARGET IN VIVO

It may seem trivial to mention that therapy with a kinase inhibitor will be successful only if the target is present and it is functional. However, it has been noted that expression of a functional target will depend on the way the cells and/or tumor are grown. Therefore, confirmation of expression and functionality in vitro may not translate into expression and functionality in vivo. Similarly, targets that are not expressed in vitro may be "switched-on" in response to the presence of growth factors in vivo. Lu et al. *(61)* have shown not only that expression of cyclin-dependent kinases is different between various cell lines but also that expression is often markedly different when the cells are grown in vitro as compared with the same cells growing as subcutaneous xenografts. The position of the tumor may also have a profound influence on the expression of genes. Uehara et al. *(62)* used immuno-histochemical detection of receptor (both native and phosphorylated) and ligand to demonstrate that expression of platelet-derived growth factor receptor (PDGFR)-β, and the ligand PDGF, can be observed in prostate cancer cells growing in close association with the bone whereas the same tumor only a few millimeters away in the adjacent muscle shows no expression of the receptor. As a consequence STI571, either alone or in combination with Taxol, was able to inhibit tumor growth in this situation. However, complete absence of PDGFR in tumor cells growing in the surrounding muscle would mean that STI571 would have no useful activity in these tumors in the absence of associated bone. These observations demonstrate the necessity to confirm the presence and functionality of the target in vivo and in the model proposed for testing a compound. Moreover, it is

---

Fig. 2. *(Continued)* time up to 24 h is shown. **(B,C)** Data and model presented in Panel A were used to simulate the tumor concentrations after 150 mg ST1571 per kg were given once per day (B) or 50 mg ST1571 per kg were given three times per day (C). Simulations were performed by TOPFIT version 2. **(D)** Tumor fragments of approx 25 mg were implanted into the left flank of each female nude mouse (initially $n = 8$/group). Treatments were started on day 8 after tumor transplantation. ST1571 was administered by mouth once per day at 50 or 150 mg/kg or three times per day at 50 mg/kg until 19 d posttumor transplantation. $*p < 0.05$ vs vehicle controls.

Table 1
Antiproliferative Activity of NVP-ADL681-NX Toward Human Cancer Cell Lines[a]

Cell line	IC50 ± SEM [μM][b]
BALB/MK (immortalized keratinocytes, EGF-dependent)	0.36 ± 0.03 (4)
NCI-H596 (lung adenocarcinoma, overexpressing EGFR)	0.77 ± 0.06 (5)
BT-474 (breast carcinoma, overexpressing EGFR and ErbB2)	0.65 ± 0.05 (4)
SK-BR-3 (breast carcinoma, overexpressing EGFR and ErbB2)	1.31 ± 0.09 (5)
T24 (bladder carcinoma, activating H-*Ras* mutation)	>10 (4)

[a]Except BALB/MK, which is a mouse-derived cell line.

[b]The dose-dependent in vitro antiproliferative activity was determined by methylene blue staining. Data are presented as Mean ± SEM of multiple experiments. Values in parentheses indicate number of independent experiments (M. Wartmann, unpublished data).

important that whenever possible the status of the pathways in the tumor be understood. If downstream members of the relevant signaling pathway are constitutively activated, then inhibition of the upstream members even if successful may have little or no influence on tumor growth.

The inhibition of protein phosphorylation represents an important measure of drug efficacy. It is clearly desirable during drug development to show that anti-tumor effects can be correlated with effects on the target kinase and PK.

4. EVALUATION OF EGFR INHIBITORS IN PRECLINICAL IN VIVO MODELS

In this section we will illustrate some of the points made earlier with reference to an inhibitor of the EGFR kinase. NVP-ADL681-NX is a potent inhibitor of the EGFR kinase and Table 1 shows that activity is dependent on the presence of functional EGFR. The compound has been profiled in a number of tumor models in mice. Efficacy has been correlated with PK and a PD end point, namely inhibition of phosphorylation of EGFR. In the following discussion, we shall also mention other compounds and models where these serve to provide further illustration.

4.1. Activity of EGFR Inhibitors in Subcutaneous Models

The activity of inhibitors of the EGFR kinase has been tested in a number of different subcutaneous models. Sirotnak (63) has reviewed many of the preclinical models used to study gefitinib (ZD1839, Iressa®) as a single agent and in combinations, as have Akita and Sliwkowski for erlotinib (64). In Table 2, a number of models used to profile EGFR kinase inhibitors are listed and the methods used and some conclusions summarized. The list is by no means exhaustive but serves to illustrate the wide range of models and endpoints that have been employed.

Table 2
Selected Examples of Tumor Models Used to Preclinically Evaluate EGFR-Family Inhibitors: Subcautaneous Xenografts

Reference	Tumor	Compound	Analysis methods	Comments
85	Colo320DM, Lovo, WiDR colon	Gefitinib	In vitro, WB, TV	Activity related to pEGFR levels
86	Fischer rat embryo fibroblasts transfected with erbB2	CP-654577 (erbB2 selective) Tarceva	In vitro, WB, TV	CP-654577 produced a dose-dependent reduction of p185(erbB2) autophosphorylation and inhibited tumor growth
67	NSCLC xenografts, LC6 and LC11 (gefitinib-sensitive), Lu116 and L27 (gefitinib-resistant)	Gefitinib	Microarray gene expression profiles in response to gefitinib (compared to vehicle)	Identified genes whose expression levels changed in sensitive tumors but not in resistant tumors
87	FaDu human squamous cell carcinoma	BIBX1382BS	TV, doubling time, BrdUrd and Ki67	BIBX1382BS significantly reduced tumor growth rates, but overall no effect on local tumor growth
88	GEO colon	ZD6474, a VEGF-R (flk-1/KDR) EGFR "dual" inhibitor	In vitro, TV, IHC	Dose-dependent reduction of neoangiogenesis and tumor growth inhibition
89	A431 epidermoid	Gefitinib	In vitro, TV, IHC	Gefitinib reduced tumor vascularity, and EGF-induced VEGF protein and mRNA

242

90	A431 epidermoid	EKB-596 (EGFR/erbB-2 inhibitor)	TV	Orally active
91	CWR22 androgen-dependent and -independent prostate cancer	Gefitinib	TV, WB	Inhibits growth of both androgen-dependent and -independent CWR22 variants
92	A431 epidermoid	Gefitinib	TV	EGFR expression level alone not predict sensitivity; no drug-resistant A431 tumors appear after long-term treatment; tumors can re-grow upon treatment cessation
93	GEO colon	Gefitinib RT	In vitro, WB, TV, IHC	Dose-dependent tumor growth inhibition with inhibition of TGF-α, VEGF, bFGF within the tumor

IHC, immunohistochemistry; RT, radiotherapy; TV, tumor volume; WB, western blot.

As mentioned previously, simple PK modeling can be used to predict doses and schedules likely to be effective. Based on PK experiments in mice bearing the human non-small-cell lung cancer tumor, NCI-H596 (Fig. 3A) and the belief that concentrations of kinase inhibitors should be held at or above $10 \times IC_{50}$ (approximates in most cases to IC_{90}) determined in vitro, tumor concentrations at a number of doses and schedules were simulated (Fig. 3,B–D). It was proposed that in this model a dose of 75 mg/kg twice daily with a single loading dose of 100 mg/kg (Fig. 3C) should produce good activity. The simulation showed that the regimen should result in rapid attainment and maintenance of tumor concentrations above $10 \times IC_{50}$. Table 3 shows the anti-tumor efficacy that was obtained with this schedule, albeit without the loading dose. Though it seems that maintaining trough levels is important to ensure adequate anti-tumor activity, it is apparent that the use of a loading dose to bring the concentration quickly to this level is not necessary. An additional advantage of this approach is that schedules can be designed to avoid excessive peak concentrations and/or total exposure (large area under concentration/time curve) that appear to have little influence on antitumor efficacy but may lead to unacceptable toxicity. In a different tumor model, A431, and using an enzyme-linked immunosorbent assay method, NVP-ADL681 was shown to reduce phosphorylation of EGFR at the same dose and schedule (Fig. 4). Up to 24 h after the last dose, phosphorylation of EGFR was still reduced whereas 48 h after the last dose, phosphorylation of the receptor was similar to that of the controls. Total levels of EGFR were not significantly affected.

As in vitro (Table 1) activity of inhibitors of EGFR kinase in vivo is also correlated with levels of EGFR phosphorylation in the untreated tumors. Using subcutaneous xenografts, Sirotnak et al. (65) demonstrated a general relationship between the activity of gefitinib and relative phosphorylated (p)EGFR expression levels: at the MTD of 150 mg/kg, by mouth, once per day, gefitinib induced partial regressions of A431 (high pEGFR expression), 70–80% inhibition among A549, SKLC-16, TSU-PR1, and PC-3 tumors (lower but highly variable pEGFR levels) and 50–55% inhibition against LX-1 tumors (lowest pEGFR expression in the panel tested). The lack of a more perfect correlation may be owing to differences in the dependency on the EGFR pathway or survival mechanisms.

One of the earlier reports relating PK to PD (both anti-tumor effects and inhibition of EGFR activation) evaluated erlotinib (OSI-7744, CP-358,774) against subcutaneous xenografts of LICR-LON-HN5 tumors, an EGF-dependent head and neck carcinoma (66). An ELISA for quantification of pEGFR was used in conjunction with high performance liquid chromotography-based determination of erlotinib in tumors obtained from treated mice. At erlotinib plasma concentrations greater than 20 μM produced approximately a 70% reduction in pEGFr levels in tumors. One hour after the administration of 100 mg/kg erlotinib, peak concentrations of 125 μM in plasma and 55 $\mu mol/kg$ in tumor were produced, but by 6 h these had rapidly declined followed by a much

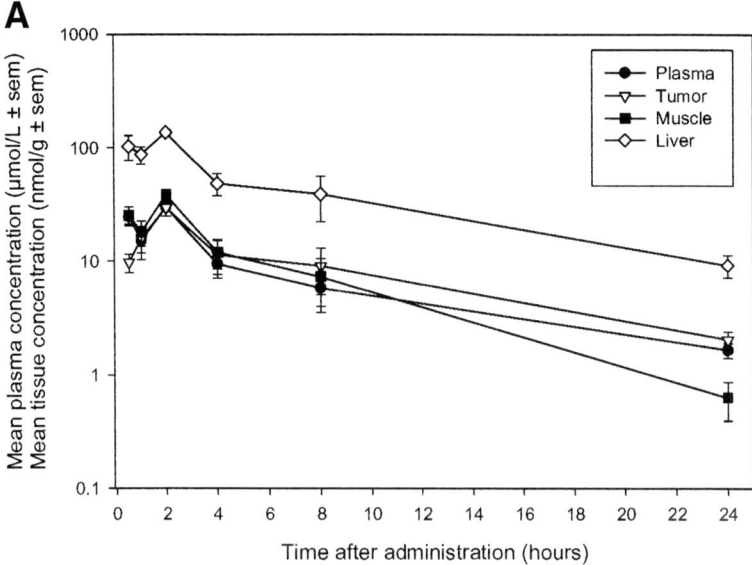

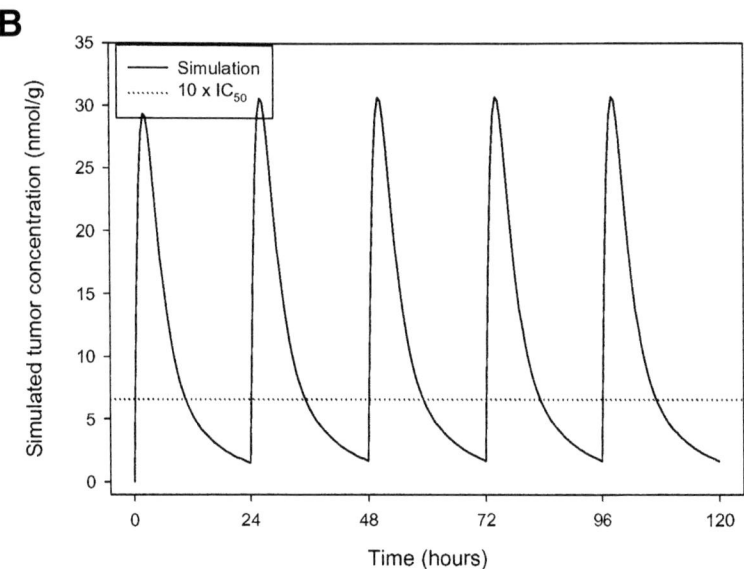

Fig. 3. Pharmacokinetics of NVP-ADL681 and pharmacokinetic simulations in nude mice bearing subcutatneous xenotransplants of the human adenosquamous lung carcinoma NCI-H596. **(A)** Experimental: plasma, tumor and tissue concentrations of NVP-ADL681 in mice after oral administration of 100 mg/kg. **(B)** Simulation: tumor concentrations of NVP-ADL681 in mice after oral dose of 100 mg/kg given once per day. **(C)** Simulation: tumor concentrations of NVP-ADL681 in mice after oral dose of 75 mg/kg given twice per day after a loading dose of 100 mg/kg. **(D)** Simulation: tumor concentrations of NVP-ADL681 in mice

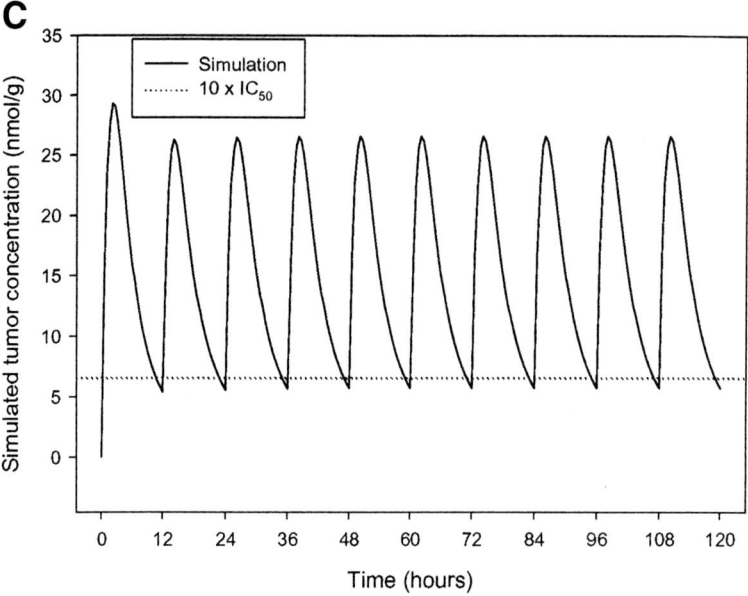

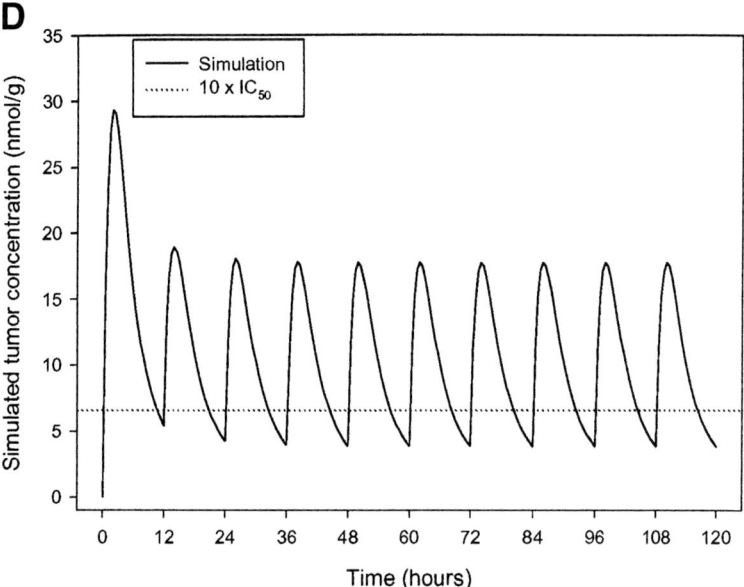

Fig. 3. *(Continued)* after oral dose of 50 mg/kg given twice per day after a loading dose of 100 mg/kg (R. Brandt and J. Brueggen, unpublished data).

slower elimination up to 24 h postadministration where the plasma and tumor concentrations were 38 µM and 4 µmol/kg, respectively. Despite declining erlotinib levels, maximal inhibition of pEGFR (75–85%) remained until approx

Table 3
Antitumor Activity of NVP-ADL681-NX and/or Taxol® Against the Human Adenosquamous Lung Carcinoma, NCI-H596, Growing in Nude Mice

| Compound | Dosage regimen | Tumor response | | Host response | |
		T/C (%)	Δ Tumor volume (mm³)	Body weight (% change)	Survival (alive/total)
Vehicle control	5 ml/kg	100	678 ± 129	4.3	9/9
Taxol	5 mg/kg i.v. q48h	116	788 ± 184	5.6	8/8
Taxol	15 mg/kg i.v. q7d	48	328 ± 127	5.1	8/8
ADL681	38 mg/kg p.o. bid	52	354 ± 64	1.2	8/8
ADL681	75 mg/kg p.o. bid	29	198 ± 62	−4.0	8/8
ADL681 + Taxol	38 mg/kg p.o. bid / 15 mg/kg i.v. q7d	39	264 ± 104	1.9	8/8
ADL681 + Taxol	38 mg/kg p.o. bid / 5 mg/kg i.v. q48h	63	430 ± 129	1.3	8/8
ADL 681 + Taxol	75 mg/kg p.o. bid / 15 mg/kg i.v. q7d	2	14 ± 38*	−1.8	7/8
ADL681 + Taxol / Taxol	75 mg/kg p.o. bid / 5 mg/kg i.v. q48h	33	223 ± 52*	7.3	5/8

Tumor fragments (ca. 40 mm³) were implanted subcutaneously. When tumor volumes had reached ca. 100 mm³ treatment started. Taxol® was given intravenously. at the dose indicated either once (q7d) or three times (q48h) weekly. NVP-ADL681-NX was given by mouth every 12 hr at the dose indicated on 5 d each week (Monday – Friday). Final tumor volumes were recorded after 22 d treatment. *$p < 0.05$ vs controls (Kruskal-Wallis ANOVA on ranks and Dunn's test). #, statistically significant change in body weight ($p < 0.05$, Paired t test). $n = 8–9$ as indicated (R. Brandt, unpublished data).

12 h postadministration; at 24 h, the inhibition was 25–40 %. This suggests that protracted inhibition of EGFR activation occurs despite apparently subeffective plasma and tumor erlotinib levels. In multiple-dose efficacy studies, there was no change in the total EGFR levels in tumors. When administered by mouth, once per day, doses greater than 12.5 mg/kg, tumor stasis was obtained, and regressions occurred at doses greater than approx 60 mg/kg. Inhibition of tumor growth was dose-dependent. Furthermore, the extent of pEGFR inhibition 1 h after single administration was dose-dependent and was strongly correlated to the degree of inhibition of tumor growth during multiple administration studies. Cessation of treatment permitted recovery of tumor growth.

These observations clearly show the desirability of correlating efficacy to PK and a PD end point. Such an approach not only will help to confirm that the activity of a compound is owing, at least in the main, to the proposed mode of

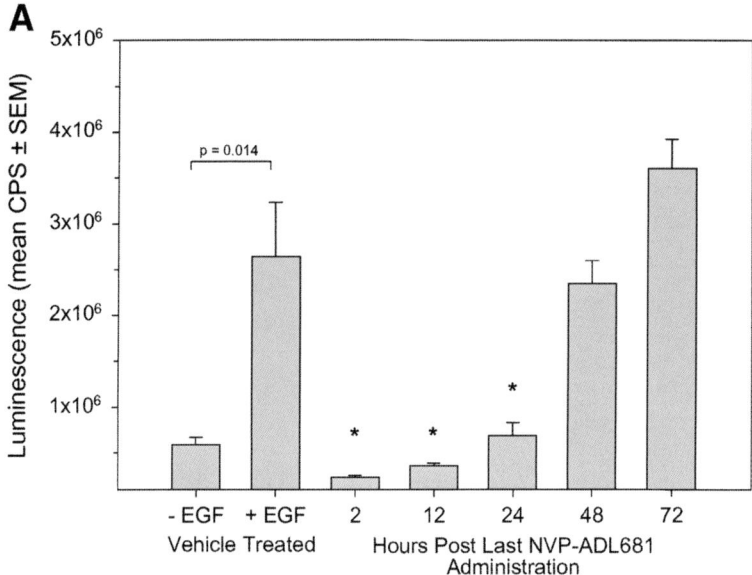

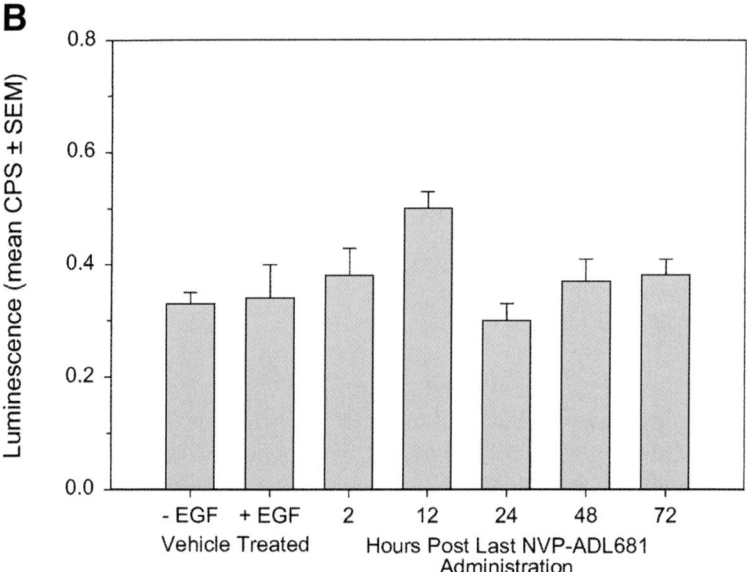

Fig. 4. Inhibition of epidermal growth factor receptor (EGFR) autophosphorylation by NVP-ADL681-NX in A431 tumors in vivo: average expression levels in tumors as determined by capture enzyme-linked immunosorbent assay (ELISA). A431 tumor bearing nude mice were treated by mouth twice per day (q12h) for seven administrations with 75 mg/kg NVP-ADL681-NX. Tumors were obtained at 2, 24, 48, or 72 h after the administration. Five minutes prior to sacrifice, mice received an intravenous injection of 0.5 µg EGFR/g body weight or 10 mL/kg saline. The tumors were extracted and subjected to capture enzyme-linked

action but will also supply information of the PK/PD relationships that may expedite clinical testing.

Subcutaneous xenografts have also been used to determine genes associated with sensitivity to EGFR inhibitors *(67)*. cDNA microarray technology was used to analyze expression profiles of 13 human tumor xenografts and 114 genes were identified whose expression levels correlated significantly with gefitinib sensitivity. Subsequently, two sensitive (LC6 and LC11) and two resistant (Lu116 and L27) nonsmall-cell lung cancers were used to determine the kinetics of changes in gene expression in response to gefitinib treatment, and a set of genes were identified whose expression differed between sensitive and resistant tumors. Such data may be useful in fully understanding the results of clinical studies in which responses to therapy are likely to be more variable than those seen in preclinical animal studies.

4.2. Activity of EGFR Inhibitors in Orthotopic Models

We have already described some data obtained with kinase inhibitors in orthotopic models. Orthotopic models seem to have advantages over subcutaneous xenografts models. They may have different expression of primary targets compared to cell culture or even the function of signaling pathways may be changed. However, although this may mean that for selected cancers in selected patients they may provide a better mimic of the clinical situation, there is no real evidence that they represent a truly better model in terms of predicting clinical outcome. In Table 4, we list a number of orthotopic models that have been used to profile inhibitors of the EGFR kinase. The table, although not exhaustive, gives an indication of the wide range of orthotopic models available and the methods that can be used to determine the outcome of therapy.

An orthotopic model of renal cell cancer *(RCC)* bone metastases has been described as being responsive to inhibition of EGFR activity *(68)*. In vitro, RBM1-IT4 RCC cells (established from a human RCC bone metastasis), respond to EGF or tumor growth factor (TGF)-α with increased cellular proliferation and EGFR autophosphorylation. When implanted into the tibia of nude mice, RBM1-IT4 RCC cells establish progressively growing lytic lesions. Immunohistochemical (IHC) analysis revealed the expression of pEGFR in both tumor cells and tumor-associated endothelial cells; PKI166 treatment decreased pEGFR by both cell types, and this was even more pronounced with mice treated with PKI166 plus Taxol®. Reduction in size and incidence of bone lesions occurred with PKI166 treatment (with or without Taxol) Furthermore, the integrity of the bone was maintained in mice treated with PKI166 or PKI166

Fig. 4. *(Continued)* immuno sorbent assay (ELISA) to determine phosphorylated EGFR **(A)** or total EGFR **(B)**. Data are the means of three to four tumors per group. *$p < 0.05$ vs EGFR-treated vehicle control group (J. Mestan and T. O'Reilly, unpublished data).

Table 4

Selected Examples of Tumor Models Used to Preclinically Evaluate EGFR-Family Inhibitors: Orthotopic Models

Reference	Tumor	Compound	Analysis methods	Comments
94	orthotopic model of oral cancer, JMAR cells were implanted into the tongues of nude mice	PKI166	Survival, IHC, TM, apoptotic fraction	Weakly active
68	Renal bone metastasis intra-tibial injection of RBM1-IT4, established from a human RCC bone metastasis	PKI166	NB and WB, histology, IHC, radiology	PKI 166 decreased pEGF-R levels in tumor cells and tumor-associated endothelial cells, decreased PNC ratios, reduced incidence and size of bone lesions
94	Murine hepatocellular carcinoma CBO140C12: orthotopic	Gefitinib	In vitro, TM	Inhibited growth of primary tumor implant and intrahepatic metastases
95	PC-3MM2 orthotopically implanted in the tiba	PKI166	TM, IHC, radiology	Reduced the incidence and size of bone tumors and destruction of bone. IHC demonstrated inhibition of phosphorylation of EGF-R on tumor and endothelial cells and induced significant apoptosis and endothelial cells within tumor lesions.
96	SN12PM6 HCC orthotopically implanted into the kidney	PKI166	TV, IHC	Inhibits growth of primary tumors and blocks formation of liver metastases

HCC, hepatocellular carcinoma; IHC, immunohistochemistry; NB, northern blot; RT, radiotherapy; TM, tumor mass; TV, tumor volume; WB, western blot.

plus Taxol, whereas extensive destruction of bone occurred in control and Taxol-treated mice. These results suggest that blockade of EGFR inhibits growth of RCC in bone by affecting both tumor-associated endothelial cells and tumor cells *(68)*.

4.3. Activity of EGFR Inhibitors in Specialized Models

Genetic manipulation of cultured cells and construction of transgenic animals have confirmed the transforming potential of the EGFR. The generation of transgenic animal models that overexpress TGF-α *(69)* or EGFR *(70)*, has also highlighted the tumorigenic potential of this receptor in vivo. Both TGF-α or hEGFR transgene expression have been shown to lead to epithelial hyperplasia and subsequently to the development of papillomas and carcinomas *(69,70)*.

A recently developed genetically engineered mammary gland tumor model (GeMag) driven by neuT overexpression in which phosphorylated, activated EGFR is present have been described *(48)*. Mammary tumors induced by neuT are characterized by a high variability in morphology, histology, and growth characteristics a situation found in breast cancer patients. Treating such tumors with Taxol resulted in a tumor response lower than in subcutaneous human xenografts, where the interassay variability is low. Based on evaluation of conventional agents, the following response categories could be proposed: nonresponder, T/C higher than 40%; partial responder T/C 40% or less; stable disease, change of tumor volume ±30%; regression decrease of tumor volume higher than 30%. Using these categories, the GeMag tumors demonstrated an overall response rate (stable disease + regression) of approx 33% for Taxol administered in an optimal regimen *(48)*. Clinical trials of Taxol in patients are also associated with variable responses *(71,72)*, with the responses in the GeMag model closely resembling those obtained clinically.

This model has been used to profile EGFR inhibitors (Fig. 5). Small-molecular-weight EGFR inhibitors could reduce the levels of pEGFR and activated neuT in vivo as determined by Western blotting of tumor extracts (not shown). In general, as compared with activity against subcutaneous NCI-H596 tumors, NVP-ADL681-NX showed much higher anti-tumor efficacy against the neuT-driven GeMag tumors producing regression. Treatment with NVP-ADL681-NX at doses of 50, 75, or 100 mg/kg twice daily produced tumor regressions ($p < 0.05$ vs controls) whereas a dose of 25 mg/kg produced essentially stable disease although this was not statistically significant (Fig. 5B). Over the range of doses used, changes in tumor volume revealed no real differences between the different dose levels. However, an alternative analysis of the data, in which the responses of individual tumors were assigned to categories of regression, stable disease, partial response, or no response (*see* preceding paragraph), while confirming the statistically significant activity, NVP-ADL681-NX revealed differences with a trend to a clear dose response (Fig. 5A). Indeed, at the lowest dose, the analysis based

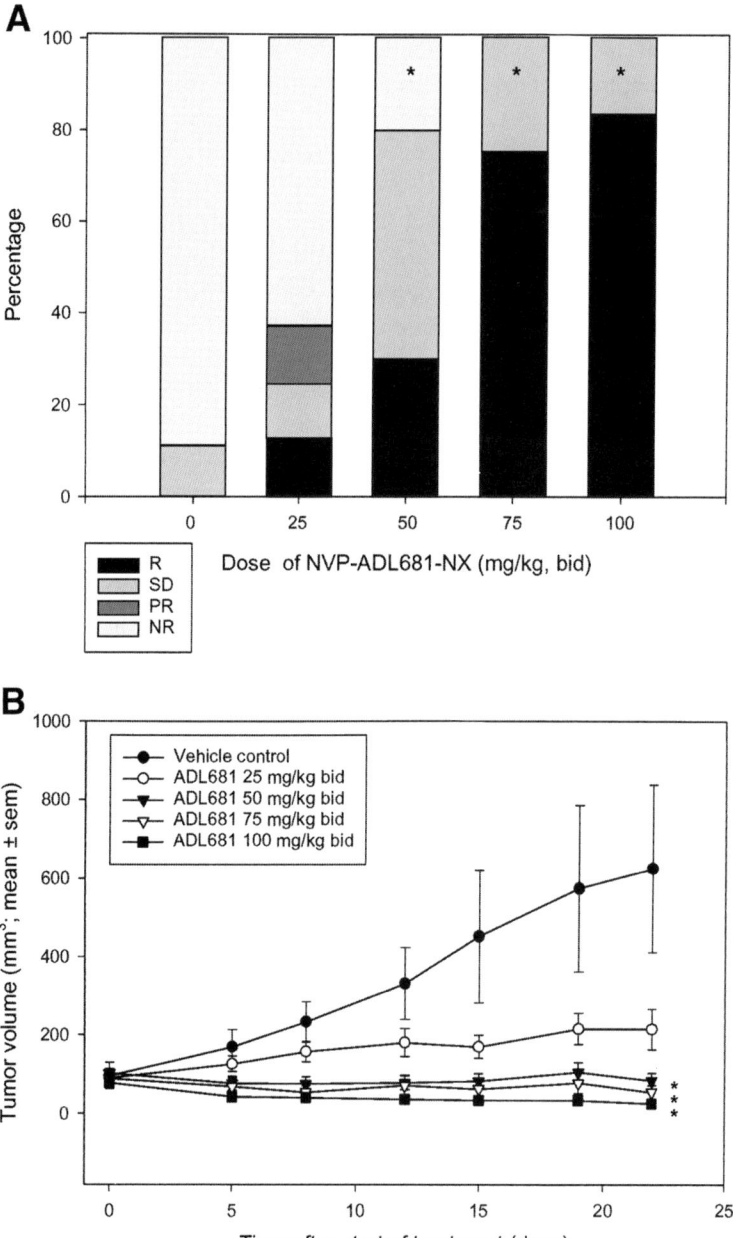

Fig. 5. Anti-tumor activity of NVP-ADL681-NX against HC11/neuT mammary gland tumors in mice. HC11 cells were transfected with *neu*T and transplanted into the cleared fat pad to female BALB/c mice. When tumors had reached a size of approx 100 mm³ treatment was started and continued for 4 wk. NVP-ADL681-NX was formulated in NMP/PEG300 (10/90 v/v) and was administered twice per day on 5 d per week (Monday–Friday). Tumor volumes

on categories revealed that only approx 40% of the tumors responded in contrast to the analysis of tumor volumes, which revealed essentially stable disease. This type of experiment indicates that alternative ways of analyzing data can lead to different conclusions. Whether this type of analysis would improve the predictive quality of a model will be revealed only when clinical data can be compared.

4.4. Activity of EGFR Inhibitors in Combination With Other Chemotherapeutic Agents

Combining treatment modalities remains the main approach for cancer therapy, but more emphasis on rational, mechanism-based combinations *(73–75)* as opposed to empirical testing *(76)* should be the aim. Although in vitro profiling provides much of the information needed to ascertain the benefits (and just as important, indicate contraindicated combinations) and provide some information on administration regimens *(77)*, animal models can be used to profile potential combinations of agents gaining information on the effect of dose and schedule on both anti-tumor effect and tolerability in the context of drug metabolism by the host *(78)*. Table 5 presents a concise review of some of the studies where the combination of an EGFR inhibitor and another form of cancer therapy has been evaluated. A variety of cell lines have been used in both subcutaneous and orthotopic locations. In general, the combinations involving the EGFR inhibitor are superior to either agent alone. However, careful inspection of the data is required in that often dose reductions are needed, and under these circumstances the anti-tumor response of the combination may not be greater than the conventional agent administered at an optimal dose.

A thorough evaluation of the EGFR inhibitor gefitinib as a combination partner for conventional agents *(65)* indicated that the anti-tumor activity of most of the conventional agents tested (cisplatin, carboplatinum, paclitaxel, docetaxel, doxorubicin, or edatrexate) was potentiated by concomitant gefitinib; vinorelbine combinations were too toxic and no potentiation was observed with gemcitabine. However, dose reduction of gefitinib was required in order to produce an acceptably tolerated combination treatment. Combination effects were observed irrespective of the extent of pEGFR expression.

Table 3 presents the activity of NVP-ADL681-NX in combination with Taxol in the NCI-H596 tumor model in nude mice. NVP-ADL681-NX was administered at its optimal dose and schedule (75 mg/kg twice daily [bid]; T/C 29%) whereas Taxol was administered at a dose and schedule (15 mg/kg once per

---

Fig. 5. *(continued)* were determined after 27 d. **(A)** The response of each individual tumor was assigned to the categories based on changes in volume over the course of the experiment. * $p < 0.025$, Crochan-Mantel-Hansel test with subsequent ranking and allowance for multiple comparisions by Holm's procedure. **(B)** Tumor volumes and body weights were determined two to three times per week. *$p < 0.05$ vs controls (analysis of variance on ranks and Dunn's test) (R. Brandt and T. O'Reilly, unpublished data).

Table 5

Selected Examples of Tumor Models Used to Preclinically Evaluate EGFR-Family Inhibitors: Combinations

Reference	Tumor	Compound	Analysis methods	Comments
85	Colo320DM, Lovo, WiDR colon	Gefitinib CPT-11	WB, TV	Combination produced supra-additive inhibitory effect on WiDR, tumor shrinkage in Lovo whereas no additive effect in COLO320DM
98	OSCC cell lines HSC2 and HSC3	Gefitinib RT	In vitro, TV, IHC	Combination caused growth inhibition and tumor regression of OSCC tumors IHC revealed gefitinib to cause decreased tumor cell proliferation when combined with RT
99	Human NSCLC A549, SK-LC-16, breast MDA-MB468, human mesothelioma (JMN)	Gefitinib RT	TV	Gefitinib significantly enhanced the antitumor action of RT, without significant adverse effects.
94	Orthotopic model of oral cancer, JMAR cells were implanted into the tongues of nude mice	PKI166 Taxol®	Survival, IHC, TM, apoptotic fraction	Combination therapy prolonged survival by increasing apoptosis
87	FaDu human squamous cell carcinoma	BIBX1382BS RT	TV, doubling time, BrdU, Ki67 staining	Combination prolongs tumor growth delay, but BIBX1382BS did not reduce the radiation dose needed to control local tumor growth
68	Renal bone metastasis intratibial injection of RBM1-IT4, established from a human RCC bone metastasis	PKI166 Taxol®	NB and WB, histology, ICC, radiology	Combination enhanced antitumor effect

254

88	GEO colon	ZD6474, a VEGF-R (flk-1/KDR) EGFR "dual" inhibitor Taxol®	In vitro, TV, IHC	Combination potentiatiate inhibition of angiogenesis, resulting in tumor regression
100	GEO colon	ZD6474, a VEGF-R (flk-1/KDR) EGFR "dual" inhibitor SC-246, a COX-1/2 inhibitor PKA-1 antisense	In vitro, TV, IHC	Triple combination produced a cooperative antitumor effect, with IHC revealed inhibition of vessel formation and expression of COX-2 and VEGF
89	A431 epidermoid	Gefitinib RT	In vitro, TV, IHC	Gefitinib potentiates the antitumor effect of single and multiple fractions of radiation; Gefitinib reduced tumor vascularity, and EGF-induced VEGF protein and mRNA
96	PC-3MM2 orthotopically implanted in the tiba	PKI166 Taxol®	TM, IHC, radiology	Combination reduced the incidence and size of bone tumors and destruction of bone, demonstrated increased inhibition of phosphorylation of EGFR on tumor and endothelial cells and induced significant apoptosis and endothelial cells within tumor lesions versus PKI166 alone

(Continued)

Table 5 (*Continued*)

Reference	Tumor	Compound	Analysis methods	Comments
91	CWR22 androgen-dependent and -independent prostate cancer	Gefitinib bicalutamide (antiandrogen) carboplatin; paclitaxel	TV, WB	Coadministration of gefitinib with a suboptimal dose of bicalutamide superior to a high-dose of bicalutamidealone. Reduced tolerability of combination. Combination increased the activity of carboplatin and paclitaxel
93	GEO colon A549 lung	Gefitinib RT	In vitro, WB, TV, IHC	Combination resulted in a significant improvement in survival, tumor growth inhibition accompanied by a significant potentiation in the inhibition of TGF-α, VEGF, bFGF within the tumor; reduction of antiangiogenisis

HCC, hepatocellular carcinoma; IHC, immunohistochemistry; NB, northern blot; RT, radiotherapy; TM, tumor mass; TV, tumor volume; WB, western blot; OSCC, oral squamous cell carcinoma.

week) that results in suboptimal anti-tumor activity (T/C 48%). The combination resulted in a clear trend (not statistically significant) to better activity (T/C 2%) than either agent alone whereas toxicity (as judged by body weights and survival) was approximately the same as that seen when NVP-ADL681-NX was given alone. Other combinations were found to be less effective (e.g., NVP-ADL681-NX 38 mg/kg bid + Taxol 15 mg/kg once per week) or even more toxic (e.g., NVP-ADL681-NX 75 mg/kg bid + Taxol 5 mg/kg three times per week). Thus it can be concluded that combination of NVP-ADL681-NX may result in improved anti-tumor activity. However, the dose and schedule used needs to be chosen with care to optimize the anti-tumor activity while preventing unwanted toxicity.

Many studies have indicated the potential of EGFR inhibitors as a combination agent for radiotherapy *(79)*, and a mechanism of action appears to be emerging: the signal cascade initiated by EGFR activation acts to facilitate survival and proliferation after ionizing radiation; consequently the action of EGFR inhibitors (inducing apoptosis, cell cycle arrest, and interfering with DNA repair) can act both on the tumor cell and also on the stromal compartment of tumors, potentiating the effects of ionizing radiation *(79)*. However, the disappointing and, based on preclinical data, unexpectedly poor anti-tumor response in combination trials in patients *(12,80)* suggests that the predictive quality of preclinical data on combination therapy may be even lower than that for monotherapy. The reasons for this discrepancy are clearly unknown, but given the extra complication of a combination partner are likely to be multiple and complex.

5. IDENTIFICATION AND PROFILING OF BIOMARKERS AND SURROGATE MARKERS

Biomarkers and surrogate markers for anticancer activity of new drugs are becoming a prerequisite for clinical testing *(81)*. Whereas surrogate markers (markers for disease progression/remission) can use well-established markers such as prostate-specific antigen for prostate cancer, biomarkers (indicators of drug activity) often must be newly developed and validated using animal models. Experiments by Bonasera et al. *(82)* suggested that use of 18F-labeled drugs in conjunction with positron emission tomography may be of limited value. Often immunohistochemical staining or Western blotting (*see* Tables 2, 4, and 5) are used to identify target activity and its inhibition in tumor tissue or other tissues. However, in a recent paper *(83)*, peripheral blood leukocytes were used as a tissue source for p70S6 kinase (p70^{s6k}), a downstream effector of mTOR, to monitor the effect of RAD001. Detailed biochemical profiling of mTOR signaling in tumors, skin, and peripheral blood mononuclear cells (PBMCs) indicated RAD001-dependent inhibition of p70^{s6k}. RAD001 demonstrated dose-dependent anti-tumor activity in the CA20948 syngeneic rat pancreatic tumors with weekly administration schedules. Comparison of an optimal (5 mg/kg) vs a suboptimal

(0.5 mg/kg) RAD001 weekly treatment schedule, demonstrated that prolonged inactivation of p70^{s6k} in PBMCs occurred only with the optimal dose. These data provide mechanistic support for the observed dose-dependent, anti-tumor efficacy of weekly treatment schedules. Supplemental experiments demonstrated that the assay could be applied to human PBMCs ex vivo, and this biomarker is currently being evaluated clinically. Taken together, these results demonstrate a correlation between the anti-tumor efficacy of intermittent RAD001 treatment schedules and prolonged p70^{s6k} inactivation in PBMCs, and suggest that monitoring of PBMC-derived p70^{s6k} activity levels could be used when assessing rapamycin treatment schedules in cancer patients.

6. PRECLINICAL TESTING OF EGFR INHIBITORS AND PREDICTION OF CLINICAL OUTCOME

Testing compounds in preclinical models is designed to provide evidence for clinical utility. In addition, the use of PK and PD end points in animal models can also allow clinical researchers to better design clinical trials. Furthermore, the development of potential biomarkers and surrogate end points in animal models, when robust and validated, will prove useful for the expedited development of new compounds in the clinic.

Despite good preclinical activity of EGFR inhibitors, the clinical activity appears lower than expected. In general, mice appear to better tolerate anticancer compounds than do other species including humans *(78)*; therefore greater compound exposure can be attained in mice as compared to humans, leading to an overestimation of pronounced anti-tumor effects in mice *(78,84)*. Targeted agents such as kinase inhibitors are assumed to be less toxic at efficacious doses than traditional oncological agents. It may therefore be considered that traditional end points such as MTD are no longer relevant with these compounds *(52)*. Data from early clinical trials have indicated that inhibitors of the EGFR kinase are indeed well tolerated at efficacious doses *(12,80)*. The consideration that they should be considered cytostatic rather than cytotoxic no longer appears valid; at least in animal experiments a number of agents have been shown to produce tumor regressions. However, the use of objective tumor response as the sole clinical end point should be challenged, particularly if the agent is given as monotherapy.

7. CONCLUSIONS AND RECOMMENDATIONS

Based on the collective experience in evaluating signal transduction inhibitors in animal models of cancer, the following can be recommended:

1. Choose the right model, one where the tumor not only expresses the target of interest, but whose growth has been demonstrated to depend on it. Subcutaneous xenografts may be needed for initial screening, but consider orthotopic models for compound profiling.

2. Perform thorough PK/PD analyses with an optimally formulated compound. Based upon this information, plan administration regimens that will provide appropriate exposure of the compound to the tumor during efficacy studies.
3. Wherever possible, incorporate biomarker and initial tolerability studies as part of the efficacy studies.
4. Fully incorporate in vivo testing in the drug discovery program with planning discussions initiated at project onset. Follow compounds being clinically tested and use clinical data to improve model and thereby increase predictive character.
5. Be fully aware of the advantages and limitations of the chosen model, and ask only appropriate questions.

REFERENCES

1. Weiss B, Shannon K. Mouse cancer models as a platform for performing preclinical therapeutic trials. *Curr Opin Genet Dev* 2003; 13:84–89.
2. Saijo N, Tamura T, Nishio K. Strategy for the development of novel anticancer drugs. *Cancer Chemother Pharmacol* 2003; 52(Suppl 1):S97–S101.
3. Fiebig HH, Dengler WA, Roth T. Human tumor xenografts: predictivity, characterization and discovery of new anticancer agents. In: Fiebig HH, Burger AM, *Relevance of Tumor Models for Anticancer Drug Development*. Contrib Oncol, vol. 54. Basel: Karger, 1999; 54:29–50.
4. Den Otter W, Steerenberg PA, Van der Laan JW. Testing therapeutic potency of anticancer drugs in animal studies: a commentary. *Regul Toxicol Pharmacol* 2002; 35(2, Pt 1):266–272.
5. Traxler P, Bold G, Buchdunger E, et al. Tyrosine kinase inhibitors: from rational design to clinical trials. *Med Res Rev* 2001; 21:499–512.
6. Kruger EA, Duray PH, Price DK, Pluda JM, Figg WD. Approaches to preclinical screening of antiangiogenic agents. *Semin Oncol* 2001; 28:570–576.
7. DeGeorge JJ, Ahn CH, Andrews PA, et al. Regulatory considerations for preclinical development of anticancer drugs. *Cancer Chemother Pharmacol* 1998; 41:173–185.
8. Curt GA. The use of animal models in cancer drug discovery and development. *Stem Cells* 1994; 12:23–29.
9. Ozawa N. Strategic proposals for avoiding toxic interactions with drugs for clinical use during development and after marketing of a new drug—proposals for designing non-clinical and clinical studies—is the non-clinical study useful? *J Toxicol Sci* 1996; 21:323–39.
10. Colombo P, Gunnarsson K, Iatropoulos M, Brughera M. Toxicological testing of cytotoxic drugs (review). *Int J Oncol* 2001; 19:1021–1028.
11. Sausville EA, Elsayed Y, Kim G. Signal transduction-directed cancer treatments. *Ann Rev Pharmacol Toxicol* 2003; 43:199–231.
12. Dancey JE, Freidlin B. Targeting epidermal growth factor receptor—are we missing the mark? *Lancet* 2003; 362:62–64.
13. Grunwald V, Hidalgo M. Developing inhibitors of the epidermal growth factor receptor for cancer treatment. *J Natl Cancer Inst* 2003; 95:851–867.
14. Normanno N, Maiello MR, De Luca A. Epidermal growth factor receptor tyrosine kinase inhibitors (EGFR-TKIs): simple drugs with a complex mechanism of action? *J Cell Physiol* 2003; 194:13–19.
15. Khalil MY, Grandis JR, Shin DM. Targeting epidermal growth factor receptor: novel therapeutics in the management of cancer. *Expert Rev Anticancer Ther* 2003; 3:367–380.
16. Herbst RS. Erlotinib (Tarceva): an update on the clinical trial program. *Semin Oncol* 2003; 30(3, Suppl 7):34–46.
17. Grunwald V, Hidalgo M. Development of the epidermal growth factor receptor inhibitor OSI-774. *Semin Oncol* 2003; 30(3 Suppl 6):23-31.

18. Herbst RS. ZD1839: targeting the epidermal growth factor receptor in cancer therapy. *Expert Opin Investig Drugs* 2002; 11:837–849.
19. Teicher BA (ed.). *Tumor Models in Cancer Research*. Totowa, NJ: Humana Press, 2002.
20. Clarke R . Human breast cancer cell line xenografts as models of breast cancer. The immuno-biologies of recipient mice and the characteristics of several tumorigenic cell lines. *Breast Cancer Res Treat* 1996; 39:69–86.
21. Auerbach R, Morrissey LW, Sidky YA. Regional differences in the incidence and growth of mouse tumors following intradermal or subcutaneous inoculation. *Cancer Res* 1978; 38:1739–1744.
22. Wagner W, Gohde W, Harle A, Bottcher HD, Histologic changes, proliferation kinetics, and growth patterns in squamous carcinomas and bone tumors transplanted to nude mice. *Strahlenther Onkol* 1989; 165:487–488.
23. Van Weerden WM, Romijn JC. Use of nude mouse xenografts in prostate cancer research. *Prostate* 2000; 43:263–271.
24. Maruo K, Ueyama Y, Hioki K, Saito M, Nomura T, Tamaoki N. Strain-dependent growth of a human carcinoma in nude mice with different genetic backgrounds: selection of nude mouse strains useful for anticancer agent screening system. *Exp Cell Biol* 1982; 50:115–119.
25. Maruo K, Emura R, Ohnishi Y, Endo S, Ueyama Y, Nomura T. Toxicity of anticancer agents, growth and chemosensitivity of human tumour xenografts in a segregating stock of AF nude mice. *Lab Anim* 1991; 25:342–347.
26. Sharkey FE, Fogh J. Considerations in the use of nude mice for cancer research. *Cancer Metastasis Rev* 1984; 3:341–360.
27. Taetle R, Rosen F, Abramson I, Venditti J, Howell S. Use of nude mouse xenografts as pre-clinical drug screens: in vivo activity of established chemotherapeutic agents against melanoma and ovarian carcinoma xenografts. *Cancer Treat Rep* 1987; 71:297–304.
28. Fodstad O, Kjonniksen I. Microenvironment revisited: time for reappraisal of some prevailing concepts of cancer metastasis. *J Cell Biochem* 1994; 56:23–28.
29. Brunner N, Boysen B, Romer J, Spang-Thomsen M. The nude mouse as an in vivo model for human breast cancer invasion and metastases. *Breast Cancer Res Treat* 1993; 24:257–264.
30. Rohan RM, Fernandez A, Udagawa T, Yuan J, D'Amato RJ. Genetic heterogeneity of angiogenesis in mice. *FASEB J* 2000; 14:871–876.
31. Hashimoto Y, Degawa M. Induction of cytochrome P450 isoforms by carcinogenic aromatic amines and carcinogenic susceptibility of rodent animals. *Pharmacogenetics* 1995; 5(Spec No):S80–S83.
32. Visser GW, Gorree GC, Peters GJ, Herscheid JD. Tissue distribution of [18F]-5-fluorouracil in mice: effects of route of administration, strain, tumour and dose. *Cancer Chemother Pharmacol* 1990; 26:205–209.
33. Kubota T. Metastatic models of human cancer xenografted in the nude mouse: the importance of orthotopic transplantation. *J Cell Biochem* 1994; 56:4–8.
34. Hoffman RM Orthotopic is orthodox: why are orthotopic-transplant metastatic models different from all other models? *J Cell Biochem* 1994; 56:1–3.
35. Manzotti C, Audisio RA, Pratesi G. Importance of orthotopic implantation for human tumor as model systems: relevance to metastasis and invasion. *Clin Exp Metastasis* 1993; 11:5–14.
36. Pocard M, Tsukui H, Salmon RJ, Dutrillaux B, Poupon MF. Efficiency of orthotopic xenograft models for human colon cancers. *In Vivo* 1996; 10:463–469.
37. Price JE. Metastasis from human breast cancer cell lines. *Breast Cancer Res Treat* 1996; 39:93–102.
38. Kuo TH, Kubota T, Watanabe M, et al. Site-specific chemosensitivity of human small-cell lung carcinoma growing orthotopically compared to subcutaneously in SCID mice: the importance of orthotopic models to obtain relevant drug evaluation data. *Anticancer Res* 1993; 13:627–630.

39. Furukawa T, Kubota T, Watanabe M, Kitajima M, Hoffman RM. A novel "patient-like" treatment model of human pancreatic cancer constructed using orthotopic transplantation of histologically intact human tumor tissue in nude mice. *Cancer Res* 1993; 53:3070–3072.

40. Furukawa T, Kubota T, Watanabe M, Kuo T-H, Kitajima M, Hoffman RM. Differential chemosensitivity of local and metastatic human gastric cancer after orthotopic transplantation of histologically intact tumor tissue in nude mice. *Int J Cancer* 1993; 54:397–401.

41. Wilmanns C, Fan D, O'Brian CA, Bucana CD, Fidler IJ. Orthotopic and ectopic organ environments differentially influence the sensitivity of murine colon carcinoma cells to doxorubicin and 5-fluorouracil. *Int J Cancer* 1992; 52:98–104.

42. Fidler IJ, Wilmanns C, Staroselsky A, Radinsky R, Dong Z, Fan D. Modulation of tumor cell response to chemotherapy by the organ environment. *Cancer Metastasis Rev* 1994; 13:209–222.

43. Dong Z, Radinsky R, Fan D, et al. Organ-specific modulation of steady-state mdr gene expression and drug resistance in murine colon cancer cells. *J Natl Cancer Inst* 1994; 86:913–920.

44. Yen-WC, Wientjes-MG, Au-JL. Differential effect of taxol in rat primary and metastatic prostate tumors: site-dependent pharmacodynamics. *Pharm Res* 1996; 13:1305–1312.

45. Jackson EL, Willis N, Mercer K. et al. Analysis of lung tumor initiation and progression using conditional expression of oncogenic K-ras. *Genes Dev* 2001; 15:3243–3248.

46. Holland EC, Hively WP, DePinho RA, Varmus HE. A constitutively active epidermal growth factor receptor cooperates with disruption of G1 cell-cycle arrest pathways to induce glioma-like lesions in mice. *Genes Dev* 1998; 12:3675–3685.

47. Bergers G, Javaherian K, Lo KM, Folkman J, Hanahan D. Effects of angiogenesis inhibitors on multistage carcinogenesis in mice. *Science* 1999; 284:808–812.

48. Brandt R, Wong AM-L, Hynes NE. Mammary glands reconstituted with neu/erbB2 transformed HC11 cells provide a novel orthotopic model for testing anticancer agents. *Oncogene* 2001; 20:5459–5465.

49. Hanahan D, Weinberg EA. The hallmarks of cancer. *Cell* 2000; 100:57–70.

50. Bohets H, Annaert P, Mannens G, et al. Strategies for absorption screening in drug discovery and development. *Curr Top Med Chem* 2001; 1:367–383.

51. Arp LH. Tumor models: assessing toxicity in efficacy studies. *Toxicol Pathol* 1999; 27(1):121–122.

52. Kerbel RS Human tumor xenografts as predictive preclinical models for anticancer drug activity in humans: better than commonly perceived—but can be improved. *Cancer Biol Ther* 2003; 2:s134–s139.

53. Cozens RM. Formulation of compounds and determination of pharmacokinetic parameters. In Zak O, Sande M, eds. *Handbook of Animal Models of Infection.* London: Academic Press, 1999: 83–92.

54. Sheiner LB, Steimer J-L. Pharmacokinetic/pharmacodynamic modeling in drug development. *Ann Rev Pharmacol Toxicol* 2000; 40:67–95.

55. Cozens R, Buchdunger E, O'Reilly T. Pharmacokinetic/pharmacodynamic relationship of the platelet-derived growth factor receptor (PDGFR) inhibitor STI571 against v-sis transformed NIH 3T3 solid tumors. 91st annual meeting of the American Association for Cancer Research 2000, abst 4579.

56. Solorazano CC, Baker CH, Tsan R, et al. Optimization for the blockade of epidermal growth factor receptor signaling for therapy of human pancreatic carcinoma. *Clin Cancer Res* 2001; 7:2563–2572.

57. Traxler P, Bold G, Buchdunger E, et al. Tyrosine kinase inhibitors: from rational design to clinical trials. *Med Res Rev* 2001; 21:499–512.

58. Rowland M, Tozer TN. *Clinical Pharmacokinetics: Concepts and Applications,* 3rd ed. Media: Williams and Wilkins, 1995.

59. Krishna DR, Klotz U. Clinical Pharmacokinetics: A Short Introduction. Berlin: Springer Verlag, 1990.

60. Blesch KS, Gieschke R, Tsukamoto Y, Reigner BG, Burger HU, Steimer JL. Clinical pharmacokinetic/pharmacodynamic and physiologically based pharmacokinetic modeling in new drug development: the capecitabine experience. *Invest New Drugs* 2003; 21:195–223.

61. Lu K, Shih C, Teicher BA. Expression of pRB, cyclin/cyclin-dependent kniases and E2F1/DP-1 in human tumor lines in cell culture and in xenografts tissues and response to cell cycle agents. *Cancer Chemother Pharmacol* 2000; 26:293–304.

62. Uehara H, Kim SJ, Karashima T, et al. Effects of blocking platelet-derived growth factor-receptor signaling in a mouse model of experimental prostate cancer bone metastases. *J Nat Cancer Inst* 2003; 95:458–470.

63. Sirotnak FM. Studies with ZD1839 in preclinical models. *Semin Oncol* 2003; 30(1, Suppl 1):12–20.

64. Akita RW, Sliwkowski MX. Preclinical studies with Erlotinib (Tarceva). *Semin Oncol* 2003; 30(3, Suppl 7):15–24.

65. Sirotnak FM, Zakowski MF, Miller VA, Scher HI, Kris MG. Efficacy of cytotoxic agents against human tumor xenografts is markedly enhanced by coadministration of ZD1839 (Iressa), an inhibitor of EGFR tyrosine kinase. *Clin Cancer Res* 2000; 6:4885–4892.

66. Pollack VA, Savage DM, Baker DA, et al. Inhibition of epidermal growth factor receptor-associated tyrosine phosphorylation in human carcinomas with CP-358,774: dynamics of receptor inhibition in situ and antitumor effects in athymic mice. *J Pharmacol Exp Ther* 1999; 291:739–748.

67. Zembutsu H, Ohnishi Y, Daigo Y, et al. Gene-expression profiles of human tumor xenografts in nude mice treated orally with the EGFR tyrosine kinase inhibitor ZD1839. *Int J Oncol* 2003; 23:29–39.

68. Weber KL, Doucet M, Price JE, Baker C, Kim SJ, Fidler IJ. Blockade of epidermal growth factor receptor signaling leads to inhibition of renal cell carcinoma growth in the bone of nude mice. *Cancer Res* 2003; 63:2940–2947.

69. Jhappan C, Stahle C, Harkins R, Fausto N, Smith G, Merlino G. TGF-α overexpression in transgenic mice induces liver neoplasia and abnormal development of the mammary gland and pancreas. *Cell* 1990; 61:1137–1146.

70. Brandt R, Eisenbrandt R, Zschiesche W, Binas B, Juergensen C, Theuring F. Mammary gland specific hEGF receptor transgene expression induces neoplasia and inhibits differentiation. *Oncogene* 2000; 19:2129–2137.

71. Cascinu S, Graziano F, Cardarelli N, Marcellini M, Giordani P, Menichetti ET et al. Phase ii study of paclitaxel in pretreated advanced gastric cancer. *Anticancer Drugs* 1998; 9:307–310.

72. Rees CN, Sinnett D, Lowdell C, English J, Coombes RC. A pilot study to evaluate paclitaxel (Taxol) as primary medical treatment for patients with inoperable stage III and IV breast carcinoma. *Eur J Cancer* 1996; 32A:2354–2356.

73. Gitler MS, Monks A, Sausville EA. Preclinical models for defining efficacy of drug combinations: mapping the road to the clinic. *Mol Cancer Ther* 2003; 2:929–932.

74. de Bono JS, Tolcher AW, Rowinsky EK. The future of cytotoxic therapy: selective cytotoxicity based on biology is the key. *Breast Cancer Res* 2003; 5:154–159.

75. Grant S, Dent P. Overview: rational integration of agents directed at novel therapeutic targets into combination chemotherapeutic regimens. *Curr Opin Investig Drugs* 2001; 2:1600–1605.

76. Damon LE, Cadman EC. Advances in rational combination chemotherapy. *Cancer Invest* 1986; 4:421–444.

77. Zoli W, Ricotti L, Tesei A, Barzanti F, Amadori D. In vitro preclinical models for a rational design of chemotherapy combinations in human tumors. *Crit Rev Oncol Hematol* 2001; 37:69–82.

78. Thompson J, Stewart CF, Houghton PJ. Animal models for studying the action of topoisomerase I targeted drugs. *Biochim Biophys Acta* 1998; 1400:301–319.
79. Sartor CI. Epidermal growth factor family receptors and inhibitors: radiation response modulators. *Semin Radiat Oncol* 2003; 13:22–30.
80. Giaconne G, Johnson DH, Manegold C, et al. A phase III clincal trial of ZD1839 («Iressa») in combination with gemcitabine and cisplatin in chemotherapy-naive patients with non-small-cell lung cancer (INTACT 1). *Ann Oncol* 2002; 13(Suppl 5):2–3.
81. Schatzkin A, Gail M. The promise and peril of surrogate end points in cancer research. *Nat Rev Cancer.* 2002; 2:19–27.
82. Bonasera TA, Ortu G, Rozen Y, et al. Potential (18)F-labeled biomarkers for epidermal growth factor receptor tyrosine kinase. *Nucl Med Biol* 2001; 28:359–374.
83. Boulay A, Zumstein-Mecker S, Stephan C, et al. The antitumor efficacy of intermittent treatment schedules with the rapamycin derivative RAD001 correlates with prolonged inactivation of p70s6k in peripheral blood mononuclear cells (PBMCs). *Cancer Research* 2004; 64:252–261.
84. Takimoto CH. Why drugs fail: Of mice and men revisited. *Clin Cancer Res* 2001; 7:229–230.
85. Koizumi F, Kanzawa F, Ueda Y, et al. Synergistic interaction between the EGFR tyrosine kinase inhibitor gefitinib ("Iressa") and the DNA topoisomerase I inhibitor CPT-11 (irinotecan) in human colorectal cancer cells. *Int J Cancer* 2004; 108:464–472.
86. Barbacci EG, Pustilnik LR, Rossi AM, et al. The biological and biochemical effects of CP-654577, a selective erbB2 kinase inhibitor, on human breast cancer cells. *Cancer Res* 2003; 63:4450–4459.
87. Baumann M, Krause M, Zips D, et al. Selective inhibition of the epidermal growth factor receptor tyrosine kinase by BIBX1382BS and the improvement of growth delay, but not local control, after fractionated irradiation in human FaDu squamous cell carcinoma in the nude mouse. *Int J Radiat Biol* 2003; 79:547–559.
88. Ciardiello F, Caputo R, Damiano V, et al. Antitumor effects of ZD6474, a small molecule vascular endothelial growth factor receptor tyrosine kinase inhibitor, with additional activity against epidermal growth factor receptor tyrosine kinase. Clin *Cancer Res* 2003; 9:1546–1556.
89. Solomon B, Hagekyriakou J, Trivett MK, Stacker SA, McArthur GA, Cullinane C. EGFR blockade with ZD1839 ("Iressa") potentiates the antitumor effects of single and multiple fractions of ionizing radiation in human A431 squamous cell carcinoma. Epidermal growth factor receptor. *Int J Radiat Oncol Biol Phys* 2003; 55(3):713–723.
90. Wissner A, Overbeek E, Reich MF, et al. Synthesis and structure-activity relationships of 6,7-disubstituted 4-anilinoquinoline-3-carbonitriles. The design of an orally active, irreversible inhibitor of the tyrosine kinase activity of the epidermal growth factor receptor (EGFR) and the human epidermal growth factor receptor-2 (HER-2). *J Med Chem* 2003; 46(1):49–63.
91. Sirotnak FM, She Y, Lee F, Chen J, Scher HI. Studies with CWR22 xenografts in nude mice suggest that ZD1839 may have a role in the treatment of both androgen-dependent and androgen-independent human prostate cancer. *Clin Cancer Res* 2002; 8:3870–3876.
92. Wakeling AE, Guy SP, Woodburn JR, et al. ZD1839 (Iressa): an orally active inhibitor of epidermal growth factor signaling with potential for cancer therapy. *Cancer Res* 2002; 62:5749–5754.
93. Bianco C, Tortora G, Bianco R, et al. Enhancement of antitumor activity of ionizing radiation by combined treatment with the selective epidermal growth factor receptor-tyrosine kinase inhibitor ZD1839 (Iressa). *Clin Cancer Res* 2002; 8:3250–3258.
94. Holsinger FC, Doan DD, Jasser SA, et al. Epidermal growth factor receptor blockade potentiates apoptosis mediated by Paclitaxel and leads to prolonged survival in a murine model of oral cancer. *Clin Cancer Res* 2003; 9:3183–3189.
95. Matsuo M, Sakurai H, Saiki I. ZD1839, a selective epidermal growth factor receptor tyrosine kinase inhibitor, shows antimetastatic activity using a hepatocellular carcinoma model. *Mol Cancer Ther* 2003; 2:557–561.

96. Kim SJ, Uehara H, Karashima T, Shepherd DL, Killion JJ, Fidler IJ. Blockade of epidermal growth factor receptor signaling in tumor cells and tumor-associated endothelial cells for therapy of androgen-independent human prostate cancer growing in the bone of nude mice. *Clin Cancer Res* 2003; 9(3):1200–1210.

97. Kedar D, Baker CH, Killion JJ, Dinney CP, Fidler IJ. Blockade of the epidermal growth factor receptor signaling inhibits angiogenesis leading to regression of human renal cell carcinoma growing orthotopically in nude mice. *Clin Cancer Res* 2002; 8(11):3592–3600.

98. Shintani S, Li C, Mihara M, et al. Enhancement of tumor radioresponse by combined treatment with gefitinib (Iressa, ZD1839), an epidermal growth factor receptor tyrosine kinase inhibitor, is accompanied by inhibition of DNA damage repair and cell growth in oral cancer. *Int J Cancer* 2003; 107:1030–1037.

99. She Y, Lee F, Chen J, et al. The epidermal growth factor receptor tyrosine kinase inhibitor ZD1839 selectively potentiates radiation response of human tumors in nude mice, with a marked improvement in therapeutic index. *Clin Cancer Res* 2003; 9:3773–3778.

100. Tortora G, Caputo R, Damiano V, et al. Combination of a selective cyclooxygenase-2 inhibitor with epidermal growth factor receptor tyrosine kinase inhibitor ZD1839 and protein kinase A antisense causes cooperative antitumor and antiangiogenic effect. *Clin Cancer Res* 2003; 9:1566–1572.

11 Phosphoproteomics in Drug Discovery and Development

Michael F. Moran, PhD, Jarrod A. Marto, PhD, Cynthia J. Brame, PhD, Olga Ornatsky, PhD, Mark M. Ross, PhD, Leticia M. Toledo-Sherman, PhD, Alfredo C. Castro, PhD, Brett Larsen, MSc, Henry Duewel, PhD, Christopher Hosfield, PhD, Christopher Orsi, MSc, Thodoros Topaloglou, PhD, Daniel Figeys, PhD, Jennifer A. Caldwell-Busby, PhD, and David R. Stover, PhD

CONTENTS

INTRODUCTION
EXPERIMENTAL PHOSPHOPROTEOMICS APPROACHES
APPLICATION EXAMPLES IN PHOSPHOPROTEOMICS
CONCLUSIONS
REFERENCES

1. INTRODUCTION

The control of enzyme activity by reversible phosphorylation and the role of protein kinases in signal transduction illustrate well the pivotal role of protein phosphorylation in cell regulation. There are numerous examples of the role of protein kinases and phosphatases in disease, and protein kinases in particular are

From: *Cancer Drug Discovery and Development:*
Protein Tyrosine Kinases: From Inhibitors to Useful Drugs
Edited by: D. Fabbro and F. McCormick © Humana Press Inc., Totowa, NJ

now established as druggable therapeutic targets *(1)*. Phosphoproteomics includes the large-scale determination of protein phosphorylation in cells and tissue, sometimes referred to as phosphoprofiling. Phosphoprofiling is a global approach that can be used to characterize biological states including therapeutic responses. It is augmented by another proteomic method, protein kinase (and protein phosphatase) interaction mapping, which together have the potential—yet to be fully realized—to provide comprehensive pictures of cellular states. A particular utility for these methods will be in drug discovery and development.

The amount of protein phosphorylation in a cell is vast. Approximately one-third of mammalian proteins are phosphorylated *(2)*. Therefore the identification of those protein modifications associated with a given pathway, cellular process, or cellular response of interest is an important goal in phosphoproteomics. A noteworthy feature of protein kinases as drug targets is the fact that their activity—and hence modulation by drugs—is frequently reflected not only in their own modification by phosphorylation (e.g., autophosphorylation), but also by the phosphorylation of their substrates, and by their protein–protein interactions. These features are the primary data of phosphoproteomics. Differential phosphoproteomics—that is, the identification of changes in protein phosphorylation against a background of largely unchanging cellular phosphorylation—is a logical but challenging approach to illuminate phosphorylated proteins of particular interest. For example, differential phosphoproteomics may reveal drug-induced changes in phosphorylation as a means to understand drug mechanisms and toxicity in preclinical models and monitor patients in clinical studies. Hence phosphoproteomics will logically find its first pharmaceutical applications in the area of protein kinase–directed therapeutics. Moreover, because phosphorylation is a universal mechanism of cell regulation, phosphoproteomics holds promise of becoming a generally applicable tool in drug discovery and development.

Phosphoprofiling is based on an ability to isolate phosphorylated peptides from an enzymatic digest of whole-cell or tissue protein extractions by using immobilized metal affinity chromatography (IMAC), followed by the identification of phosphopeptides and their corresponding proteins by mass spectrometry *(3)*. Such samples contain many thousands of phosphopeptides and their identification and characterization by mass spectrometry and data analysis have benefited from the development of new mass spectrometry instrumentation and software solutions, as described below. Imbedded in the data are two basic pieces of information: the sequence context of the phosphorylation sites, and, in a differential experiment, a ratio of the abundance of a given peptide in two samples. The former can be used to identify the protein from which the peptide was derived by searching protein databases, whereas the latter can be used to quantify the peptide, at least in a relative or comparative sense. In addition, the total data set itself is likely to contain information in the form of patterns (e.g., drug or disease signatures) that reflect the physiological state of the cell.

The importance of "global"—in a total cellular context—attempts at protein phosphorylation analysis predate current proteomics trends and terminology. For example, early practitioners of two-dimensional gel electrophoretic analysis of ^{32}P metabolically labeled cell-derived protein extracts appreciated the potential information content inherent to cellular phosphorylation patterns *(4)*. Although greatly facilitating our understanding of signal transduction and oncogene function, these methods did not readily provide insight into the identity of phosphorylated proteins or sites of phosphorylation. The introduction of protein phospho-specific antibodies and other tools such as protein phosphotyrosine-binding SH2 domains allowed the identification, by nongenetic means, of the substrates and interacting proteins in kinase-mediated signal transduction *(5,6)*. Although various sequence-specific antibodies are available and being put to use in array format to characterize disease states *(7)*, a limitation exists in that the antibody reagents available for these applications are based on the limited set of known phosphorylations discovered largely in experimental systems that may not reflect precisely those that may occur in animals. A goal for phosphoproteomics is therefore the global analysis of protein phosphorylation in tissue that will provide both detail at the level of individual protein regulation, and phosphoprofile patterns that reflect cellular states as a means to recognize and understand developmental and disease progression, and therapeutic responses, just to name a few.

In this chapter, we briefly review advances in measuring protein interactions of protein kinases, and follow with a more focused review of new approaches for global protein phosphorylation mapping by application of advanced proteomics methods based largely on mass spectrometry-based analytical methods. A significant bioinformatics infrastructure and tools for acquiring and analyzing mass spectrometry-based phosphopeptide data are required to support phosphoproteomics, but are not the subject of this chapter.

2. EXPERIMENTAL PHOSPHOPROTEOMICS APPROACHES

2.1. Capture of Phosphopeptides

Protein phosphorylation sites are labile. Protein phosphorylation can occur at ice temperatures, and phosphatases can remain active in cell lysates. Therefore, our experimental strategy is to process samples rapidly, include phosphatase and protease inhibitors, and employ conditions that denature proteins. The goal is to preserve the in vivo spectrum of phosphorylation, and not introduce bias toward any particular class of phosphoprotein, phosphopeptide, or phospho amino acid. Total protein preparations from tissue or tissue culture samples are made by using Trizol reagent (Invitrogen), a solution that contains chaotropes and phenol. Cellular protein is recovered from the organic phase following the subtraction of nucleic acids, and proteins and their phosphate ester bonds are stabilized by using these denaturing conditions essentially as described in

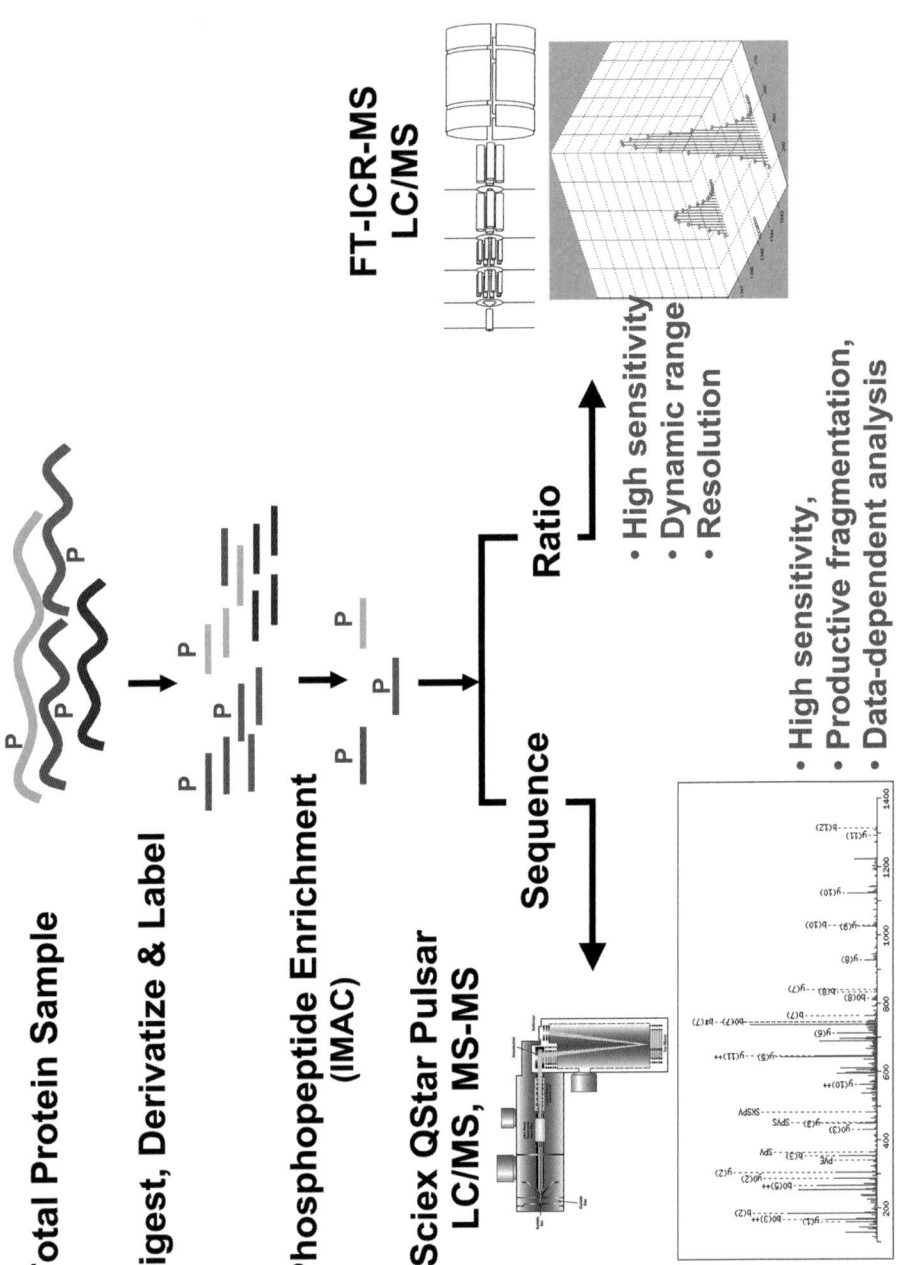

Total Protein Sample

Digest, Derivatize & Label

Phosphopeptide Enrichment (IMAC)

Sciex QStar Pulsar LC/MS, MS-MS

Ratio

Sequence

FT-ICR-MS LC/MS

- High sensitivity
- Dynamic range
- Resolution

- High sensitivity,
- Productive fragmentation,
- Data-dependent analysis

Fig. 1.

268

Ficarro et al. *(3)*. Following enzymatic digestion of total cellular protein, IMAC is used to bind phosphate-containing peptides. Peptides will bind the IMAC matrix through carboxylate and phosphate groups. To eliminate the former, peptide carboxylate groups are converted to methyl esters prior to IMAC. Details of this methodology and a description of controls verifying that by this approach peptide retention to the column is a function of phosphate groups (and therefore sensitive to phosphatase treatment) were reported *(3)*. A recent publication by Jensen and co-workers describes an alternative approach for phosphopeptide isolation (applied to isolate Arabidopsis membrane proteins) that omits chemical derivatization prior to IMAC *(8)*. By this approach, a greater recovery of singly phosphorylated phosphopeptides may be achieved. Another purpose of the chemical derivatization of peptides prior to IMAC is that it allows for the differential isotopic labeling of two peptide mixtures prior to their comparative analysis. In this approach, each sample is carboxy methylated by using d0- or d3-containing methanol such that corresponding peptides will differ in mass by 3 Da in the two samples. The differentially modified samples are then mixed prior to IMAC and mass spectrometric analysis of eluted phosphopeptides. The mass spectrometry intensity ratio of identical peptides that differ in mass by 3 Da is used to quantify the relative abundance of the peptide in the two samples being compared.

An advantage to the chemical derivatization-IMAC approach is its relative simplicity. There are few steps from the tissue sample to the enriched phosphopeptide mixture. Proteins in the initial sample (whole-cell lysates) are rendered denatured at the first step to stabilize phosphate ester bonds, and the chemical derivatization reaction is directed against carboxylate groups, not the more rare, and therefore less concentrated phosphoryl groups in the initial protein digest. Other successful approaches to enrich for phosphate-containing peptides/proteins prior to mass spectrometric analysis include the application of a series of chemical reactions directed at sites of phosphorylation to convert them into affinity tags *(9,10)*.

2.2. Mass Spectrometry and Data Analysis

Two different mass spectrometry strategies are used in concert to analyze phosphopeptide mixtures enriched by IMAC (Fig. 1). Phosphopeptide sequencing involves LC-MS/MS analysis with an MDS-Sciex QStar Pulsar mass spectrometer followed by a Mascot-based search of a protein database to identify phosphopeptides and the proteins from which they are derived. For quantification

---

Fig. 1. *(opposite page)* Phosphoproteomics allows both qualitative and quantitative analysis. Proteins are digested and then the peptides are derivatized to form either d0 or d3 methyl esters. Immobilized metal affinity chromatography (IMAC) is used to enrich for phosphorylated peptides, which are then analyzed by LC/MS and MS-MS on the QStar for sequence identifications and by LC/MS with the Fourier transform mass spectrometer for peptide abundance quantification (d0/d3 labeled pairs).

a custom-built 7-Tesla ion cyclotron resonance-based Fourier transform mass
spectrometer (FTMS) is used to analyze mixtures of differentially mass
labeled (as described above) peptide samples. This allows direct comparison
of relative quantities of phosphopeptides in combined samples. Proprietary
software is used to align FTMS and QStar data to assign peptide identities
with their cognate differentially mass-labeled pairs for quantification.

Identification of peptides by LC-MS/MS follows a data-dependent approach.
The identification of peptide pairs by using FTMS requires data processing
including the grouping of charged states and creation of monoisotopic peaks,
the assembly of chromatographic peaks corresponding to derivatized (±mass
label) peptide pairs, and determination of their intensity ratio as an expression
of their relative abundance in the two samples under comparison.

3. APPLICATION EXAMPLES IN PHOSPHOPROTEOMICS

Protein–protein interactions play a significant role in intracellular signal
transduction processes. Protein kinases as a class of proteins appear relatively
more interactive than proteins in general. That is to say they frequently are found
engaged in protein–protein interactions (11). Indeed protein kinases and phosphatases are often multidomain proteins containing modules that direct their
assembly into protein complexes and affect their subcellular localization (12).
For example, intracellular signaling proteins containing Src homology (SH)2
domains are often known to become tyrosine phosphorylated as a consequence
of their SH2-mediated interactions with activated growth factor receptor tyrosine
kinases (6), and mitogen-activated protein kinase cascades including various
kinases and their substrate proteins are known to associate directly and through
specialized scaffolding proteins (13). Hence, the mapping of the interactions of
protein kinases provides considerable insight into their function.

3.1. Protein Kinase, Phosphatase, and Substrate Interactions in Yeast

Protein kinases must physically associate with their substrate proteins at least
transiently to effect their covalent modification. In many instances these interactions can be captured by the method of co-immunoprecipitation, a process that has
been industrialized and scaled as one type of proteomics approach to identify interacting proteins (14,15). We recently described such a method for protein interaction mapping employing immunopurification and mass spectrometry as an
effective means to reveal interactions between a variety of protein types including
protein kinases and phosphatases (14,16). Figure 2 depicts the protein–protein
interactions of protein kinases and protein phosphatases in budding yeast. Within
this data set are verified examples of kinase-substrate pairs, and we conclude that
protein interaction mapping is an effective means to rapidly generate a shell of
knowledge around a given protein kinase or pathway of interest (14).

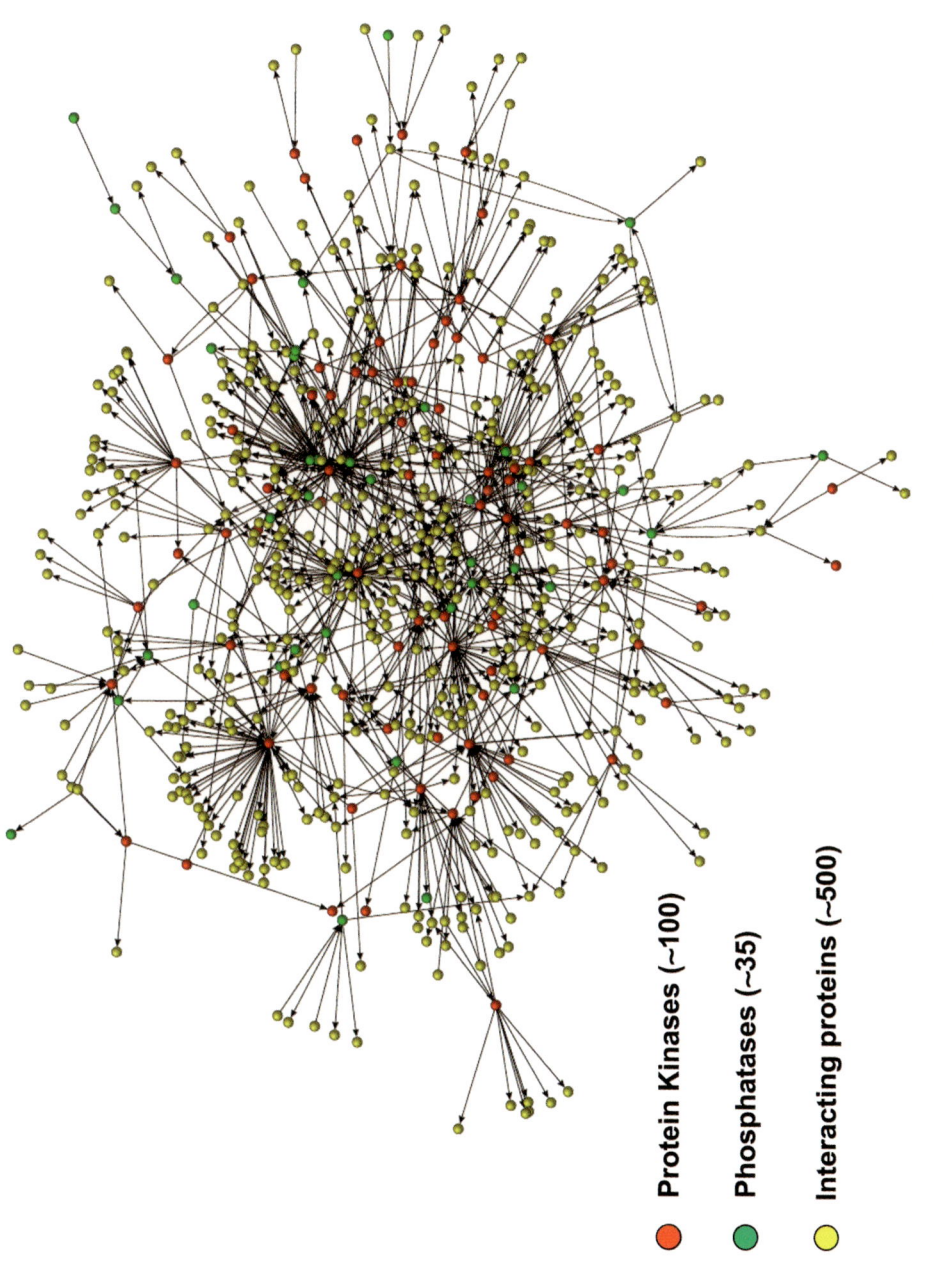

Fig. 2.

Protein Kinases (~100)

Phosphatases (~35)

Interacting proteins (~500)

An application of this approach is in biomarker discovery and assay development. For example, protein interactions of a protein kinase drug target are defined in a relevant model system(s), and then tested systematically to determine if the interacting proteins are phosphorylated by the kinase. If these phosphorylation events can be monitored in tissue samples (by mass spectrometry or enzyme-linked immunosorbent assay-type assays, for example) and display a dose–response relationship with drug treatment, they may serve as biomarkers of kinase modulation by inhibitory drugs.

3.2. Gleevec™/STI571 K562 Example

Human leukemia cells (K562) derived from a pleural effusion associated with chronic myelogenous leukemia, and known to express the activated protein tyrosine kinase BCR-ABL, were studied by the methods described above for the effects of BCR-ABL inhibition by the drug STI571/Gleevec (17,18). Cultured cells (10^8 cells) were treated for 24 h with or without 1 μM STI571 and processed as depicted in Fig. 3. Table 1 presents a partial listing of phosphopeptides found in lysates from these cultures, and their ratios measured in two such experiments. From analysis of total cell lysates from K562 cells in these experiments, some 274 phosphorylation sites were mapped from 137 different peptides derived from 93 distinct proteins. The majority of sites are phosphoserine, followed by phosphothreonine, and then phosphotyrosine, each less abundant by roughly an order of magnitude. We presume this to reflect the true in vivo abundances of these three phosphoamino acid species. The average of two phosphorylated hydroxyamino acids per peptide may reflect a bias for doubly phosphorylated peptides in the IMAC procedure, but this remains to be determined.

Immunoprecipitation of BCR-ABL protein from K562 cells afforded a direct measure of its phosphorylation and protein–protein interactions. As summarized in Table 2 (A and B), treatment of K562 cells for 24 h with STI571 caused changes in the phosphorylation and associated proteins of BCR-ABL. Most notable is the STI571-induced loss of tyrosine phosphorylation at BCR residue 177. Phosphotyrosine 177 is a known binding site for the Ras-linked adaptor protein Grb2, and this association is known to be required for BCR-ABL to transform various cells in culture and to cause myeloproliferative disease in mouse models (19,20). Curiously we did not observe tyrosine 393-containing peptides in our analysis. This residue resides in the activation loop of the Abl kinase domain. In addition to BCR residue Y177 as noted above, STI571 treatment effected the loss of phosphorylation at Abl residues Y185, Y232, and S569. Curiously, S459 was found to become phosphorylated in response to STI571 treatment. We did not observe phosphorylation at Abl Y257, which was reported to be phosphorylated in K562 cells in response to STI571 (21). The Grb2-related adaptor protein GRAP-2, the phosphoprotein p62 DOK, and the lipid phosphatase SHIP2 were found to co-immunoprecipitate with BCR-ABL recovered from K562 cells. As

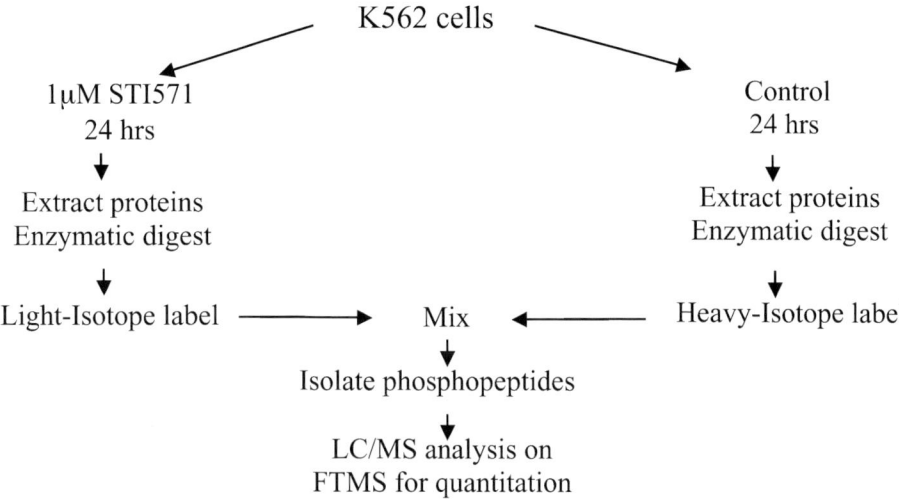

K562 cells

1 μM STI571   Control
24 hrs   24 hrs

Extract proteins   Extract proteins
Enzymatic digest   Enzymatic digest

Light-Isotope label ⟶ Mix ⟵ Heavy-Isotope label

Isolate phosphopeptides

LC/MS analysis on
FTMS for quantitation

Fig. 3. Global protein phosphorylation analysis. Schematic outline of the experimental approach to global phosphoprofiling. Total cell/tissue extracts are denatured and digested, typically by using trypsin. Carboxylate groups are converted to methyl esters, and during this derivatization process a 3-Da mass tag may be added such that comparative analysis can be conducted as described in the text. Phosphate-containing peptides are enriched by using immobilized metal affinity chromatography (IMAC). Phosphopeptide mixtures are subjected to LC-MS/MS analysis for peptide identification and phosphorylation mapping. Differential analysis by Fourier transform mass spectrometer is used to quantify relative changes in the abundance of phosphopeptides. A variety of software tools are used to align the protein differential and phosphorylation/identification data sets to enable the identification of phosphopeptides modulated by the drug treatment.

expected from the loss of phosphorylation at BCR Y177, the GRAP-2 interaction, and that of DOK were abolished by STI571 treatment (Table 2). The phosphoprotein p62 (Dok) is a strong marker of Ph1-positive BCR-ABL-expressing blast cells in crisis phase *(22)*, and is in fact a most prominent phosphotyrosine-containing protein substrate of a variety of activated and oncogenic tyrosine kinases *(6)*. These findings illustrate the potential of a systematic phosphoproteomics approach to discover pharmacodynamic biomarkers to assist in the discovery and clinical development of protein kinase inhibitor drugs.

Fig. 2. Protein interactions in *Saccharomyces cerevisiae*. Protein interactions of protein kinases (red dots), and protein phosphatases (green dots) are represented. Protein interactions were measured by co-immunoprecipitation from yeast lysates epitope-tagged kinases and phosphatases, followed by LC-MS/MS analysis of recovered proteins. Yellow dots represent interacting proteins. Arrows point from bait proteins to the prey proteins they were found to interact with *(14)*.

Table 1
Phosphoprofile of STI571/Gleevec™ in Human K562 Cells

Peptide sequence	Protein name	Acc. #	Ratio ± Gleevec	
SSpTPLHpSPSPIR	BAG-FAMILY MOLECULAR CHAPERONE REGULATOR–3	O95817	2.23:1	2.5:1
EAETKpSPLVpSPSK	BETA ADDUCIN	P35612	0:7.4e6	0:4.1e6
SEDARpSpSPSQR	DNA REPLICATION LICENSING FACTOR MCM4	P 33991	1.7:1	1.5:1
TGpSESpSQTGTSTTSSR	EUKARYOTIC TRANSLATION INITIATION FACTOR 4B			
pSRpSRpTPPAIR	hypothetical protein KIAA0324 - human	P23588	4.7:1	4.5:1
pSRpTPPVTR	hypothetical protein KIAA0324 - human	T02345	2.3:1	2.2:1
HGpTPDPpSPR	hypothetical protein MGC13125	T02345	1.3:1	1.6:1
YpSPSQNpSPIHHIPSR	KIAA0164 gene product	NP_1 16114.1	4.8:1	5.4:1
LRLpSPpSPTSQR	LAMIN A/C	NP_055554.1	1.6:1	1.5:1
LKLpSPpSPSSR	LAMIN B1	P02545	1.2:1	1.4:1
YRpTPpSRpSR	peptidyl-propyl isomerase G (cyclophilin G)	P20700	12:1	1.2:1
GPPpSPPAPVMHpSPSR	NUCLEAR PROTEIN SKIP (SKI-INTERACTING PROTEIN)	NP_004783.1	8.5:1	1.4e 7:0
YHGHpSMSDPGVpSYR	PYRUVATE DEHYDROGENASE E1 COMPONENT ALPHA SUBUNIT	Q13573	1.7:1	1.9:1
SRSpSSpSPPPK	RNA binding protein	P08559	1.46:1	1.35:1
pSPpSPKPTK	RNA-binding protein S1, serine-rich domain	BAA83718	2.2:1	1.95:1
RQpSPSPSpTRPIR	SER/ARG -RELATED NUCLEAR MATRIX PROTEIN	NP_006702.1	2.9:1	1.7:1
TRHpSPpTPQQSNR	SER/ARG-RELATED NUCLEAR MATRIX PROTEIN	NP_005830	1.75:1	1.5:1
	SER/ARG-RELATED NUCLEAR MATRIX PROTEIN	NP_005830	2.2:1	2.5:1

(Continued)

Table 1 *(Continued)*

Peptide sequence	Protein name	Acc. #	Ratio ± Gleevec	
RRpSFpSIpSPVR	SON_HUMAN SON PROTEIN (SON3)	P18583	2.4:1	3.3:1
pSRpSGpSIKGSR	SPLICING FACTOR, ARGININE/SERINE-RICH 7	Q16629	1.4:1	1.4:1
KHpSPpSPPPPTPTESR	SWI/SNF RELATED ACTIN DEPENDENT REGULATOR OF CHROMATIN	XP_006719	1.35:1	1.3:1
KRpSPSPpSPTPEAK	SWI/SNF RELATED ACTIN DEPENDENT REGULATOR OF CHROMATIN	XP_006719	1.4:1	1.4:1
pSVpSRpSPVPEK	SPLICING FACTOR, ARGININE/SERINE-RICH 5	Q13243	0:1.4e7	0:8e6
pSHpSPMSNR	TR2H_HUMAN TRANSFORMER-2 PROTEIN HOMOLOG	Q13595	1.1:1	1.8:1
pSRpSYpTPEYR	TR2H_HUMAN TRANSFORMER-2 PROTEIN HOMOLOG	Q13595	2.2:1	2.6:1
pSGpSNPNR	XRC1_HUMAN DNA-REPAIR PROTEIN XRCC1	P18887	2.0:1	1.9:1

Phosphoproteins were characterized in K562 cells treated with or without STI571/Gleevec as indicated. One hundred thirty-seven phosphopeptides were characterized from 93 different proteins. A total of 274 sites of phosphorylation were mapped.

Table 2
Phosphoprofile of BCR-ABL and BCR-ABL-Interacting Proteins as a Function of STI571/Gleevec™ in Human K562 Cells

	Panel A	
Site	*–STI571*	*+STI571*
BCRpY177	+	–
BCRpS459	–	+
BCRpY644	+	+
AblpY185	+	–
AblpY226	+	+
AblpY232	+	–
AblpS569	+	–
	Panel B	
Protein	*–STI571*	*+STI571*
GRAP2	+	–
p62 Dok1	+	–
SHIP2	+	+

BCR-ABL was immunoprecipitated from K562 cells treated with or without STI571/Gleevec as indicated. Immuniprecipitates were analyzed by mass spectrometry to identify BCR-ABL phosphorylation (panel A), and to identify associated, co-immunoprecipitated proteins associated with BCR-ABL (panel B).

4. CONCLUSIONS

Phosphoproteomics is an emerging technology and holds great promise for facilitating a better understanding of cellular states. The large amounts of information generated by phosphoproteomics represent yet another "omics" data set. When combined with data sets arising from expression microarray analyses and protein interaction mapping a deeper understanding of underlying biology is expected to emerge than would be gleaned from any single data set. Perhaps the most promising application of phosphoproteomics is in the areas of drug development, and diagnostics. Antibody-based phosphoproteomics for profiling clinical samples has been demonstrated *(7)*.

Mass spectrometry-based global profiling of tissue phosphorylation holds great promise as a generally applicable means to identify early disease markers, and biomarkers to assist in the selection, dosing, and monitoring of patients in the clinic. In the case of protein kinase inhibitors, the monitoring of drug-affected phosphopeptides may serve as pharmacodynamic biomarkers. In addition, phosphoprofiling of tissue may facilitate lead profiling in drug discovery. Another attraction to phosphoproteomics in drug discovery and development is its potential to become applied as a universal type of assay. The preparation of cell and tissue samples for analysis involves rather simple denaturing conditions that stabilize protein phosphorylation, and may be applied with a variety of starting materials representing discovery, preclinical, and clinical stages of drug development. For example, the identification of drug-associated phosphopeptides during early drug discovery may form the basis of cell- and animal-based assays through preclinical studies, and pharmacodynamic biomarkers during clinical development. Given the universal role of protein phosphorylation in cell regulation, phosphoproteomics should find utility in the development of drugs and diagnostics in many indications.

As presented in this chapter, feasibility is now established for phosphoproteomics and phosphoprofiling. However, advances are required for these methods to achieve better utility in drug discovery and development. Challenges that must be met include the industrialization of the technology such that acceptable levels of reproducibility and confidence in the data are achieved, and reducing the amount of starting materials required so that the approach is more amenable to the direct analysis of clinical specimens such as needle biopsies and microdissected tumors. Mass spectrometry hardware and software, and bioinformatics are advancing at a rapid pace, and this will greatly facilitate phosphoproteomics.

The human genome is estimated to encode more than 500 protein kinases. However, we have only a very limited knowledge of the individual sites of phosphorylation on proteins and the function and temporal dynamics of each of these modifications. Indeed, inspection of thousands of newly sequenced phosphorylation sites in a variety of tissues and cell types at MDS Proteomics indicates that only a minor proportion of these sites match known consensus sequences (unpublished observations). These data, in the form of understood individual phosphorylation sites, and complex patterns of phosphorylation are a rich source of information that may serve to define cellular states. Augmentation of the methods described in this chapter with approaches such as subcellular fractionation, antibody-based enrichment, and chemi-proteomics (e.g., the identification of kinase inhibitor-binding proteins) will improve phosphoproteomics by allowing more focused investigations into subpopulations of proteins and pathways of interest in the phosphoproteome. Feasibility is clearly established for the systematic identification and monitoring of protein phosphorylation as a measure of cellular states. Translating this technology into useful applications in oncology such as early disease detection, personalized medicine, and clinical drug development remain important goals for phosphoproteomics.

ACKNOWLEDGMENTS

We thank Forest White, Adrian Heilbut, Yuen Ho, and Mike Tyers for their contributions reviewed in this chapter. We are indebted to Dr. Don Hunt for guidance and access to FTMS technology essential to this study.

REFERENCES

1. Fabbro D, Parkinson D, Matter A. Protein tyrosine kinase inhibitors: new treatment modalities? *Curr Opin Pharmacol* 2002; 2(4):374–381.
2. Hubbard MJ, Cohen P. On target with a new mechanism for the regulation of protein phosphorylation. *Trends Biochem Sci* 1993; 18(5):172–177.
3. Ficarro SB, McCleland ML, Stukenberg PT, et al. Phosphoproteome analysis by mass spectrometry and its application to *Saccharomyces cerevisiae*. *Nat Biotechnol* 2002; 20(3):301–305.
4. Cooper JA, Hunter T. Changes in protein phosphorylation in Rous sarcoma virus–transformed chicken embryo cells. *Mol Cell Biol* 1981; 1(2):165–178.
5. Kamps MP, Sefton BM. Identification of multiple novel polypeptide substrates of the v-src, v-yes, v-fps, v-ros, and v-erb-B oncogenic tyrosine protein kinases utilizing antisera against phosphotyrosine. *Oncogene* 1988; 2(4):305–315.
6. Moran MF, Koch CA, Anderson D, et al. Src homology region 2 domains direct protein-protein interactions in signal transduction. *Proc Natl Acad Sci USA* 1990; 87(21):8622–8626.
7. Wulfkuhle JD, Aquino JA, Calvert VS, et al. Signal pathway profiling of ovarian cancer from human tissue specimens using reverse-phase protein microarrays. *Proteomics* 2003; 3(11):2085–2090.
8. Nuhse TS, Stensballe A, Jensen ON, Peck SC. Large-scale analysis of in vivo phosphorylated membrane proteins by immobilized metal ion affinity chromatography and mass spectrometry. *Mol Cell Proteomics* 2003; 2(11):1234–1243.
9. Zhou H, Watts JD, Aebersold R. A systematic approach to the analysis of protein phosphorylation. *Nat Biotechnol* 2001; 19(4):317–318.
10. Oda Y, Nagasu T, Chait BT. Enrichment analysis of phosphorylated proteins as a tool for probing the phosphoproteome. *Nat Biotechnol* 2001; 19(4):379–382.
11. Bader GD, Heilbut A, Andrews B, Tyers M, Hughes T, Boone C. Functional genomics and proteomics: charting a multidimensional map of the yeast cell. *Trends Cell Biol* 2003; 13(7):344–356.
12. Pawson T, Scott JD. Signaling through scaffold, anchoring, and adaptor proteins. *Science* 1997; 278(5346):2075–2080.
13. Morrison DK, Davis RJ. Regulation of map kinase signaling modules by scaffold proteins in mammals. *Annu Rev Cell Dev Biol* 2003; 19:91–118.
14. Ho Y, Gruhler A, Heilbut A, et al. Systematic identification of protein complexes in *Saccharomyces cerevisiae* by mass spectrometry. *Nature* 2002; 415(6868):180–183.
15. Gavin AC, Bosche M, Krause R, et al. Functional organization of the yeast proteome by systematic analysis of protein complexes. *Nature* 2002; 415(6868):141–147.
16. Figeys D, McBroom LD, Moran MF. Mass spectrometry for the study of protein-protein interactions. *Methods* 2001; 24(3):230–239.
17. Druker BJ, Tamura S, Buchdunger E, et al. Effects of a selective inhibitor of the Abl tyrosine kinase on the growth of Bcr-Abl positive cells. *Nat Med* 1996; 2(5):561–566.
18. Capdeville R, Buchdunger E, Zimmermann J, Matter A. Glivec (STI571, imatinib), a rationally developed, targeted anticancer drug. *Nat Rev Drug Discov* 2002; 1(7):493–502.
19. Pendergast AM, Quilliam LA, Cripe LD, et al. BCR-ABL-induced oncogenesis is mediated by direct interaction with the SH2 domain of the GRB-2 adaptor protein. *Cell* 1993; 75(1):175–185.
20. Million RP, Van Etten RA. The Grb2 binding site is required for the induction of chronic myeloid leukemia-like disease in mice by the Bcr/Abl tyrosine kinase. *Blood* 2000; 96(2):664–670.
21. Salomon AR, Ficarro SB, Brill LM, et al. Profiling of tyrosine phosphorylation pathways in human cells using mass spectrometry. *Proc Natl Acad Sci USA* 2003; 100(2):443–448.
22. Wisniewski D, Strife A, Wojciechowicz D, Lambek C, Clarkson B. A 62-kilodalton tyrosine phosphoprotein constitutively present in primary chronic phase chronic myelogenous leukemia enriched lineage negative blast populations. *Leukemia*. 1994; 8(4):688–693.

Index

4-Hydroperoxycyclophosphamide
and Glivec, 154
5-Fluorouracil, 233
11q23 translocations in leukemia/lymphoma, 123, 125, 132

A

A12 in IGF-IR inhibition, 15
Abl (Abelson). *See also* Bcr-Abl protein
Glivec and, 147, 205, 222, 223
inhibition of, 3–5, 20, 89
phosphorylation problems,
preventing, 194
structure of, 192, 203–205, 219
ABX-EGF described, 9
Activity of kinases, inhibiting, 3
Acute lymphoblastic leukemia (ALL)
Bcr-Abl protein in, 3–5
FLT3 in, 97
JAK inhibition in, 131
Acute myeloid leukemia (AML)
described, 93–94, 100, 103
FLT3 in, 97, 98, 103
point mutations in, 6
SOCS in, 127
SU5416 for, 101
tandem duplications in, 98–99
Acute promyelocytic leukemia (APL), 94
Adult T-cell leukemia/lymphoma
(ATLL), 126
Affinity chromatography, 197
AG490
features of, 129, 130
and Glivec, 155
and IL-6, 132

in JAK inhibition, 18, 131, 132, 134
structure of, 131
AG1295 in FLT3 inhibition, 101
AG1296
in FLT3 inhibition, 101
in Glivec inhibition, 169, 171
AG1801 in apoptosis induction, 132
Agrin, 211
Akt. *See* PKB and PI3-K enzymes
Allylamino-17-demethoxygeldanamycin
(17-AAG) and Glivec, 154, 155
AML. *See* Acute myeloid leukemia
(AML)
Anemia, 87
Angiogenesis, 11, 61
Angiozyme in clinical trials, 13
Animal models
of cancer, 242–243, 249
compounds
administration/pharmacokinetics
of, 235–240
preselection in, 234–235
of disease, 231–235, 258–259
of Glivec, 150, 169
Antagonists. *See also individual compound/pathway by name*
depeptidation of, 34
design of, 38
identifying, 35
Antibodies in development, 9
Apoptosis
Fas-induced, 131
inducing
via Glivec, 149
via JAK, 129, 132, 134

via Src, 89
and PI3-K, 60
Arsenic trioxide and Glivec, 154c
Arteriosclerosis, allograft-associated, 6
Asterriquinone E in Src SH2 antago-
nism, 45
Atherosclerosis and PDGF, 170–171
ATP analogs in Src inhibition, 74
Avastin. *See* Bevacizumab (Avastin)

B

Baculovirus system in protein sequencing, 191, 194–195, 197, 223
B-cell lymphoma, 125, 127
B-cells, development of and FLT3, 95, 97
Bcr-Abl protein
cell cycle control and, 149
cells, characteristics of, 149
in CML, 3–6, 145
formation of, 205
inhibition of, 89–90, 155–156
levels, decreasing, 154
phosphorylation of, 272–276
Bevacizumab (Avastin)
described, 9
in RPTK inhibition, 3
in VEGFR inhibition, 12
Binding pockets, mapping, 32, 33
Biomarkers, identification/profiling of,
257–259, 276
Bone loss models of Src activity in vivo,
80–83
Bone metastases
animal models of, 249
cortactin and, 73
Breast cancer
characteristics of, 251
Her2 in, 19
PP60^{c-src} in, 32
and Src activity/expression, 73
Btk (Bruton's tyrosine kinase), structure
of, 192, 206
Busulfan and Glivec, 154

C

C225 in cancer therapy, 9–10
Caffeine in PI3-K inhibition, 56, 57
Cancer
animal models for pharmacology,
232–234, 242–243
described, 234
EGFR in, 8, 242–243
genetically engineered models of, 234
PDGF in, 172, 173
PI3-K inhibition in, 62–63
point mutations in, 18
progression and JAKs, 17–18
Src inhibition in, 83–86
stroma, formation of, 176
tyrosine kinase targets in, 2
Canertinib (CI-1033)
in cancer therapy, 9
in EGFR/Her2 inhibition, 11
side effects of, 11
structure of, 8
Carboplatin/carboplatinum, clinical
testing of, 154, 253, 256
Carboxy-terminal Src kinase (Csk), structure of, 192, 195, 196, 206–207
CDP-860 described, 5, 9
Cell cycle control
in Bcr-Abl cells, 149
JAK in, 132, 134
and PI3-K, 60–61
PI3-Ks in, 60–61
signal strength, modulating, 165
CEP-701 as FLT3 inhibitor, 6, 7, 102
CEP-5214 as FLT3 inhibitor, 102
Cetuximab (Erbitux, IMC-225), 3, 9
CGP 41251 (PKC412), inhibition by,
102–103
CGP76030
activity
bone loss models of Src inhibition
in vivo, 80–83
in cancer, 83–85, 89–90
in vitro, 76–77

mechanism of action, 77–78, 84, 85
side effects of, 87, 88, 89
structure of, 75
CGP77675
  activity
    in cancer, 86, 89, 90
    in vitro, 76–77
  mechanism of action, 77–78
CGP79833, bone loss models of Src
    activity in vivo, 80–83
CGP79883
  pharmacokinetic parameters, 78
  side effects of, 87, 88
  structure of, 75
CGP81699
  bone loss models of Src activity in
    vivo, 80–83
  pharmacokinetic parameters, 78
  side effects of, 87
  structure of, 75
Chemotaxis and PDGF, 165
Chemotherapy
  AG490, 131, 132
  and Glivec, 153–155
  in leukemia, 101
  in NCSLC treatment, 11
Chromatography in protein sequencing,
    196–198
Chronic myeloid leukemia (CML)
  Bcr-Abl protein in, 3–6, 145
  described, 145–146
  in Glivec development, 150–153
  JAK activation in, 120, 125
  point mutations in, 6, 18, 205
  treatment of, 18, 103, 205, 219
Chronic myelomonocytic leukemia
    (CMML), 174–175
Circumin (Diferuloymethane) in JAK
    inhibition, 132, 133
Cisplatin, clinical testing of, 233, 253
c-Kit
  expression of, 95
  Glivec and, 147, 212–213

inhibition of, 3–5, 101, 102
structure of, 193, 211–214, 220
Cladribine and Glivec, 154
Clathrin and PI3-K enzymes, 55
Cloning/expression in protein sequencing,
    194–195
c-Met
  overview, 16–17
  and Src, 73
  structure of, 193, 214
COLIA1-PDGFB, expression of, 174
Colorectal cancer
  PP60^{c-src} in, 32
  and Src activity/expression, 73
  Src inhibition in, 83, 84, 85
  treatment, 12, 13, 233
  VEGFR inhibition in, 13
Compound 1 in Src SH2 antagonism, 33
Compound 2 in Src SH2 antagonism, 33
Compound 3 in Src SH2 antagonism, 33
Compound 4 in Src SH2 antagonism, 33, 34
Compound 5 in Src SH2 antagonism, 33
Compound 6 in Src SH2 antagonism, 33, 34
Compound 7 in Src SH2 antagonism,
    33, 34, 35
Compound 8 in Src SH2 antagonism, 33, 35
Compound 9 in Src SH2 antagonism, 33, 35
Compound 10 in Src SH2 antagonism, 33, 35
Compound 11 in Src SH2 antagonism, 34, 35
Compound 12 in Src SH2 antagonism, 34, 35
Compound 13 in Src SH2 antagonism, 34, 35
Compound 14 in Src SH2 antagonism, 34, 35
Compound 15 in Src SH2 antagonism, 34, 35
Compound 16 in Src SH2 antagonism, 36
Compound 17 in Src SH2 antagonism, 36, 37
Compound 18 in Src SH2 antagonism, 36, 37
Compound 19 in Src SH2 antagonism, 38, 39
Compound 20 in Src SH2 antagonism, 38, 39
Compound 21 in Src SH2 antagonism, 38, 39
Compound 22 in Src SH2 antagonism, 38, 39
Compound 23 in Src SH2 antagonism, 39, 40
Compound 24 in Src SH2 antagonism, 39, 40
Compound 25 in Src SH2 antagonism, 39, 40
Compound 26 in Src SH2 antagonism, 39, 40

Compound 27 in Src SH2 antagonism, 39, 40
Compound 28 in Src SH2 antagonism, 39, 40
Compound 29 in Src SH2 antagonism, 41–43
Compound 30 in Src SH2 antagonism, 42, 43
Compound 31a, structure of, 75
Compound 31 in Src SH2 antagonism, 42, 43
Compound 32 in Src SH2 antagonism, 42, 43
Compound 33 in Src SH2 antagonism, 42, 43
Compound 34 in Src SH2 antagonism, 42, 43, 45
Compound 35 in Src SH2 antagonism, 42, 43
Compound 36 in Src SH2 antagonism, 44
Compound 37 in Src SH2 antagonism, 44
Compound 38 in Src SH2 antagonism, 44, 45
Compound 39 in Src SH2 antagonism, 44, 45
Compound 40 in Src SH2 antagonism, 44, 45
Compound 41 in Src SH2 antagonism, 45
Compound 42 in Src SH2 antagonism, 45
Construct definition in protein sequencing, 191, 194
Cortactin and bone metastases, 73
CP-690-440 in JAK inhibition, 16, 17
CP-69050 in JAK inhibition, 133
CP-358,774. *See* Erlotinib (Tarceva)
CR4, features of, 130, 133
C-Src. *See* Src homology 2 (SH2) domain
CT52923 in Glivec inhibition, 169
CT53518 (MLN518)
    in Glivec inhibition, 169
    inhibition by, 102
Cucurbitacin I in JAK inhibition, 133
Cysteine-188 in Src SH2 antagonism, 36, 37
Cytarabine (Ara-C) and Glivec, 152, 153, 154
Cytokine receptor complexes, composition of, 119
Cytomegalovirus, 123
Cytopenia, leukemia-induced, 101

D

D835 mutations in leukemia prognosis, 99
D-64406, inhibition by, 102
D-65476, inhibition by, 102

Daunorubicin and Glivec, 154
Decitabine and Glivec, 154
Dendritic cells and FL, 94, 96, 97
Dermatofibrosarcoma protuberans (DFSP), 174
DFG motif in Abl kinases, 204, 205
Diabetes, drug-induced, 208
Diabetic retinopathy treatment, 14
Differential phosphoproteomics, 266
Differential scanning calorimetry, 198
Dihydroxyquinolones in Src SH2 antagonism, 45
Disulfide formation, preventing, 194
Docetaxel, clinical testing of, 253
Dose, predicting, 235–237, 244
Doxorubicin
    clinical testing of, 253
    for colon cancer, 233
    and Glivec, 154
Drug design, 218–219
Dual kinase inhibition approach, effectiveness of, 86

E

Edatrexate, clinical testing of, 253
EGFR. *See* Epidermal growth factor receptor (EGFR)
EKB-569
    in EGFR/Her2 inhibition, 9, 11
    structure of, 8
EKB-596, clinical testing of, 243
Enzymes, selectivity of in Src inhibition, 88
EphA2/B2, structure of, 193, 195, 214–215
Epidermal growth factor receptor (EGFR)
    in cancer, 8, 242–243
    clinical testing of, 10–11, 232, 241–258
    inhibitors
        activity in combination therapy, 253–257
        activity in orthotopic models, 249–251
        activity in specialized models, 251, 253

small molecule, 10–11, 19
overview, 7–10
and p100α, 60
and PDGF, 172
phosphorylation, inhibition of, 237, 244
and Src, 73
structure of, 193, 215–216
Erbitux. *See* Cetuximab (Erbitux, IMC-225)
Erk kinase in Src inhibition, 80
Erk MAPK pathway and PDGF, 165, 168
Erlotinib (Tarceva)
in cancer therapy, 9, 18
clinical testing of, 232, 241–244,
246–247
in EGFR/Her2 inhibition, 10–11, 247
structure of, 8
Ethylene glycol in protein sequencing, 198
Etoposide and Glivec, 154
Ewing family sarcoma, 172

F

FAK (Focal adhesion kinase), structure
of, 192, 207–208
Farnesyltransferases and Glivec, 153–155
Fermentation technology in protein
sequencing, 195
FGFRs. *See* Fibroblast growth factor
receptors (FGFRs)
Fibroblast growth factor receptors
(FGFRs), 14, 193
Fibrotic diseases and PDGF, 171–172
Flavopiridol and Glivec, 155
FLT3
in B-cell development, 95, 97
expression in leukemias, 97, 212
and HSCs, 94, 95, 96
inhibition of, 6, 7, 20, 101–103
internal tandem duplications in, 98–99
kinase inhibitors, 100–103
ligand described, 95–97
mutations, 6–7, 97–99
receptor described, 94–95
signaling, biological consequences
of, 99–100

structure of, 193, 211–212
TKD mutations in, 99
FLT3 gene, human, 94, 98
Fluoroquinolones in PI3-K inhibition,
56, 57
Focal adhesion kinase (FAK), structure
of, 192, 207–208
Food and compound PK, 237, 240
Fourier transform mass spectrometer
(FTMS), 270
Fragment-based screening, 224
Fusion proteins. *See also individual
protein by name*
in CMML, 174–175
in protein sequencing, 197

G

Gastric cancer, 233
Gefitinib (Iressa)
in cancer therapy, 3, 9, 18
clinical testing of, 232, 241–243,
250, 253, 254, 256
in EGFR/Her2 inhibition, 10–11, 244
structure of, 8
Geldanamycin and Glivec, 154
Gemcitabine
and Glivec, 154
in tumor inhibition, 63
Gene amplification in JAK expression, 125
Gene targeting
of EGFR inhibitors, 249
of PI3-K isoforms, 58–59
Gene transfer
in lung fibrosis, 172
retroviral-mediated, 96
γ-Irradiation and Glivec, 154
Gleevec. *See* Imatinib mesylate (Gleevec,
Glivec, STI-571)
Glioblastoma
EGFR in, 8
Glivec for, 172
and PDGF, 176, 177
VEGFR inhibition in, 13
Glivec. *See* Imatinib mesylate (Gleevec,
Glivec, STI-571)

Glomerulonephritis and PDGF, 171
Glycerol in protein sequencing, 198
Gp130/JAK signaling pathway, 128, 129
Grb2. *See* Growth-factor receptor-bound
    protein 2 (Grb2)
Growth-factor receptor-bound protein 2
    (Grb2)
    and JAK proteins, 120
    overview, 40–46
GST fusion proteins in protein sequenc-
    ing, 197
Guanosine triphosphate-14564
    described, 102

H

Hck, structure of, 192
hEGRF expression in hyperplasia, 251
Hematopoiesis, regulation of, 96
Hematopoietic stem cells (HSCs)
    and FLT3, 94, 95, 96
    and JAK proteins, 120
Her2
    in breast cancer, 19
    inhibition of, 9–11, 244, 247
    small molecule inhibitors of, 10–11
Herbimycin-A, 89, 101
Herceptin. *See* Trastuzumab (Herceptin)
HES. *See* Hypereosinophilic syndrome (HES)
HIF-1 (Hypoxia-inducible factor-1), 61
Hodgkin's lymphoma, 125
Homoharringtonine and Glivec, 153, 154
h-R3 described, 9
HSCs. *See* Hematopoietic stem cells (HSCs)
HT29 xenografts, Src inhibition in, 84, 85, 89
Human papilloma virus, 123
Hydroxylapatite chromatography, 197
Hydroxyurea and Glivec, 154
Hypereosinophilic syndrome (HES)
    mutations in, 5, 19
    PDGFR signaling in, 175
Hyperplasia, 251
Hypoxia-inducible factor-1 (HIF-1) in
    PI3-K inhibition, 61

I

IC87114 in PI3-K inhibition, 56, 57
IGF-IR. *See* Insulin-like growth factor 1
    receptor (IGF-IR)
IMAC (Immobilized metal affinity
    chromatography), 266, 269
Imatinib mesylate (Gleevec, Glivec,
    STI-571)
    and Abl (Abelson), 147, 205, 222, 223
    animal models of, 150, 169
    in cancer therapy, 3–5, 18, 116, 134,
        172, 174, 175
    development of, 146–147, 150–153, 198
    dose–response relationship in CML, 151
    hematological/cytogenic response in
        CML, 152
    indications for, 6
    and paclitaxel, 176
    pharmacological profile, 147–150,
        205, 237–239, 272–276
    protein kinase inhibition by, 148
    resistance to, 5, 20, 153–155, 219
    side effects of, 151
    in Src inhibition, 90
IMC-225. *See* Cetuximab (Erbitux,
    IMC-225)
IMC-IC11, 9, 12
Imidazoles
    in protein sequencing, 196
    in Src SH2 inhibition, 35
Imidazopyridines in PI3-K inhibition,
    56, 57
Indolinones in Src inhibition, 74
Inhibition. *See individual drug/gene/*
    pathway by name
Insect cells in protein sequencing,
    195–197, 223
Insulin-like growth factor 1 receptor
    (IGF-IR)
    overview, 15, 16
    structure of, 193, 198, 208–211
Insulin receptor kinase (IRK) domain
    sequencing of, 191, 194, 196, 199, 200

structure, 208–211, 220
Interferons
and Glivec, 152, 153, 154
and JAK, 119, 121
Interleukin-6 (IL-6)
and AG490, 132
in multiple myeloma, 127–129
Interleukins and JAK, 119, 121, 122
Ion-exchange chromatography, 197–198
Iressa. *See* Gefitinib (Iressa)
Irinotecan in cancer therapy, 9–10
IRIS study on Glivec, 152
Isoxazoles in NMR screening, 225

J

JAB proteins. *See* Suppressor
of cytokine signaling family
(SOCS) proteins
JAK (Just another kinase) family
activation/structure of, 116–120, 126, 133
in cellular transformation, 125
in hematological diseases, 123–130
inhibition, pharmacological, 130–134
inhibitors, features of, 130, 131
in leukemia/lymphoma, 123–127
in multiple myeloma, 125, 127–131
negative regulation of, 120–122
overview, 17–18, 115–116
src homology 2 (SH2) domain and,
120, 121, 126–127
Janex-1 in JAK inhibition, 18

K

K-252a in c-Met inhibition, 16, 17
K562 cells, Glivec in, 272–276
Kidney development and PDGF, 167
Kinase insertion domains (KIDs), 211, 217

L

L-744832 and Glivec, 153–155
Lapatinib (GW-572016), 8, 9
Lck, structure of, 192, 201
LCK-SH2 domain, antagonists of, 37–40
Leflunomide in JAK inhibition, 132

Leukemia
11q23 translocations in, 123, 125, 132
abnormalities, inducing, 98, 123
in chemotherapy, 101
FLT3 in, 97, 212
JAKs in, 123–127
Philadelphia chromosome translocations in, 123, 125, 132, 133,
145, 205
treatment, 90
Libraries
in protein structure determination, 220
in Src SH2 antagonism, 40
Liver cancer, 233
Lucentis. *See* Ranibizumab (Lucentis)
Lung cancer
and Src activity/expression, 73
treatment of, 233
VEGFR inhibition in, 13
Lung fibrosis and PDGF, 171–172
LY294002
and Glivec, 153, 155
in PI3-K inhibition, 56, 57, 62–63
Lymphoma
11q23 translocations in, 123, 125, 132
JAKs in, 123–127
Philadelphia chromosome translocations in, 123, 125, 132, 133,
145, 205
T-cell, 126, 127

M

Macular degeneration treatment, 14
Mafosfamide and Glivec, 154
Mammalian target of rapamycin
(mTOR), 61, 257
MAPK pathway and PDGF, 165, 168
Markers, identification/profiling of,
257–258
Mass spectrometry in protein sequenc-
ing, 269–270, 276, 277
MDS. *See* Myelodysplastic syndrome
(MDS)
MDX-210 in cancer therapy, 9, 10

MDX 447 in cancer therapy, 9, 10
Met. *See* c-Met
Methotrexate and Glivec, 154
Methylxanthines in PI3-K inhibition, 56, 57
Midostaurin/PKC412 as FLT3 inhibitor, 6, 7, 20
Mitomycin-C for lung cancer, 233
Mitoxantrone and Glivec, 154
MLN-518 as FLT3 inhibitor, 6, 7
MLN518 (CT53518), inhibition by, 102
MMAC1. *See* PTEN and PI3-K enzymes
Monoclonal antibodies and PDGF, 169
mTOR (Mammalian target of rapamycin), 61, 257
Mubritinib (TAK-165)
    in cancer therapy, 9
    structure of, 8
Multiple myeloma
    IL-6 in, 127–129
    JAK in, 125, 127–131
    SOCS in, 127
Murine melanoma, 176
MuSK (Muscle-specific kinase), structure of, 193, 211
Mutations
    in AML, 94
    and drug resistance, 19–20, 103
    FLT3, 6–7, 97–99
    in JAK proteins, 123
    point
        in CML, 6, 18, 205
        in leukemia, 98, 123
        in MDS, 6, 7
        in PI3-K inhibition, 58
    screening, 177
    in Src, 73
Mycosis fungoides, 126
Myelodysplastic syndrome (MDS)
    point mutations in, 6, 7

N

Neovastat in VEGFR inhibition, 14
Neurofibromatosis type I, 89

Nimustine (ACNU) and Glivec, 154
non-Hodgkin's lymphoma, 125
Non-small-cell lung cancer (NCSLC) and EGFR, 8, 10, 11
Nuclear magnetic resonance spectroscopy (NMS), 187, 191, 195, 219–225
NVP-AAK980
    activity
        bone loss models of Src inhibition in vivo, 80–83
        in cancer, 83, 85, 90
        in vitro, 77
    pharmacokinetic parameters, 78
    side effects of, 87, 88
    structure of, 75
NVP-ADL681, 244–246
NVP-ADL681-NX, 241, 247, 248, 251–253, 257, 258
NVP-ADW742 in IGF-IR inhibition, 15, 16
NVP-AEW541 in IGF-IR inhibition, 15

O

Olomucines, purine-derived
    activity in vitro, 76–77
    bone loss models of Src activity in vivo, 80–83
    pharmacokinetic parameters, 78
    side effects of, 87
    in Src inhibition, 74
    structure of, 75
OSI-7744. *See* Erlotinib (Tarceva)
Osteoclast models in Src inhibition, 77–80
Osteoporosis, 73, 86–87
Ovarian cancer
    EGFR/Her2 inhibition in, 11
    VEGFR inhibition in, 13

P

p38 kinase in Src inhibition, 80
p55α/γ described, 53, 59
P56^{lck}. *See* LCK-SH2 domain, antagonists of
p65 in PI3-K inhibition, 59
p70^{S6k} kinase, 257, 258

p85α described, 53, 59
p100α/β/δ/γ
  in cell migration, 62
  described, 53, 54
  inhibition of, 56–59
Paclitaxel (Taxol)
  clinical testing of, 247, 249, 253,
      255–257
  in EGFR inhibition, 251, 253–255
  and Glivec, 176
  PK of, 233
Pancreatic cancer, 233
PD-166285, structure of, 75
PD184352 and Glivec, 153, 155
PDGFR. See Platelet-derived growth
    factor receptor (PDGFR)
Pegaptanib in VEGFR inhibition, 14
Pegylated interferons and Glivec, 154
Pericyte recruitment by PDGF, 167, 176
PHA-665752 in c-Met inhibition, 16, 17
Pharmacology, animal tumor models
    for, 232–240
Philadelphia chromosome translocations
    in leukemia/lymphoma, 123, 125,
    132, 133, 145, 205
Phosphoinositide 3-kinases (PI3-Ks)
  in the cell cycle, 60–61
  and cell migration, 61–62
  downstream effectors of, 55–56
  drug development prospects, 63–64
  and FLT3, 100
  and IL-6, 129
  inhibitors of, 56–61
  isoforms, gene targeting of, 58–59
  and JAK proteins, 120
  overview, 53–55
  and PDGF, 165, 167, 168
  in tumor inhibition, 62–64
Phosphoinositide phosphatases and PI3-
    K enzymes, 55
Phosphoprofiling, 266
Phosphoproteomics
  approaches to, 267, 269, 270
  overview, 265–269, 276–277

Phosphorylation
  Abl problems, preventing, 194
  Bcr-Abl protein, 272–276
  inhibition in EGFR, 237, 244
  of proteins, 266, 267, 277
  of Src homology 2 (SH2) domain, 270
PI3-Ks. See Phosphoinositide 3-kinases
    (PI3-Ks)
Piceatannol in JAK inhibition, 132
PIK3CA gene in cancer expression, 59, 63
PK1166, 235–237, 249, 250, 254, 255
PKB and PI3-K enzymes, 56, 60
PKC412 (CGP 41251), inhibition by,
    102–103, 169
Placenta, PDGF in, 167
Platelet-derived growth factor (PDGF)
  antagonists
    mechanism of action, 164, 165
    overview, 168–169
  autocrine/paracrine function of, 173,
      175–177
  described, 161–162
  developmental/physiological roles of,
      166–168
  Glivec and, 147, 172, 174, 175
  and interstitial fluid pressure, 176–177
  isoforms/receptors, 162–165
  JM domain in, 212
  lung fibrosis and, 171–172
  MAPK pathway and, 165, 168
  mutations, screening for, 177
  and p100α, 60
  PI3-Ks and, 165, 167, 168
  receptors in disease, 170–177
  Src homology 2 (SH2) domain, 163, 165
Platelet-derived growth factor receptor
    (PDGFR)
  activation of, 164, 165
  inhibition of, 3–5, 105
PLC-γ1 and PDGF, 165, 167
PNU 156804, 131, 133
Point mutations. See Mutations, point
PP-1
  in Src inhibition, 89, 90

structure of, 75
PP60c-src domain, antagonists of, 32–37
PreScission in protein purification, 196, 197
Prostate cancer, Src inhibition in, 86, 89
Proteins
    fusion, 174–175, 197
    phosphorylation of, 266, 267, 277
    structure, determining
        baculovirus system in, 191, 194–
            195, 197, 223
        chromatography, 196–198
        cloning/expression in, 194–195
        construct definition, 191, 194
        fermentation technology, 195
        insect cells in, 195–197, 223
        methods of determination, 187,
            191, 219–225
        purification, 195–196
Protein tyrosine kinases (PTKs)
    interactions, mapping, 270, 271
    preparation for sequencing methods,
        191, 194–198
    sequence alignment, structure-based,
        188–190
    small-molecule inhibitors of, 4
    structural biology of, 187–191
    structures of, 199–219
    as targets for therapy, 1–3, 265–267
    taxonomic relationships of, 72
Protein tyrosine phosphatases (PTPs) in
    JAK activation, 120–121
Proteosomal degradation
    and Glivec, 155
    and JAK proteins, 122
PS341 and Glivec, 155
PTEN and PI3-K enzymes, 55, 59–60,
    63, 64
PTKs. See Protein tyrosine kinases
    (PTKs)
Pyk2, inhibition of, 78, 79
Pyrazolopyrimidines
    in Src inhibition, 74, 89, 90
    structure of, 75
Pyridolopyrimidines

in PI3-K inhibition, 56, 57
in Src inhibition, 74
structure of, 75
Pyridone 6
    features of, 130, 133
    structure of, 131
Pyrrolopyrimidines
    activity in vitro, 76–77
    bone loss models of Src activity in
        vivo, 80–83
    enzymatic selectivity of, 76
    mechanism of action, 77–80
    pharmacokinetic parameters, 78
    side effects of, 87
    in Src inhibition, 74
    structure of, 75

Q

Quinolinecarbonitriles, 74, 75

R

RAD001, 257–258
Radiotherapy and EGFR inhibitors, 257
Ranibizumab (Lucentis) in VEGFR
    inhibition, 14
Rapamycin in cancer treatment, 61
Ras signaling and PDGF, 165
Rat models of Src inhibition, 87
Receptor protein tyrosine kinases
    (RPTKs)
    in malignancy, 3
Renal cell cancer, 249
Resistance
    to imatinib mesylate (Gleevec,
        Glivec, STI-571), 5, 20, 153–
        155, 219
    via mutation, 19–20, 103
Resonance assignment defined, 221, 222
Restenosis and PDGF, 170
Rheumatoid arthritis treatment, 14
RPR101511A in Glivec inhibition, 169
RPTKs. See Receptor protein tyrosine
    kinases (RPTKs)

S

Salz regimen described, 12, 13
SCH66336 and Glivec, 153–155
SCID. *See* Severe combined immunodeficiency (SCID) treatment
Screening, NMR based, 223–225
SELEX aptimers and PDGF, 169
Semaxanib, 12, 13
Severe combined immunodeficiency (SCID) treatment, 17, 62, 123, 134
SH2. *See* Src homology 2 (SH2) domain
SHAPES screening strategy, 224, 225
Signal transducer and activator of transcription. *See* STAT proteins, activation of
SKI-606 in Src inhibition, 90
SOCS proteins. See Suppressor of cytokine signaling family (SOCS) proteins
Sodium chloride in protein sequencing, 197
Src homology 2 (SH2) domain
    antagonists of, 33, 35, 36, 39, 42, 44
    c-Met and, 73
    and FLT3 signaling, 100
    inhibitors
        activity in cancer models, 83–86
        activity in vitro, 76–77
        activity in vivo, 80–83
        limitations of, 86–89
        mechanism of action, 77–80
        overview, 74–76
    and JAK proteins, 120, 121, 126–127
    in negative regulation, 201
    and PDGF, 163, 165
    phosphorylation of, 270
    structural/functional relationships, 72, 73, 192
    structure of, 201–203, 219
    substrates of, 73
    as treatment target, 31–32, 71–73, 89–90
SSI-1 proteins. *See* Suppressor of cytokine signaling family (SOCS) proteins

STAT proteins, activation of, 119–120, 127–129, 132, 134, 149
Staurosporine
    in Csk structure, 207
    in JAK inhibition, 132
    in ZAP70 structure, 207
STI-571. *See* Imatinib mesylate (Gleevec, Glivec, STI-571)
SU5416 in FLT3 inhibition, 101
SU5614 in FLT3 inhibition, 101
SU-6668, 12, 13
SU9518 in Glivec inhibition, 169
SU-11248
    as FLT3 inhibitor, 6, 7
    in Glivec inhibition, 169
    in VEGFR inhibition, 13
SU11657, inhibition by, 102
Suberoylanilide hydroxamic acid and Glivec, 155
Superdex 75 gel in protein sequencing, 198
Suppressor of cytokine signaling family (SOCS) proteins, 121–122
Survivin, 129, 132

T

Tarceva. *See* Erlotinib (Tarceva)
Target selection in clinical studies, 18–19
Taxol. *See* Paclitaxel (Taxol)
Taxotere and Glivec, 154
T-cell lymphoma, 126, 127
T-cells
    activation of, 207
    FL release in, 96–97
TEL-JAK translocations in malignant disease, 125, 126
TEL-PDGFRB protein, expression in CMML, 175
TGF-α expression in hyperplasia, 251
Theophylline in PI3-K inhibition, 56, 57
Thiotepa and Glivec, 154
Tie-2
    in cancer, 14
    regulation of, 217
    structure of, 218

Tobacco etch virus (TEV) in protein purification, 196
Topotecan and Glivec, 154
TRAIL and Glivec, 155
Translocations in malignant disease, 123, 125
Trastuzumab (Herceptin)
    described, 9
    in EGFR/Her2 inhibition, 10
    in RPTK inhibition, 3
Treosulfan and Glivec, 154
Trizol reagent, 267
Tumors. *See* Cancer
TYK2 described, 116, 117, 119
Tyr416, 203
Tyr527, structure of, 201, 202
Tyr759 in gp130/JAK signaling, 129
Tyr1162, 209
Tyrphostins
    in FLT3 inhibition, 101
    in JAK inhibition, 130, 134

U

Ubiquitin, degradation of, 122

V

Vascular endothelial growth factor receptor (VEGFR), 11–14, 193
Vascular endothelial growth factor (VEGF)
    cysteine residue spacing in, 163
    inhibition of, 101, 102, 176
    and PI3-K inhibition, 61
    and Src inhibition, 90
Vatalinib
    in rheumatoid arthritis treatment, 14
    structure of, 12

in VEGFR inhibition, 12, 13
VEGF. *See* Vascular endothelial growth factor (VEGF)
VEGFR. *See* Vascular endothelial growth factor receptor (VEGFR)
Vincristine and Glivec, 154
Vinorelbine, clinical testing of, 253

W

WHI-P97 in JAK inhibition, 133
WHI-P131
    features of, 130, 133
    structure of, 131
WHI-P154 in JAK inhibition, 133
Wortmannin
    and Glivec, 153, 155
    in PI3-K inhibition, 56, 57, 62
    side effects of, 62
WP1066 in apoptosis induction, 132

X

Xenografts, placement of, 233, 242, 243

Y

Y759 in gp130/JAK signaling, 129
Yeast, interactions in, 270, 272, 273

Z

ZAP70, structure of, 192, 207
ZD1839. *See* Gefitinib (Iressa)
ZD-6474
    clinical testing of, 242, 255
    structure of, 12
    in VEGFR inhibition, 14
ZM39923, 130, 131
ZM449829, 130, 131